Fundamentals of Molecular Symmetry

Series in Chemical Physics

The Series in Chemical Physics is an international series that meets the need for up-to-date texts on theoretical and experimental aspects in this rapidly developing field. Books in the series range in level from introductory monographs and practical handbooks to more advanced expositions of current research. The books are written at a level suitable for senior undergraduates and graduate students and will also be useful for practising chemists, physicists and chemical physicists who wish to refresh their knowledge of a particular field.

Series Editors: **J H Moore**, University Of Maryland-College Park, USA
N D Spencer, ETH-Zurich, Switzerland

Series in Chemical Physics

Fundamentals of Molecular Symmetry

Philip R Bunker

Steacie Institute for Molecular Sciences
National Research Council
Ottawa, Canada

Per Jensen

Theoretical Chemistry
Bergische Universität Wuppertal, Germany

Institute of Physics Publishing
Bristol and Philadelphia

British Library Cataloguing-in-Publication Data

A catalogue record for this book is available from the British Library.

ISBN 0 7503 0941 5

Library of Congress Cataloging-in-Publication Data are available

Commissioning Editor: John Navas
Editorial Assistant: Leah Fielding
Production Editor: Simon Laurenson
Production Control: Sarah Plenty
Cover Design: Victoria Le Billon
Marketing: Louise Higham and Ben Thomas

Published by Institute of Physics Publishing, wholly owned by The Institute of Physics, London

Institute of Physics Publishing, Dirac House, Temple Back, Bristol BS1 6BE, UK

US Office: Institute of Physics Publishing, The Public Ledger Building, Suite 929, 150 South Independence Mall West, Philadelphia, PA 19106, USA

Printed in the UK

Contents

Preface

In 1998 we wrote the second edition of the research-level text *Molecular Symmetry and Spectroscopy*. The present book is on the broader subject of molecular symmetry and it is at the student level. It is designed to explain the basis for what is called 'symmetry' in chemistry and to show how symmetry helps in the solution of problems in spectroscopy and in molecular orbital theory. A crucial part of the book is concerned with explaining the relationship between the geometrical symmetry of a molecule, as expressed using the point group symmetry of its equilibrium structure, and the true symmetry of a molecule, as expressed using the *molecular symmetry* group. The elements of the molecular symmetry group involve nuclear permutations and the space-fixed inversion operation called E^*. We aim at giving a balanced account of molecular symmetry using both point groups and molecular symmetry groups.

The book is organized into four parts. Part 2 introduces geometrical (point group) symmetry and true symmetry, and discusses how point group symmetry derives by approximation from true symmetry. Part 3 shows how these two symmetries are used in solving problems. These two parts could be a book in themselves but we felt it appropriate to add the introductory part 1 in order to provide the reader with a brief account of spectroscopy, quantum mechanics and the derivation of molecular wavefunctions. In the final part 4, we develop more advanced ideas and discuss current research on symmetry; the latter focuses on the attempts that are being made, using atomic and molecular spectroscopy experiments, to determine the extent of the breakdown of each of the symmetries that are invoked in describing matter.

> Throughout the text, we introduce 'shadow boxes' such as this to focus attention on a particularly significant statement.

At the end of each chapter in parts 1, 2 and 3, we have included problems involving the application of the ideas developed in the chapter. The answers to selected problems are given in appendix A at the end of the book; the problems that have answers are marked with an asterisk. In appendix B, we give character tables and, in appendix C, we give a short list of books for further reading.

We are grateful to those at the IOP who suggested that we write this student text and we appreciate their encouragement in the completion of the project. PRB thanks the Alexander von Humboldt Foundation whose award allowed him to spend time at the University of Wuppertal during which part of the book was written.

Many colleagues and friends have given us advice and help for which we are very grateful, and we list their names here: O Baum, S Brünken, G W Fuchs, T F Giesen, S G Kukolich, F Lewen, M Litz, P Neubauer-Guenther, S Patchkovskii, R D Poshusta, A Ruoff, A Stolow, J Tennyson, J K G Watson, G Winnewisser and S N Yurchenko.

Line drawings were produced with xfig, figures involving numerical data were produced with idl, and figures showing 3D objects were drawn with MAPLE. Published spectra were initially digitized to a set of numerical (x, y) points which were used as input for idl. The complete text was produced using $\text{\LaTeX}\,2_\varepsilon$, and the figures were inlined as eps files; our text was then copy-edited by the IOP.

Figure acknowledgments

We have included adapted versions of the following published figures, and we thank the authors and publishers for their permission to do this.

Figure 1.1. From figure 1 of Maillard J-P *et al* 1990 *Astrophys. J.* **363** L37. Published by the American Astronomical Society.

Figure 1.3. From figure 2 of Owyoung A *et al* 1978 *Chem. Phys. Lett.* **59** 156. Published by Elsevier.

Figure 1.4. From figure 1 of McKellar A R W and Watson J K G 1998 *J. Mol. Spectrosc.* **191** 215. Published by Academic Press.

Figure 13.7. From figure 2 of Fellers R S *et al* 1999 *J. Chem. Phys.* **110** 6306. Published by the American Institute of Physics.

Figure 13.8. From figure 3 of Fellers R S *et al* 1999 *J. Chem. Phys.* **110** 6306. Published by the American Institute of Physics.

Figure 13.10. From figure 2b of Liu K *et al* 1994 *J. Am. Chem. Soc.* **116** 3507. Published by the American Chemical Society.

Figure 15.2. From figure 2 of Arnold R *et al* 1999 *Eur. Phys. J.* A **6** 361. Published by Springer.

Table acknowledgment

The character tables in appendix B are largely from appendix A of P R Bunker and Per Jensen, *Molecular Symmetry and Spectroscopy*, 2nd edn, NRC Research Press. Published by the National Research Council of Canada.

Philip R Bunker
Per Jensen
22 July 2004

PART 1

SPECTROSCOPY AND THE QUANTUM STATES OF MOLECULES

Chapter 1

Molecular spectroscopy

The study of the extent of the absorption, emission and scattering of electromagnetic radiation by matter, as a function of the wavelength of the radiation and of the nature of the matter, is the subject of spectroscopy. We concentrate on situations involving weak electromagnetic radiation[1] and gas phase molecular samples. In these circumstances, classical theory is used to describe the radiation and quantum mechanics to describe the molecules and their interaction with the radiation.

The classical theory of electromagnetic radiation is based on Maxwell's 1860 theory of the electromagnetic field. Electromagnetic radiation consists of oscillating electric and magnetic fields by virtue of which it carries electric and magnetic energy from a source to a detector. The electric and magnetic fields that constitute the radiation oscillate at the same frequency ν [in units of cycles s^{-1} or hertz (Hz)]; these fields oscillate perpendicular to each other and to the direction of propagation of the radiation. In a vacuum, radiation propagates at the speed of light c (= 299 792 458 m s^{-1}) and the distance between adjacent field oscillation crests is the wavelength λ, where

$$\lambda = c/\nu. \tag{1.1}$$

1.1 Molecular spectra

Electromagnetic radiation emitted from a region of auroral activity in the upper atmosphere of Jupiter can be dispersed to yield an *emission spectrum* such as shown in figure 1.1. An emission spectrum is a plot of the intensity of the radiation emitted from a source as a function of its wavelength, frequency or wavenumber $\tilde{\nu}$ ($\tilde{\nu} = 1/\lambda$; invariably quoted in cm^{-1} units). Figure 1.2 is part of the *absorption spectrum* of carbon monoxide (CO) at 300 K plotted as transmittance (see below) versus wavenumber. Figure 1.3 is part of the Raman spectrum of methane CH_4. A Raman spectrum is obtained by illuminating a

---

[1] 'Weak' is defined at the end of section 1.5, where we discuss power broadening.

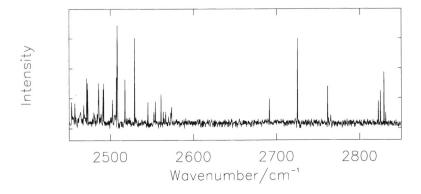

Figure 1.1. Part of the emission spectrum from a region of auroral activity in the upper atmosphere of Jupiter. Adapted from Maillard J-P *et al* 1990 *Astrophys. J.* **363** L37.

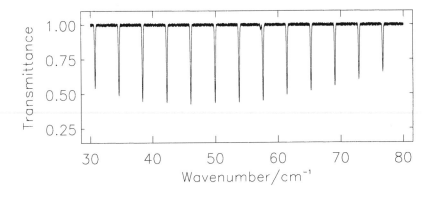

Figure 1.2. Part of the absorption spectrum of carbon monoxide at 300 K.

sample with monochromatic radiation (the *exciting* radiation) and measuring the intensity of the dispersed scattered radiation as a function of its difference (or shift) in frequency or wavenumber from that of the exciting radiation. These are the three most common types of molecular spectra and they consist of spectral lines each having a position, intensity and shape.

Absorption spectra involve a measurement of transmittance as a function of frequency, wavelength or wavenumber; an example is given in figure 1.2. Transmittance is defined with the help of the Lambert–Beer law, which states: If a monochromatic and parallel beam of electromagnetic radiation at wavenumber $\tilde{\nu}$ with intensity $I_0(\tilde{\nu})$ passes through a length l of gas at a concentration c, the transmitted radiation has intensity $I_{\mathrm{tr}}(\tilde{\nu})$ given by

$$I_{\mathrm{tr}}(\tilde{\nu}) = I_0(\tilde{\nu}) \exp[-lc\epsilon(\tilde{\nu})] \tag{1.2}$$

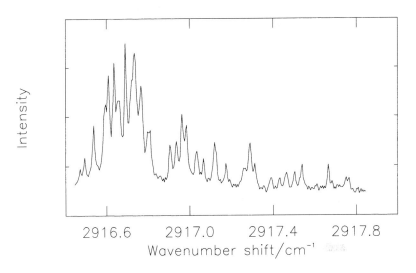

Figure 1.3. Part of the Raman spectrum of methane. Adapted from Owyoung A *et al* 1978 *Chem. Phys. Lett.* **59** 156.

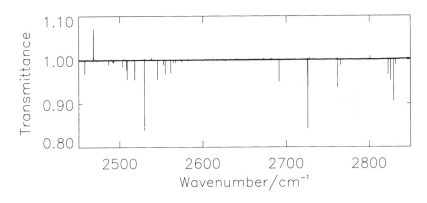

Figure 1.4. A laboratory absorption spectrum of H_3^+ in an electrical discharge through H_2. Taken from McKellar A R W and Watson J K G 1998 *J. Mol. Spectrosc.* **191** 215. The emission feature at 2469 cm^{-1} is caused by H atoms, and the other weaker emission features are caused by H_2.

where the function $\epsilon(\tilde{\nu})$ is the absorption coefficient. The transmittance τ is defined as the ratio

$$\tau = \frac{I_{\mathrm{tr}}(\tilde{\nu})}{I_0(\tilde{\nu})} = \exp[-lc\epsilon(\tilde{\nu})]. \tag{1.3}$$

Figure 1.4 is a laboratory absorption spectrum of H_3^+ in an electrical discharge through hydrogen gas. Comparing the Jupiter emission spectrum, in figure 1.1 with this laboratory spectrum it is clear that H_3^+ ions are present in the atmosphere of Jupiter during an aurora[2].

> The analytical use of spectroscopy is easy to understand and is based on the fact that each molecule has a unique spectrum that characterizes it. Molecular spectra also tell us the temperature of the sample and its concentration. Using quantum mechanics, the spectrum of a molecule can be interpreted to give the structure, bond strengths and other properties of the molecule involved.

1.2   The energies of molecules in the gas phase

We think of a molecule in a gas sample at a particular instant in time as moving with a certain speed and as having a certain amount of internal energy. The internal energy can be approximately separated as the sum of the rotational energy, the vibrational energy and the electronic energy. The rotational energy is the kinetic energy of the overall rotational motion of the molecule, the vibrational energy results from the relative motions of the nuclei and the electronic energy is the energy of the electrons as they orbit the nuclear framework. The internal energy is called the rotation–vibration–electronic energy or the *rovibronic* energy for short.

A molecule with mass M and speed v moving in an unconstrained way in free space has translational energy $Mv^2/2$; this energy can assume any value between zero and infinity. In contrast, the internal energy is *quantized*, i.e. only certain values of the internal energy occur, characteristic of finite motions. The pattern of the discrete internal energies is a unique characteristic of a molecule, and each molecule has a 'fingerprint' of internal energy levels. In figure 1.5, the lowest rotational energy levels for the CO, H_2O, CH_3D and CH_4 molecules are shown. In order to be able to relate molecular energy level separations directly to the wavenumber positions of the related spectral lines [see equation (1.7) and the discussion after it], the energies in figure 1.5 are divided by hc, where h $(= 6.626\,069\,3 \times 10^{-34}$ J s$)$ is Planck's constant, and they are quoted in cm^{-1} units.

For small strongly bound molecules like those in figure 1.5, the rotational energy level spacings divided by hc are about 1 to 50 cm^{-1} and the vibrational

---

[2] The rotational temperature of the H_3^+ that emits the spectrum shown in figure 1.1 is determined from the spectrum to be about 1000 K, whereas that of the laboratory spectrum is 287 K. Because of this the line intensities below 2600 cm^{-1} in figure 1.1 are very different from what they are in the laboratory spectrum.

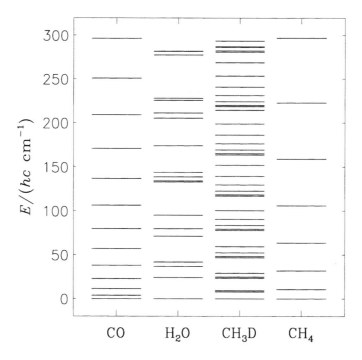

Figure 1.5. The possible rotational energy levels below 300 cm^{-1} for several simple molecules.

energy level spacings divided by hc are about 1000 to 4000 cm^{-1}. For closed-shell molecules, the electronic energy level spacings divided by hc are about 20 000 to 100 000 cm^{-1}. For larger, heavier, weakly bound or open-shell molecules, smaller energy level spacings occur.

As an aside here, the division of the electromagnetic spectrum into three main regions roughly reflects the division of molecular energies into rotational, vibrational and electronic energies. These three regions are the microwave region (wavenumbers from 0.1 to 1 cm^{-1}), the infrared region (wavenumbers from 10^2 to 10^4 cm^{-1}) and the visible/ultraviolet region (wavenumbers from 10^4 to 10^6 cm^{-1}). Other regions are the radiofrequency region (below 0.1 cm^{-1}), the millimetrewave region (from 1 to 10^2 cm^{-1}), and the x-ray and γ-ray regions (above 10^6 cm^{-1}).

In a gas sample, the speed and, hence, translational energy of each individual molecule is continually changing as a result of collisions with other molecules and with the walls of the containing vessel. However, because of the large number of molecules in a gas sample, the distribution of speeds remains constant for an isolated sample at thermal equilibrium. For example, in an isolated sample of carbon monoxide gas at thermal equilibrium at 300 K, at any instant in time,

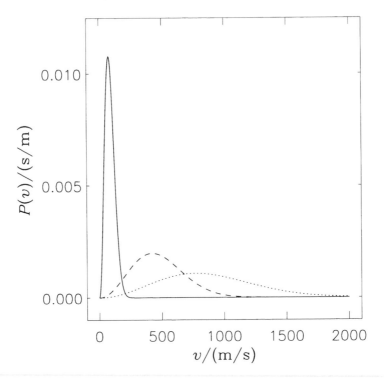

Figure 1.6. The distribution of translational speeds for the CO molecule at temperatures of 10 (——), 300 (- - - -) and 1000 K (······).

19.3% of the molecules will have speeds between 400 and 500 m s^{-1}, whereas 1.0% of the molecules will have speeds between 0 and 100 m s^{-1}, and 1.8% of the molecules will have speeds between 900 and 1000 m s^{-1}.

The expression for the distribution of speeds in an isolated ideal gas sample at thermal equilibrium can be calculated using the methods of statistical mechanics and it is called the *Maxwell distribution of speeds*. This distribution is such that the probability of a molecule having speed between v and $v + dv$ is given by $P(v)\, dv$, where

$$P(v) = 4\pi (M/2\pi kT)^{3/2} v^2 e^{-Mv^2/2kT}. \qquad (1.4)$$

In this equation, $k\ (= 1.380\,650\,5 \times 10^{-23}\ \mathrm{J\ K^{-1}})$ is the Boltzmann constant, M is the mass of the molecule and T is the absolute temperature. Equation (1.4) was used to calculate the percentages given in the preceding paragraph for the distribution of speeds of CO molecules at 300 K.

In figure 1.6, the distribution of translational speeds at temperatures of 10, 300 and 1000 K for the CO molecule is plotted. Because of the v^2 factor in equation (1.4) no molecule has zero speed, and because of the exponential factor no molecule has infinite speed. In between, there is a maximum at the most probable speed given by $(2kT/M)^{1/2}$. For higher temperatures, or lower mass, the most probable speed increases and the whole distribution spreads out and moves to higher speeds.

When a molecule suffers an *inelastic* collision, it changes its internal (rovibronic) energy as well as its speed. For a large number of molecules at thermal equilibrium, the collisions between the molecules distribute the molecules among their internal energy states in a way that reflects the temperature and the Maxwell distribution of speeds so that, at any instant, the fraction $F(E_i)$ of the molecules in the internal energy level E_i is given by the *Maxwell–Boltzmann distribution law*:

$$F(E_i) = \frac{g_i e^{-E_i/kT}}{\sum_j g_j e^{-E_j/kT}} \tag{1.5}$$

where the sum in the denominator (the denominator is called the *partition function*) runs over all the discrete possible energies E_j; each E_j is only counted once in the sum. The value of g_i is the number of states having energy E_i; this is called the *degeneracy* of the energy level E_i [see equation (2.72), and the discussion after it, for an example of a state that has $g_i > 1$]. Figure 1.7 shows the fraction $F(E_i)$ of CO molecules, at thermal equilibrium, in each of its rotational states (see figure 1.5), for temperatures of 10, 300 and 1000 K. At low temperatures, very few rotational energy states are populated.

1.3   The positions of spectral lines

An isolated molecule in an initial internal energy state E_i can absorb energy from a weak electromagnetic radiation field and change its internal energy state to a final one with energy E_f; from the conservation of energy, the radiation absorbed has frequency ν_{if} satisfying the *Bohr frequency condition*

$$h\nu_{if} = E_f - E_i = \Delta E_{if} \tag{1.6}$$

where h is Planck's constant. Put another way, a photon can be absorbed if the energy of the photon $h\nu$ is in *resonance* with the molecular internal energy difference ΔE concerned. Such resonant absorption causes the molecule to make a *transition* from one energy *level* to another. Dividing both sides of equation (1.6) by hc gives the wavenumber version as

$$\tilde{\nu}_{if} = \nu_{if}/c = E_f/hc - E_i/hc = \Delta E_{if}/hc \tag{1.7}$$

where $\tilde{\nu}_{if}$ is the wavenumber of the radiation.

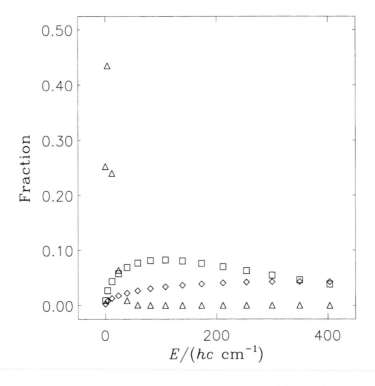

Figure 1.7. The fraction $F(E_i)$ of CO molecules in each of its rotational states for temperatures of 10 ($\triangle$), 300 ($\square$) and 1000 K ($\diamond$), at thermal equilibrium; see equation (1.5).

Thus, in an absorption spectrum, the wavenumber of every spectral line gives the difference between two internal molecular energies divided by hc. Internal energies divided by hc, and quoted in cm^{-1} units, are called *term values*; their separations can be directly related to the wavenumber positions of spectral lines in cm^{-1}. For this reason, spectroscopists generally quote the term values rather than the energies of molecular states.

The *assignment* of the upper and lower energy levels of each transition to their position in the ladder of energy levels of the molecule under investigation is one of the tasks of experimental molecular spectroscopy. The ultimate goal here is the determination of the term values of all internal energy levels for the molecule. The separations of the energy levels in a molecule can be analysed to yield molecular properties such as structure and bond strengths. The discrete line structure of a spectrum shows directly that the internal energy states are quantized. If there were no restrictions on the values of the internal energy, then the 'spectrum' would exhibit continuous absorption at all wavenumbers with no lines.

1.4   The intensities of spectral lines

The intensity of a spectral line in absorption is proportional to the fraction of the molecules $F(E_i)$ in the initial energy state E_i of the transition and, at thermal equilibrium, this fraction can be varied, according to equation (1.5), by changing the temperature. A low-temperature sample has just a few of the lowest levels populated and many molecules are in these levels, so the spectrum will consist of fewer stronger lines than that obtained for a high-temperature sample. The variation of a spectrum with temperature will clearly help in its assignment since transitions originating in highly excited levels (so-called *hot* transitions) will be stronger at higher temperatures in a predictable way.

     In addition to absorbing resonant radiation, molecules also undergo resonant stimulated emission. In stimulated emission, radiation of frequency ν_{if} stimulates a molecule in an excited energy level E_f to emit radiation of the same frequency ν_{if} and to drop into a lower energy level E_i. For this to occur, the energy $h\nu_{if}$ must be in resonance with the energy difference ($E_f - E_i$). This process competes with the absorption process and reduces the amount of absorption by the multiplicative factor

$$R_{\text{stim}}(f \rightarrow i) = 1 - \exp(-h\nu_{if}/kT). \tag{1.8}$$

At low frequencies in the GHz region, this is an important cause of reduced absorption intensity.

     In a process that is the complete opposite of resonant absorption, the stimulated emission process can cause the intensity of radiation at frequency ν_{if} to be amplified, and this process is used in a laser (Light Amplification by Stimulated Emission of Radiation). The successful operation of a laser requires that the excited level at E_f be continually repopulated using energy from an electrical discharge or other means and that the nature of the molecule, and its energy level ladder, be such that the lower level at E_i be rapidly depopulated so that it does not absorb the lasing radiation.

     Apart from the dependence on initial state population and extent of stimulated emission, the intensity of a line has an intrinsic value called the *line strength* $S(f \leftarrow i)$, see equation (2.87), and this depends on the specific properties of the two energy states involved. In fact, some transitions have zero line strength. The absorption spectrum of CO shown in figure 1.2 only involves transitions between adjacent rotational energy levels in figure 1.5; transitions between non-adjacent rotational energy levels here have zero line strength and are said to be *forbidden*. Of all possible transitions, only a selection are allowed and symmetry is used to determine the *selection rules* that govern this behaviour. The H_3^+ spectra shown in figures 1.1 and 1.4 involve transitions between different vibrational states; transitions between its rotational energy levels are forbidden by the simplest selection rules. However, by studying the symmetry properties of the levels the possibility emerges that very weak transitions can occur between some of the rotational levels of H_3^+. Some transitions are less forbidden than others and symmetry can help us understand whether small effects that are normally neglected can come into play to make a forbidden transition observable. Symmetry selection rules, and the spectra of CO and H_3^+ illustrated here, are discussed further in chapter 12.

We also show in chapter 12, how the quantitative value of the line strength depends on molecular properties. For example, the line strength of a transition between different rotational energy levels depends on the value of the molecular electric dipole moment [see equation (2.88)] and the line strength of a transition between different vibrational energy levels depends on how the value of the dipole moment changes with molecular deformation. All this information is important in building a complete understanding of the properties of a molecule.

By integrating the absorption coefficient over the line one obtains the expression

$$I(f \leftarrow i) = \frac{8\pi^3 N_A \nu_{if}}{(4\pi\epsilon_0)3hc^2} \frac{e^{-E_i/kT}}{\sum_i g_i e^{-E_i/kT}}[1 - \exp(-h\nu_{if}/kT)]S(f \leftarrow i)$$

(1.9)

for the intensity of the absorption line for the transition from the state i with energy E_i, in thermal equilibrium at the temperature T, to the state f with energy E_f, where $h\nu_{if} = E_f - E_i$, N_A ($= 6.022\,141\,5 \times 10^{23}$ mol^{-1}) is the Avogadro constant, and ϵ_0 [$= 10^7/(4\pi c^2)$ F m^{-1}, where c is in m s^{-1}] is the permittivity of free space (also called the electric constant).

1.5 The shapes of spectral lines

Spectral lines have a finite width and a characteristic shape. Important causes of line broadening are the Doppler effect, the finite lifetime of molecular energy states and the power of the radiation.

The molecules in a gas sample are not at rest but have a distribution of speeds, given by equation (1.4), and the frequency that a molecule 'feels' as it moves with speed v relative to the direction of propagation of radiation having frequency ν_0 is shifted by $\nu_0(v/c)$ because of the Doppler effect. Molecules moving towards the radiation source will absorb on the low-frequency side of the line centre and molecules moving away will absorb on the high-frequency side. Making use of equation (1.4), one can determine that the lineshape function arising from the Doppler shifts of all the molecules in a gas sample is

$$S(v) = S(v_0) \exp\left[-\frac{Mc^2}{2kT}\left(\frac{v - v_0}{v_0}\right)^2 \right] \tag{1.10}$$

for a line centred at v_0. The function in equation (1.10) is a Gaussian function. It has a full width at half height FWHH (the frequency width of the line at half the maximum intensity) given by

$$\text{FWHH} = \frac{2v_0}{c}\left(\frac{2kT}{M}\ln 2\right)^{1/2}$$

$$\approx 7.15 \times 10^{-7}\left(\frac{T}{M/u}\right)^{1/2} v_0 \tag{1.11}$$

where u is the unified atomic mass unit[3].

Around equation (1.8), the process of resonant stimulated emission was introduced. Molecules also *spontaneously* emit resonant radiation and drop into a lower energy level; any transitions down that are allowed by the selection rules can occur. As a result, molecular energy levels have a finite *natural radiative lifetime*. This has the effect of broadening the energy levels and spectral lines; the FWHH is related to the lifetime τ in ps by the relation

$$\text{FWHH/cm}^{-1} \approx \frac{5.3}{\tau/\text{ps}} \quad \text{or} \quad \text{FWHH/GHz} \approx \frac{159}{\tau/\text{ps}}. \tag{1.12}$$

Spontaneous emission from an upper level having energy E_f to a lower level having energy E_i occurs with the emission of radiation having frequency v_{if} that satisfies equation (1.6). The rate of spontaneous emission is proportional to v_{if}^3 and so this lifetime is shorter for highly excited levels. The lifetime of a state can also be reduced by *predissociation*, which is a process whereby a molecule falls apart after a certain time. This process can occur only if the state has an energy greater than the dissociation energy of the molecule. Predissociation leads to the appearance of very broad *diffuse* lines in a spectrum.

Collisions that change the internal energy reduce the lifetime of a state. The collisional lifetime (the mean time between collisions) is reduced, and the linewidth increased, by raising the gas pressure; this cause of broadening is, thus, referred to as *pressure broadening*. At low pressures (less than about 10 Torr[4]), pressure broadening (or natural radiative lifetime broadening) gives rise to a Lorentzian lineshape function

$$S(v, v_0) = \frac{1}{\pi}\left[\frac{\Delta v}{(v_0 - v)^2 + (\Delta v)^2}\right] \tag{1.13}$$

---

[3] $1\,u = 1.660\,538\,86 \times 10^{-27}$ kg; also called the dalton or the atomic mass constant.
[4] 1 Torr ≈ 133.322 Pa.

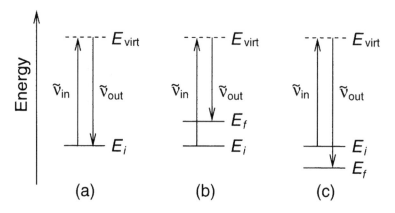

Figure 1.8. Energy-level diagrams of (a) Rayleigh scattering, (b) Stokes Raman scattering, and (c) anti-Stokes Raman scattering.

where $2\Delta\nu = 1/(2\pi\tau)$, τ is the lifetime, and ν_0 is the central frequency. The FWHH $2\Delta\nu$ is $1/(2\pi\tau)$ which is given by the second of equations (1.12).

The simultaneous occurrence of Doppler and lifetime broadening leads to a lineshape that is a convolution of the two lineshape functions called a Voigt function. A detailed treatment of pressure broadening for higher pressures leads to a more complicated unsymmetrical lineshape function and a small shift in the centre frequency, that can both be related to the nature of the intermolecular forces.

This discussion of lineshapes, and the development of equation (1.9) for the intensity of an absorption line, assume that the radiation is weak. By definition, the intensity of the radiation is weak if the absorption is a linear function of that intensity, i.e. if the transmittance is independent of the intensity of the radiation. As the power of the radiation is increased, the molecules absorbing in the centre of the absorption line (where the absorption is the greatest) will start to absorb radiation at a greater rate than that at which they can return to the lower level of the transition to achieve thermal equilibrium. As a result, the centre of the absorption line will start to saturate and the line will broaden; this is called *power broadening*.

1.6   Raman spectra

In figure 1.3, we show part of a Raman spectrum obtained by illuminating CH_4 with monochromatic radiation. In Raman spectroscopy, the exciting radiation, whose wavenumber we denote $\tilde{\nu}_{in}$, is normally visible laser light and the sample absorbs and scatters photons from this light beam. As indicated in figure 1.8, we can think of the absorption as transferring molecules from an initial state with energy E_i, to a highly excited, so-called *virtual state*, with energy E_{virt}. In the vast majority of cases, the molecules return from the virtual state to the initial state as shown in figure 1.8(a) and photons with wavenumber $\tilde{\nu}_{out} = \tilde{\nu}_{in}$ are emitted (or scattered). This process is known as *Rayleigh scattering*. However, a tiny fraction of the molecules (about 1 in 10^7) transfer from the virtual state to a final state different from the initial state. When this happens, there is a *Raman shift* and the scattered radiation has wavenumber $\tilde{\nu}_{out} \neq \tilde{\nu}_{in}$. The energy of the final state is $E_f \neq E_i$. We can have $E_f > E_i$ [*Stokes Raman scattering*, figure 1.8(b)] so that $\tilde{\nu}_{out} < \tilde{\nu}_{in}$, or $E_f < E_i$ [*anti-Stokes Raman scattering*, figure 1.8(c)] with $\tilde{\nu}_{out} > \tilde{\nu}_{in}$. Figure 1.8 shows that the Raman shift

$$\tilde{\nu} = \tilde{\nu}_{in} - \tilde{\nu}_{out} = (E_f - E_i)/(hc) \tag{1.14}$$

corresponds to the energy difference between the final and initial states; such energy differences can be obtained from Raman experiments.

The intensity of Raman-scattered light is proportional to the Raman line strength $S_{Raman}(f \leftarrow i)$, which is analogous to the line strength of an absorption or emission transition. The calculation of $S_{Raman}(f \leftarrow i)$ is discussed in section 12.5. The Raman intensity is also proportional to the intensity of the exciting light, to the concentration of molecules in the initial state, to $\tilde{\nu}_{out}^4$, and to the solid angle of observation. In addition, the intensities observed in a Raman experiment depend on the angle between the electric field vector of the exciting light and that of the scattered light; this angle is determined by the experimental set-up.

1.7 Problems

1.1 What are the wavenumber, frequency, and energy per photon of (a) visible radiation having a wavelength of 500 nm, (b) infrared radiation having a wavelength of 5 μm, and (c) microwave radiation having a wavelength of 5 cm? Quote the frequency in appropriate units such as THz or GHz, and the wavenumber in cm^{-1}.

1.2 The successive rotational energy levels of the $^{12}C^{16}O$ molecule depicted in figure 1.5 are labelled $J = 0, 1, 2, \ldots$ and, to three significant figures, their energy divided by hc is $E_i = 1.92J(J + 1)$ cm^{-1}; the degeneracy[5] of each level is $g_i = (2J + 1)$. Calculate the numerator in the Maxwell–Boltzmann distribution function equation (1.5) for a range of values of J for temperatures T of 10, 30 and 1000 K to confirm the J-value at which the population is a maximum according to figure 1.7; in the appropriate

---

[5] This is the *m*-degeneracy that will be introduced in section 2.7.

units $k \approx 0.695$ cm^{-1} K^{-1}. By setting equal to zero the differential of the numerator with respect to J, determine an expression as a function of T for the value of J at which the population is a maximum. Check that this leads to the correct result for $T = 10$, 300 and 1000 K (J has to be an integer).

1.3   Use equation (1.11) to calculate the Doppler widths (FWHH) of infrared absorption lines of the H$_3^+$ molecule ($M/u \approx 3$) and CH$_4$ molecule ($M/u \approx 16$) at around 3000 cm^{-1} at 1000 K and at 10 K. What would the Doppler widths be for lines of these molecules at these temperatures at wavelengths around 1 cm (in the microwave region) or 100 nm (in the ultraviolet region)? The appropriate units for linewidth are cm^{-1} in the infrared and ultraviolet and kHz in the microwave.

1.4   In a recent experiment[6], NH molecules in the excited (A) electronic state were generated by first forming ground (X) electronic state molecules using an ArF laser to photolyse NH$_3$, and then using a tunable dye laser to pump NH from the X-state up to the A-state. NH molecules in the A-state then emit *fluorescent* radiation as they drop back down to the X-state, and the intensity of this radiation at thermal equilibrium at room temperature was found to decay according to

$$I_F(t) \propto n(t) = n(0)e^{-t/\tau_{\text{eff}}} \tag{1.15}$$

where t is time. By making measurements at five different NH$_3$ pressures between 2–10 Pa, the decay rate was found to fit the expression

$$1/\tau_{\text{eff}} = 1/\tau + kn_{\text{NH}_3} \tag{1.16}$$

where τ is the radiative lifetime, k is a constant and n_{NH_3} is the number density of NH$_3$ molecules. It was found, for a particular rotation–vibration level of the A-state, that $\tau = 438$ ns and $k = 6.7 \times 10^{-10}$ cm^3 s^{-1}. Plot this lifetime and the linewidth of the emission line, as a function of NH$_3$ pressure. At room temperature, $n_{\text{NH}_3} \approx 2.4 \times 10^{15}$ cm^{-3} for a pressure of 10 Pa.

[6] Hake A and Stuhl F 2002 *J. Chem. Phys.* **117** 2513.

Chapter 2

Quantum mechanics

2.1　The Schrödinger equation

The equation

$$\frac{d}{dx}e^{a_n x} = a_n e^{a_n x} \tag{2.1}$$

is an example of an *eigenvalue equation*, in which the operator d/dx acting on the function $e^{a_n x}$ gives, as result, a constant 'a_n' times the function. In general, an eigenvalue equation has the form

$$\hat{O}\psi_n = O_n \psi_n \tag{2.2}$$

where $\hat{O}$ is a differential operator, ψ_n is a function, and O_n is a constant. A function ψ_n that satisfies this equation is an *eigenfunction* of the operator $\hat{O}$, and the constant numerical factor O_n is the *eigenvalue* of the operator $\hat{O}$ appropriate for the eigenfunction ψ_n; the subscript n (= 1, 2, 3, etc) labels the different solutions. Restricting the eigenfunctions of an eigenvalue equation so that they have certain properties can lead to the eigenvalues having discrete values like molecular energies.

> Schrödinger postulated the way of setting up an eigenvalue equation for a molecule and the way of interpreting and restricting the eigenfunctions, so that the eigenvalues are the molecular energies.

This special eigenvalue equation has come to be called the *Schrödinger equation*.

2.2   The postulates of quantum mechanics

Quantum mechanics is used to describe nature at the atomic and molecular level, in which there is quantization of the values of *observables* O such as energy and angular momentum. The theory of quantum mechanics leads to quantization by introducing rules or *postulates* concerning the way operators are set up to represent observables and the way the eigenfunctions are restricted and interpreted.

Using classical mechanics for an atom or molecule that consists of l particles (electrons and nuclei), labelled $r = 1, 2, \ldots, l$ with mass m_r, most observables can be written as a function of Cartesian coordinates X_r, Y_r, and Z_r and momenta P_{Xr} ($= m_r\dot{X}_r = m_r\,dX_r/dt$, where t is time), P_{Yr}, and P_{Zr}. The first postulate of quantum mechanics states that if one replaces the momenta P_{Xr}, P_{Yr} and P_{Zr} in the classical expression for an observable O, by the partial differential operators $\hat{P}_{Xr}$, $\hat{P}_{Yr}$ and $\hat{P}_{Zr}$ according to the rules

$$\hat{P}_{Xr} = -i\hbar\partial/\partial X_r \tag{2.3}$$
$$\hat{P}_{Yr} = -i\hbar\partial/\partial Y_r \tag{2.4}$$

and

$$\hat{P}_{Zr} = -i\hbar\partial/\partial Z_r \tag{2.5}$$

where $i = \sqrt{-1}$ and $\hbar = h/2\pi$, then the resultant differential operator $\hat{O}$ *represents* the observable O from which it was derived. If the observable depends only on coordinates, and not on momenta, then the expression for the quantum mechanical operator is identical to the classical expression for the observable. The operator that represents the energy is called *the Hamiltonian operator* or, simply, *the Hamiltonian*.

The second postulate of quantum mechanics states that the only possible values for an observable are the eigenvalues of the operator that represents it when the eigenfunctions ψ_n are *restricted* to be single valued and when they are *interpreted* to be such that

$$P_n\,d\tau = \psi_n^*\psi_n\,d\tau = |\psi_n|^2\,d\tau \tag{2.6}$$

is the probability that ψ_n has its coordinates in the *volume element* $d\tau$. The volume element for integration, $d\tau$, is a short hand notation in which, for example, if the functions ψ_n were expressed in an l-particle Cartesian coordinate space $(X_1, Y_1, Z_1, \ldots, X_l, Y_l, Z_l)$,

$$d\tau = dX_1\,dY_1\,dZ_1 \ldots dX_l\,dY_l\,dZ_l. \tag{2.7}$$

By appropriate transformations, $d\tau$ can be expressed in any coordinate system.

It is often asserted that a postulate of quantum mechanics is that the eigenfunctions be square integrable, i.e. that

$$\int |\psi_n|^2\,d\tau < \infty \tag{2.8}$$

but this is only true for bound (and quantized) states; see, for example, the eigenfunctions given in equation (2.68) for a case in which there is no quantization. When there is quantization, we *normalize* the eigenfunctions so that

$$\int P_n \, d\tau = \int \psi_n^* \psi_n \, d\tau = \int |\psi_n|^2 \, d\tau = 1 \tag{2.9}$$

in order that the total probability over all space be unity. When the eigenfunctions are not square integrable, the ratio of $|\psi_n|^2$ at two different points is the ratio of their probabilities.

With this restriction and interpretation of the eigenfunctions, the possible values of the energy are the eigenvalues E_n of the Hamiltonian operator $\hat{H}$:

$$\hat{H}\psi_n = E_n \psi_n. \tag{2.10}$$

This equation is the (time independent) Schrödinger equation or wave equation, and the eigenfunctions ψ_n are the *wavefunctions* that represent (or simply *are*) the state of the system. The E_n are the possible *stationary state* energies of the system. The state having the lowest energy is called the *ground* state and all other states are called *excited* states.

The second postulate still allows us to multiply a wavefunction by $\exp(i\theta)$, where θ is any real constant, without changing $P_n = |\psi_n|^2$, since $|\exp(i\theta)|^2 = 1$. The factor $\exp(i\theta)$ is called a *phase factor* and a wavefunction multiplied by a phase factor is the same state. However, it is necessary to define the phase factor used and to be consistent, because relative phase factors of wavefunctions can be significant.

A state ϕ_n might not be an eigenfunction of a particular operator $\hat{O}$. The third postulate of quantum mechanics concerns such a situation and it states that the expected (or most probable) value of an observable O for a system in a state ϕ_n is the integral of the product $\phi_n^* \hat{O} \phi_n$, where $\hat{O}$ is the operator that represents O; such an integral is called an *expectation value* and we write it as

$$O_{nn} = \langle n|\hat{O}|n \rangle = \int \phi_n^* \hat{O} \phi_n \, d\tau. \tag{2.11}$$

The fourth postulate, concerning the time dependence of wavefunctions, is given in section 14.1 and the fifth, concerning the symmetry of wavefunctions under the effect of the permutation of identical particles, is given in section 9.1.

2.2.1 Operators and eigenfunctions

We summarize here some important facts and definitions. An operator $\hat{O}$ is *Hermitian* if

$$\int \psi_m^* \hat{O} \psi_n \, d\tau = \int (\hat{O}\psi_m)^* \psi_n \, d\tau = \int (\psi_n^* \hat{O} \psi_m)^* \, d\tau. \tag{2.12}$$

The eigenvalues of a Hermitian operator are real and operators corresponding to real physical observables must be Hermitian. It can be proved that the eigenfunctions of a Hermitian operator that have different eigenvalues are orthogonal, i.e. they are such that

$$\int \psi_m^* \psi_n \, d\tau = 0 \qquad (2.13)$$

unless $m = n$. Equations (2.9) and (2.13) are summarized by saying that for a set of *orthonormal* functions ψ_1, ψ_2, ψ_3, etc.

$$\int \psi_m^* \psi_n \, d\tau = \delta_{mn} \qquad (2.14)$$

where δ_{mn} is the Kronecker delta.

Acting on a function of X, $\phi(X)$ say, with the operator difference $(X\hat{P}_X - \hat{P}_X X)$ gives

$$
\begin{aligned}
(X\hat{P}_X - \hat{P}_X X)\phi(X) &= [X(-i\hbar\partial/\partial X) - (-i\hbar\partial/\partial X)X]\phi(X) \\
&= -i\hbar X \frac{\partial\phi(X)}{\partial X} + i\hbar \frac{\partial}{\partial X}[X\phi(X)] \\
&= -i\hbar X \frac{\partial\phi(X)}{\partial X} + i\hbar\left[\phi(X) + X\frac{\partial\phi(X)}{\partial X}\right] \\
&= i\hbar\phi(X).
\end{aligned}
\qquad (2.15)
$$

We can write

$$(X\hat{P}_X - \hat{P}_X X)\phi(X) = i\hbar\phi(X) \qquad (2.16)$$

and formally 'cancel' out the $\phi(X)$ from each side of the equation to obtain the *operator equation*:

$$X\hat{P}_X - \hat{P}_X X = i\hbar. \qquad (2.17)$$

An operator equation means that each side of the equation produces the same result when it acts on a function. Two operators $\hat{O}_1$ and $\hat{O}_2$ commute if the following operator equation is true:

$$(\hat{O}_1\hat{O}_2 - \hat{O}_2\hat{O}_1) = 0. \qquad (2.18)$$

We introduce the notation

$$[\hat{O}_1, \hat{O}_2] = (\hat{O}_1\hat{O}_2 - \hat{O}_2\hat{O}_1) \qquad (2.19)$$

where $[\hat{O}_1, \hat{O}_2]$ is called the *commutator* of $\hat{O}_1$ and $\hat{O}_2$. Thus, the commutator of X and $\hat{P}_X$ is not zero and these two operators do not commute.

Observables are represented by Hermitian operators and special care must sometimes be exercised when using the first postulate to set up an operator. For example, an expression such as $X\hat{P}_X$ is not Hermitian because X and $\hat{P}_X$ do not commute. In such a case, it is necessary to properly symmetrize the classical expression before converting it to quantum mechanical form. For example, instead of $X P_X$, one must write $(X P_X + P_X X)/2$.

If a stationary state energy level E_n is k-fold degenerate, then there are k linearly independent[1] and orthogonal eigenfunctions $\psi_{n1}, \psi_{n2}, \ldots, \psi_{nk}$ having the same energy eigenvalue E_n of the Hamiltonian operator $\hat{H}$. Any other eigenfunction ψ_p having eigenvalue E_n can only be a linear combination of this *complete set* of k functions, i.e.

$$\psi_p = \sum_{j=1}^{k} c_{pj} \psi_{nj} \qquad (2.20)$$

where the c_{pj} are constants. For the level E_n, we would have $g_i = k$ in equation (1.5). Stationary states can be non-degenerate or they can be degenerate. If the level E_n is non-degenerate and ψ_n is an eigenfunction with eigenvalue E_n, then the only other functions that can be eigenfunctions having eigenvalue E_n are of the form $c\psi_n$ where c is a constant. If $\psi_1, \psi_2, \ldots$ are eigenfunctions of $\hat{H}$, then linear combinations of them can be chosen that are also simultaneously the eigenfunctions of any operator that commutes with $\hat{H}$. Conversely if two operators (such as X and $\hat{P}_X$) do not commute then there are no non-trivial functions that are simultaneously the eigenfunctions of both.

2.3 Diagonalizing the Hamiltonian matrix

It frequently happens that we know a set of functions ψ_n^0 say, that are approximately the eigenfunctions of a Hamiltonian $\hat{H}$, i.e. $\hat{H} \psi_n^0 = E_n^0 \psi_n^0 + X$, where $X \ll E_n^0 \psi_n^0$. Such a set of known functions are called *basis functions* and one can use them to determine the true eigenfunctions and eigenvalues by setting up and diagonalizing the Hamiltonian matrix; we now explain this procedure using some definitions and results from matrix algebra that are collected together in section 2.9.

Consider a Hamiltonian $\hat{H}$ and a set of orthonormal basis functions ψ_n^0 (where $n = 1, 2, 3, \ldots$). We introduce the integrals

$$H_{mn} = \langle m | \hat{H} | n \rangle = \int (\psi_m^0)^* \hat{H} \psi_n^0 \, d\tau \qquad (2.21)$$

which can be arranged in a matrix where m and n are the row and column indices, respectively; we call such integrals *matrix elements* of the Hamiltonian, and the entire matrix is called *the Hamiltonian matrix*. The diagonal matrix elements are the expectation values of the energy for the basis functions and we write

$$H_{nn} = \langle n | \hat{H} | n \rangle = E_n^0. \qquad (2.22)$$

---

[1] These k functions are linearly independent if there is no relation of the type $c_1 \psi_{n1} + c_2 \psi_{n2} + \cdots + c_k \psi_{nk} = 0$ (apart from the trivial one with all $c_i = 0$) connecting them.

For convenience, we organize the matrix so that $E_1^0 \leq E_2^0 \leq E_3^0 \ldots$.

The values of the matrix elements would change if we used different basis functions. For example, if we set up the Hamiltonian matrix using the normalized eigenfunctions ψ_n of $\hat{H}$, the diagonal matrix elements would be the eigenvalues and the off-diagonal matrix elements would vanish. The Hamiltonian matrix would then be said to be *diagonal*.

It is the presence of non-vanishing off-diagonal matrix elements that spoils the functions ψ_n^0 as eigenfunctions. However, the degree to which the functions are spoiled does not only depend on the magnitudes of the off-diagonal matrix elements between them but it also depends on the differences between their diagonal matrix elements. To explain this, we will first show how the eigenfunctions and eigenvalues can be determined from the values of the Hamiltonian matrix elements and then focus on a simple 2×2 example. The 2×2 example is extremely important since it is used to analyse interactions or *perturbations* between energy levels caused by a previously neglected part of the Hamiltonian which gives rise to a non-vanishing off-diagonal matrix element. In such circumstances, the states ψ_n^0 would be called *zero-order* states.

We wish to determine the eigenfunctions and eigenvalues, ψ_j and E_j ($j = 1, 2, 3, \ldots$), of the Hamiltonian $\hat{H}$, using the complete set of basis functions ψ_n^0. Since the basis set is complete, by definition we can write the unknown eigenfunctions ψ_j in terms of them as:

$$\psi_j = \sum_n C_{jn} \psi_n^0 \tag{2.23}$$

where the C_{jn} are the eigenfunction coefficients that remain to be determined. Since $\hat{H}\psi_j = E_j\psi_j$,

$$\hat{H}\left[\sum_n C_{jn}\psi_n^0\right] = E_j\left[\sum_n C_{jn}\psi_n^0\right]. \tag{2.24}$$

To determine the ψ_j and E_j, we proceed as follows: Multiply each side of equation (2.24) on the left by $(\psi_m^0)^*$, make use of the fact that $\hat{H}$ and the C_{jn} commute with each other, and finally integrate each side over all space. Using equations (2.21) and (2.14), this gives

$$\sum_n C_{jn} H_{mn} = E_j \sum_n C_{jn}\delta_{mn} \tag{2.25}$$

which can be rewritten as the matrix product

$$\sum_n (H_{mn} - \delta_{mn}E_j)\tilde{C}_{nj} = 0 \tag{2.26}$$

where $\tilde{C}$ is the transpose of C, see section 2.9. Apart from the useless solution that all elements of $\tilde{C}_{nj} = 0$, the solution obtained is the following *secular* equation for the eigenvalues E_j:

$$|H_{mn} - \delta_{mn} E| = 0 \qquad \text{if } E = E_j. \tag{2.27}$$

This states that the determinant[2] of the matrix $(H_{mn} - \delta_{mn} E)$ vanishes if E is an eigenvalue. This enables us to determine the eigenvalues of $\hat{H}$ and an l-dimensional Hamiltonian matrix leads to a secular equation with l eigenvalues. The E_j are the eigenvalues both of the operator $\hat{H}$ and of the matrix H that represents it using a basis set.

By substituting the eigenvalues E_j one at a time into equation (2.26), we obtain l simultaneous equations (as $m = 1$ to l) for the $\tilde{C}_{nj}$ and we obtain the elements in the jth column of the matrix $\tilde{C}$. Since $\tilde{C}_{nj} = C_{jn}$ these coefficients form the jth row of the matrix C; these are the coefficients of the basis functions ψ_n^0 in the eigenfunction ψ_j and we can appreciate how well (or badly) ψ_j^0 represents ψ_j from their values. The orthonormality of the functions demands that the elements of C satisfy

$$\sum_n C_{jn}^* C_{kn} = \delta_{jk} \tag{2.28}$$

which means that the matrix C is unitary, see section 2.9.

Using matrix notation, it can be shown that the elements of the matrix C are such that

$$CHC^{-1} = \Lambda \tag{2.29}$$

where H is the Hamiltonian matrix and Λ is a matrix having non-vanishing diagonal elements $\Lambda_{jj} = E_j$ (the eigenvalues) and $\Lambda_{ij} = 0$ if $i \neq j$. We say that the *similarity transformation* of the Hamiltonian matrix H by the matrix of eigenfunction coefficients C in equation (2.29) diagonalizes H to produce the diagonal matrix Λ of eigenvalues. The process of *diagonalizing* a Hamiltonian matrix in a basis set is a routine procedure, once we have determined the elements of the Hamiltonian matrix, and standard computer routines are available.

As a simple example, we consider a two-dimensional (Hermitian) Hamiltonian matrix for the zero-order states ψ_1^0 and ψ_2^0:

$$\begin{bmatrix} E_1^0 & H_{12} \\ H_{12}^* & E_2^0 \end{bmatrix}. \tag{2.30}$$

Assuming, for simplicity, that H_{12} is real, equation (2.27) reduces to

$$(E_1^0 - E)(E_2^0 - E) - H_{12}^2 = 0 \tag{2.31}$$

and solving for E, one obtains the two roots:

$$E_1 = E_1^0 - S \tag{2.32}$$

---

[2] See equation (2.95) for the definition of a determinant.

and

$$E_2 = E_2^0 + S \tag{2.33}$$

where $E_1^0 \leq E_2^0$, the energy shift

$$S = \tfrac{1}{2}\left(\sqrt{4H_{12}^2 + \Delta^2} - \Delta\right) \tag{2.34}$$

and $\Delta = E_2^0 - E_1^0$ is the difference between the diagonal matrix elements. Using equations (2.26) and (2.28) to obtain the eigenfunction coefficients, one gets

$$\psi_1 = c^+ \psi_1^0 - c^- \psi_2^0 \tag{2.35}$$

and

$$\psi_2 = c^+ \psi_2^0 + c^- \psi_1^0 \tag{2.36}$$

where

$$c^{\pm} = \frac{1}{\sqrt{2}}\left[1 \pm \frac{\Delta}{\sqrt{4H_{12}^2 + \Delta^2}}\right]^{1/2}. \tag{2.37}$$

The expressions for S and $c^{\pm}$ depend on both the off-diagonal matrix element H_{12} and the diagonal matrix element difference Δ.

For situations when $|H_{12}| \ll \Delta$, the leading terms in the binomial expansion of $\sqrt{4H_{12}^2 + \Delta^2} = \Delta(1 + 4H_{12}^2/\Delta^2)^{1/2}$ can be used to give the approximate solution

$$S \approx \frac{H_{12}^2}{\Delta} \tag{2.38}$$

with

$$c^+ \approx 1 - \frac{H_{12}^2}{2\Delta^2} \tag{2.39}$$

and

$$c^- \approx \frac{H_{12}}{\Delta}. \tag{2.40}$$

Equations (2.38)–(2.40) are also obtained using second-order perturbation theory (see below) for the 2×2 case.

We can now quantitatively represent the very commonly occurring phenomenon of a 2×2 energy level interaction or perturbation. The functions ψ_1^0 and ψ_2^0 would be eigenfunctions of the Hamiltonian if the off-diagonal matrix element H_{12} were zero; as H_{12} grows, the two states perturb, repel, or interact with, each other. The lower level E_1 moves down, the higher level E_2 moves up, and the wavefunctions of the two states ψ_1^0 and ψ_2^0 gradually become more and more mixed in the eigenfunctions ψ_1 and ψ_2. The extent to which the zero-order functions are mixed, and the amount by which the levels repel each other depend on the size of the off-diagonal matrix element H_{12} and on the zero-order energy separation Δ. The above equations allow us to calculate these effects for any values of H_{12} and Δ. For $\Delta = 0$, we obtain $c^+ = c^- = 1/\sqrt{2}$; we have a 50:50 mixing of the two zero-order states and a maximal energy shift of H_{12}. However, as $H_{12} \to 0$, c^+ approaches 1 while c^- approaches 0; now $\psi_1 \to \psi_1^0$ and $\psi_2 \to \psi_2^0$, with energy shifts that go to zero.

Perturbations can involve more than two levels interacting simultaneously and a matrix larger than 2×2 will then have to be diagonalized. It often happens that we have to consider the complete basis set consisting of an infinite number of functions. Obviously, we can only set up and diagonalize the Hamiltonian matrix in a finite number of basis functions, so we have to *truncate* the matrix. If the size of the truncated matrix is $n^{\text{trunc}} \times n^{\text{trunc}}$, then its diagonalization will yield n^{trunc} approximate eigenvalues and eigenfunctions. If we are only interested in the lowest p states, say, then we must choose n^{trunc} to be much larger than p in order that this approximate approach leads to satisfactory results for the p states of interest. This approximate approach is called the *variational* approach. In practice, the size of n^{trunc} is increased until a further increase has negligible effect on the p eigenvalues of interest; the calculation is then said to have *converged*. The lowest eigenvalue of the truncated Hamiltonian matrix $E_{\text{lowest}}^{\text{trunc}}$ will not be precisely equal to the lowest eigenvalue $E_{\text{lowest}}^{\text{exact}}$ of the Hamiltonian. The *variational theorem* states that $E_{\text{lowest}}^{\text{trunc}}$ is always above $E_{\text{lowest}}^{\text{exact}}$, i.e.

$$E_{\text{lowest}}^{\text{trunc}} - E_{\text{lowest}}^{\text{exact}} = \Delta E > 0. \tag{2.41}$$

In a well-behaved problem, increasing the number of basis functions reduces ΔE and $E_{\text{lowest}}^{\text{trunc}}$ converges to $E_{\text{lowest}}^{\text{exact}}$ as the number of basis functions is increased. The existence of very efficient numerical computer routines for diagonalizing large matrices make the variational procedure of practical and general use.

In situations where the off-diagonal matrix elements are small compared to the differences in diagonal matrix elements, an alternative procedure for determining eigenfunctions and eigenvalues called *perturbation theory* can be used. In this procedure, we write the Hamiltonian operator as

$$\hat{H} = \hat{H}^0 + \lambda \hat{H}' \tag{2.42}$$

where the eigenfunctions of $\hat{H}^0$ are the known basis functions ψ_n^0 and where $\hat{H}'$ (called *the perturbation*) has non-vanishing off-diagonal matrix elements. By changing the expansion constant λ from zero to one, the perturbation is switched on. In this approach, analytical expressions for the eigenfunctions and eigenvalues of $\hat{H}$ are obtained as power series in λ involving the matrix elements of $\hat{H}'$ in the basis functions ψ_n^0 and differences between diagonal matrix elements. Truncating these expressions at the terms quadratic in λ produces the results of *second-order* perturbation theory and this is normally where the procedure is terminated. The results obtained using perturbation theory are useful if $\hat{H}'$ has a small effect and one can gain an understanding of the effects of $\hat{H}'$ from the analytical expressions obtained. Perturbation theory is used in the development of the *effective rotational*

Hamiltonian (see section 11.5). This Hamiltonian is crucially important in the practical analysis of spectra.

2.4   The molecular Schrödinger equation

The classical expression for the total energy of a molecule consisting of l particles, N nuclei and $l - N$ electrons, is

$$E_{\text{total}} = T + V. \tag{2.43}$$

The kinetic energy is given by

$$T = \tfrac{1}{2} \sum_{r=1}^{l} m_r (\dot{X}_r{}^2 + \dot{Y}_r{}^2 + \dot{Z}_r{}^2) \tag{2.44}$$

where particle r has mass m_r (the mass of an electron being m_e) and $\dot{X}_r$, $\dot{Y}_r$ and $\dot{Z}_r$ are the components of its velocity in a space-fixed XYZ axis system. In SI units, the electrostatic potential energy that results from the repulsions and attractions between the particles is

$$V = \sum_{r<s=1}^{l} \frac{C_r C_s e^2}{4\pi \epsilon_0 R_{rs}} \tag{2.45}$$

where $C_r e$ is the charge[3] of particle r (the charge on an electron is $-e$), R_{rs} is the interparticle distance given by

$$R_{rs} = [(X_r - X_s)^2 + (Y_r - Y_s)^2 + (Z_r - Z_s)^2]^{1/2} \tag{2.46}$$

and ϵ_0 is the permittivity of free space (introduced on page 12).

The energy expression $T + V$ can be separated into two parts: The translational energy and the internal (rovibronic) energy. In chapters 3, 4 and 5, we discuss the separation of the rovibronic energy into electronic, vibrational and rotational parts. Such separations of variables are of central importance in making the equations that occur easier to handle and to understand; they always involves making coordinate changes.

2.5   The separation of translational energy

Whenever we make a change of coordinates, there are two ways of proceeding:

(I)   First to set up the quantum mechanical molecular Hamiltonian in the initial coordinates and then to change to the new coordinates in the resultant differential equation that is the Schrödinger equation using the *chain rule*,

---

[3] The elementary charge $e = 1.602\,176\,53 \times 10^{-19}$ C; this is the charge on a proton.

(II) First to change coordinates in the classical expression to obtain the classical energy expression in the new coordinates and then to set up the quantum mechanical Hamiltonian, and Schrödinger equation, in the new coordinates.

We use method (II) here.

We know that the translational energy is the kinetic energy $Mv^2/2$, where $M = \sum_r m_r$ is the molecular mass and v is the speed of the molecular centre of mass through space. To change coordinates so that the molecular kinetic energy T involves this, we must explicitly introduce the coordinates of the centre of mass, which we call (X_0, Y_0, Z_0). Thus, the XYZ coordinates of the particles are written as

$$X_r = X_r + X_0 \tag{2.47}$$
$$Y_r = Y_r + Y_0 \tag{2.48}$$

and

$$Z_r = Z_r + Z_0 \tag{2.49}$$

where (X_r, Y_r, Z_r) are the coordinates of particle r in an XYZ axis system that is parallel to the space-fixed XYZ axis system[4] but which has origin at the molecular centre of mass (X_0, Y_0, X_0). We must write the kinetic energy T in terms of the new set of $3l$ coordinates

$$X_0, Y_0, Z_0, X_2, Y_2, Z_2, \ldots, X_l, Y_l, Z_l \tag{2.50}$$

and their velocities $\dot{X}_0, \ldots, \dot{Z}_l$, where we have eliminated X_1, Y_1 and Z_1 using

$$X_1 = -\frac{1}{m_1} \sum_{r=2}^{l} m_r X_r \tag{2.51}$$

with similar equations for Y_1 and Z_1.   The $(3l - 3)$ coordinates $X_2, Y_2, Z_2, \ldots, X_l, Y_l, Z_l$ are the *internal* coordinates; they specify the positions of the particles relative to the centre of mass of the molecule.

Using equations (2.47)–(2.49) for $r = 2$ through l gives

$$\frac{1}{2} \sum_{r=2}^{l} m_r (\dot{X}_r^2 + \dot{Y}_r^2 + \dot{Z}_r^2) = \frac{1}{2} \sum_{r=2}^{l} m_r (\dot{X}_r^2 + \dot{Y}_r^2 + \dot{Z}_r^2)$$

$$+ \sum_{r=2}^{l} m_r (\dot{X}_r \dot{X}_0 + \dot{Y}_r \dot{Y}_0 + \dot{Z}_r \dot{Z}_0)$$

$$+ \frac{1}{2} \left(\sum_{r=2}^{l} m_r \right) (\dot{X}_0^2 + \dot{Y}_0^2 + \dot{Z}_0^2). \tag{2.52}$$

[4] Be careful to notice the font distinction here. We use upright font for the XYZ axes; they have space-fixed origin and space-fixed orientation. We use italic font for the XYZ axes that are parallel to them and which, therefore, have space-fixed orientation, but which have their origin at the molecular centre of mass.

Using equation (2.51), and the similar equations for Y_1 and Z_1, in equations (2.47)–(2.49), we obtain

$$\frac{1}{2}m_1(\dot{X}_1{}^2 + \dot{Y}_1{}^2 + \dot{Z}_1{}^2) = \frac{1}{2m_1}\sum_{r,s=2}^{l} m_r m_s(\dot{X}_r\dot{X}_s + \dot{Y}_r\dot{Y}_s + \dot{Z}_r\dot{Z}_s)$$

$$-\sum_{r=2}^{l} m_r(\dot{X}_r\dot{X}_0 + \dot{Y}_r\dot{Y}_0 + \dot{Z}_r\dot{Z}_0)$$

$$+\tfrac{1}{2}m_1(\dot{X}_0{}^2 + \dot{Y}_0{}^2 + \dot{Z}_0{}^2). \tag{2.53}$$

Adding equations (2.45), (2.52) and (2.53) gives

$$E_{\text{total}} = T_{\text{trans}} + T_{\text{rve}} + V \tag{2.54}$$

where V is expressed in terms of the coordinates $X_2, \ldots, Z_l$. In equation (2.54), the translational kinetic energy is

$$T_{\text{trans}} = \tfrac{1}{2}M(\dot{X}_0{}^2 + \dot{Y}_0{}^2 + \dot{Z}_0{}^2) \tag{2.55}$$

and the internal (rovibronic) kinetic energy that results from the motion of the particles in the molecule relative to the molecular centre of mass is

$$T_{\text{rve}} = \tfrac{1}{2}\sum_{r=2}^{l} m_r(\dot{X}_r{}^2 + \dot{Y}_r{}^2 + \dot{Z}_r{}^2)$$

$$+ \frac{1}{2m_1}\sum_{r,s=2}^{l} m_r m_s(\dot{X}_r\dot{X}_s + \dot{Y}_r\dot{Y}_s + \dot{Z}_r\dot{Z}_s). \tag{2.56}$$

T_{rve} and V do not involve the coordinates or velocities of the centre of mass, and T_{trans} does not involve the internal coordinates or velocities.

> There is a complete separation of the internal and translational degrees of freedom in the energy expression.

We can write the total energy as

$$E_{\text{total}} = E_{\text{trans}} + E_{\text{rve}} \tag{2.57}$$

where the translational energy E_{trans} is the pure kinetic energy term T_{trans} given in equation (2.55); there is no potential energy contribution to the translational energy for a molecule moving in an unconstrained way in field free space. The rovibronic energy is

$$E_{\text{rve}} = T_{\text{rve}} + V. \tag{2.58}$$

2.5.1 The translational Schrödinger equation

To follow the procedure outlined above [see equations (2.3)–(2.10)] for obtaining the translational Schrödinger equation, we begin by expressing the classical energy, given in equation (2.55), in terms of momenta P_α rather than velocities in order to obtain it in Hamiltonian form:

$$H_{\text{trans}} = \frac{1}{2M}(P_{X0}{}^2 + P_{Y0}{}^2 + P_{Z0}{}^2) \qquad (2.59)$$

where $P_{X0} = M\dot{X}_0$ etc. To obtain the Schrödinger equation, we replace the momenta P_{X0}, P_{Y0} and P_{Z0} by the partial differential operators $\hat{P}_{X0} = -i\hbar\partial/\partial X_0$, $\hat{P}_{Y0} = -i\hbar\partial/\partial Y_0$, $\hat{P}_{Z0} = -i\hbar\partial/\partial Z_0$, in H_{trans} to yield the Hamiltonian operator for the translational motion:

$$\begin{aligned}
\hat{H}_{\text{trans}} &= \frac{1}{2M}(\hat{P}_{X0}{}^2 + \hat{P}_{Y0}{}^2 + \hat{P}_{Z0}{}^2) \\
&= -\frac{\hbar^2}{2M}\left(\frac{\partial^2}{\partial X_0{}^2} + \frac{\partial^2}{\partial Y_0{}^2} + \frac{\partial^2}{\partial Z_0{}^2}\right)
\end{aligned} \qquad (2.60)$$

and we set up the eigenvalue equation

$$\hat{H}_{\text{trans}}\Phi_{\text{trans}}^{(n)}(X_0, Y_0, Z_0) = E_{\text{trans}}^{(n)}\Phi_{\text{trans}}^{(n)}(X_0, Y_0, Z_0). \qquad (2.61)$$

Equation (2.61) gives the translational Schrödinger equation for a molecule moving in an unconstrained way in free space. From the second quantum mechanical postulate, the eigenfunctions have to be single valued and the relative probabilities must be given by $|\Phi_{\text{trans}}^{(n)}(X_0, Y_0, Z_0)|^2$; they are then the translational wavefunctions of the molecule and the eigenvalue $E_{\text{trans}}^{(n)}$ is the translational energy of the molecule when it is in the state $\Phi_{\text{trans}}^{(n)}(X_0, Y_0, Z_0)$.

Since $\hat{H}_{\text{trans}}$ is the sum of three independent terms in X_0, Y_0 and Z_0, we can separate the translational Schrödinger equation into three by writing

$$E_{\text{trans}} = E_{\text{transX}}^{(n_X)} + E_{\text{transY}}^{(n_Y)} + E_{\text{transZ}}^{(n_Z)} \qquad (2.62)$$

and

$$\Phi_{\text{trans}}(X_0, Y_0, Z_0) = \psi_{\text{transX}}^{(n_X)}(X_0)\psi_{\text{transY}}^{(n_Y)}(Y_0)\psi_{\text{transZ}}^{(n_Z)}(Z_0). \qquad (2.63)$$

We substitute these two equations into equation (2.61). Making use of the fact that

$$\frac{\partial^2}{\partial X_0{}^2}\Phi_{\text{trans}}(X_0, Y_0, Z_0) = \psi_{\text{transY}}^{(n_Y)}(Y_0)\psi_{\text{transZ}}^{(n_Z)}(Z_0)\frac{\partial^2}{\partial X_0{}^2}\psi_{\text{transX}}^{(n_X)}(X_0) \qquad (2.64)$$

with similar equations for the effects of $\partial^2/\partial Y_0{}^2$ and $\partial^2/\partial Z_0{}^2$, we can divide the resultant equation through by $\Phi_{\text{trans}}(X_0, Y_0, Z_0)$ to obtain the three independent equations

$$-\frac{\hbar^2}{2M}\frac{d^2}{dX_0^2}\psi_{transX}^{(n_X)}(X_0) = E_{transX}^{(n_X)}\psi_{transX}^{(n_X)}(X_0) \tag{2.65}$$

$$-\frac{\hbar^2}{2M}\frac{d^2}{dY_0^2}\psi_{transY}^{(n_Y)}(Y_0) = E_{transY}^{(n_Y)}\psi_{transY}^{(n_Y)}(Y_0) \tag{2.66}$$

and

$$-\frac{\hbar^2}{2M}\frac{d^2}{dZ_0^2}\psi_{transZ}^{(n_Z)}(Z_0) = E_{transZ}^{(n_Z)}\psi_{transZ}^{(n_Z)}(Z_0). \tag{2.67}$$

The translational energy $E_{transX}^{(n_X)}$ has to be positive and real and the most general solution we can write for the eigenfunction of equation (2.65) is

$$\psi_{transX}^{(n_X)}(X_0) = A\cos(k_X X_0) + B\sin(k_X X_0) \tag{2.68}$$

or, equivalently,

$$\psi_{transX}^{(n_X)}(X_0) = C\exp(ik_X X_0) + D\exp(-ik_X X_0) \tag{2.69}$$

where $k_X = (2ME_{transX}^{(n_X)})^{1/2}/\hbar$, A and B are arbitrary constants, $C = (A - iB)/2$ and $D = (A + iB)/2$. Similar equations can be written for the eigenfunctions of equations (2.66) and (2.67) so that we have

$$E_{trans} = \frac{\hbar^2}{2M}(k_X^2 + k_Y^2 + k_Z^2). \tag{2.70}$$

> There is no quantization of the translational states of an unconfined molecule moving in free space. The translational wavefunctions are plane waves and the eigenvalues (energies) can be any positive real number.

This provides a very simple example of what happens when we separate the coordinates in a Hamiltonian. Here, since the Hamiltonian can be written as the sum of three independent parts, we have reduced a three-dimensional Schrödinger equation (2.61) to three separate one-dimensional Schrödinger equations (2.65)–(2.67). The eigenvalues are obtained as the sum of the eigenvalues of the one-dimensional Schrödinger equations in equation (2.62) and the eigenfunctions are obtained as the product of the one-dimensional eigenfunctions in equation (2.63).

For a molecule confined to remain within a sample cell (which has finite dimensions), there is quantization of the translational states. To show how this comes about, we consider the situation in which the molecule is confined within a cube-shaped box with side L that has one corner at the point $(X_0, Y_0, Z_0) =$

$(0, 0, 0)$ and which lies in the positive octant of the (X_0, Y_0, Z_0) axis system. In this circumstance, the eigenfunctions satisfy equations (2.63)–(2.69) within the box (i.e. when the X_0, Y_0 and Z_0 coordinates are between 0 and L). However, with the probability interpretation of the eigenfunctions they must be zero outside the box and must go smoothly to zero at the walls of the box. From equation (2.68) for $\psi_{transX}^{(n_X)}(X_0)$, we see that functions that go smoothly to zero at $X_0 = 0$ must have $A = 0$ and for them also to go smoothly to zero at $X_0 = L$, they must also have $k_X L = n_X \pi$ where n_X is a positive integer. Thus, for a molecule confined within this box, the translational wavefunctions [from equations (2.63) and (2.68)] within the box are given by

$$\Phi_{trans}^{(n_X, n_Y, n_Z)}(X_0, Y_0, Z_0) = N \sin\left(\frac{n_X \pi}{L} X_0\right) \sin\left(\frac{n_Y \pi}{L} Y_0\right) \sin\left(\frac{n_Z \pi}{L} Z_0\right)$$
(2.71)

where n_X, n_Y and n_Z must be positive integers and N is a constant. Outside the box the translational wavefunctions vanish. The normalizing constant N is determined to be $(8/L^3)^{1/2}$ by setting the the integral of $|\Phi_{trans}|^2 \, d\tau$ within the box to be unity since that is the probability of finding the molecule within the box [see equation (2.9)]. Since $k_X L = n_X \pi$ etc, we have the result that

the translational energies of a molecule of mass M constrained to move within a cube of side L are quantized. They are given by

$$E_{trans}^{(n_X, n_Y, n_Z)} = \frac{h^2}{8ML^2}(n_X^2 + n_Y^2 + n_Z^2)$$
(2.72)

where the quantum numbers n_X, n_Y and n_Z are positive integers.

The lowest state has $n_X = n_Y = n_Z = 1$ with energy $3h^2/(8ML^2)$ and the first excited translational state has energy $6h^2/(8ML^2)$. The energy separation is $3h^2/(8ML^2)$. This lowest excited state is actually three states with quantum numbers $(n_X, n_Y, n_Z) = (2, 1, 1)$, $(1, 2, 1)$ or $(1, 1, 2)$ which all have the same energy; such a state is said to be three-fold degenerate or to have a degeneracy of three. The separation in energy $[3h^2/(8ML^2)]$ between the two lowest states for a $^{12}C^{16}O$ molecule (mass $M \approx 4.65 \times 10^{-26}$ kg) constrained to move in a cubic box with side $L = 10^{-2}$ m is 3.5×10^{-38} J. Dividing by hc, and quoting in cm^{-1}, the wavenumber separation is obtained as 1.8×10^{-15} cm^{-1}. This incredibly small energy separation shows how the quantization that results from using the rules of quantum mechanics disappears for all practical purposes for systems having macroscopic dimensions when the rules of classical mechanics are satisfactory.

We can use this analysis of the energy levels of a particle in a box to get an approximate estimate of the energy separations involved when electrons move about within the limits of molecular dimensions or when nuclei vibrate in bonds. For an electron (mass $M \approx 9.11 \times 10^{-31}$ kg) constrained to move within a cubic box of side $L = 0.3$ nm (which gives a box that has roughly the volume over which an outer electron moves in a small molecule), the particle in a box analysis leads to a wavenumber separation between the two lowest states of 1.01×10^5 cm^{-1}. For a particle constrained to move in one dimension within a length L, the energy is given by $h^2 n^2/(8ML^2)$ where $n = 1, 2, 3, \ldots$ (obtained by just considering the X_0 motion, for example, in the three-dimensional analysis) and the energy separation between the two lowest states is given by $3h^2/(8ML^2)$ just as for motion within a three-dimensional box. For a proton (mass $M \approx 1.67 \times 10^{-27}$ kg) constrained to move in one dimension within a length of 0.03 nm (which roughly equals the stretching vibrational amplitude in a molecule), the wavenumber separation between the lowest two energy levels is 5500 cm^{-1}, a factor of about 1/20th of the electronic wavenumber (or energy) separation.

> The first excited electronic state has an energy much larger than that of the first excited vibrational state because the electron mass is so much less than any nuclear mass; this more than compensates for the fact that electronic motions are less constrained than nuclear vibrational motions in molecules.

2.6 The rovibronic Schrödinger equation

After separating out the translation, the classical expression for the rovibronic (internal) energy of a molecule that consists of l particles (nuclei and electrons) is obtained from equations (2.45), (2.56) and (2.58) as

$$E_{\text{rve}} = \tfrac{1}{2} \sum_{r=2}^{l} m_r (\dot{X}_r{}^2 + \dot{Y}_r{}^2 + \dot{Z}_r{}^2)$$

$$+ \frac{1}{2m_1} \sum_{r,s=2}^{l} m_r m_s (\dot{X}_r \dot{X}_s + \dot{Y}_r \dot{Y}_s + \dot{Z}_r \dot{Z}_s)$$

$$+ \sum_{r<s=1}^{l} \frac{C_r C_s e^2}{4\pi \epsilon_0 R_{rs}}. \qquad (2.73)$$

Starting with this classical rovibronic energy expression and using the first postulate, we obtain the rovibronic Schrödinger equation.

In the rovibronic energy expression E_{rve}, the motion of the particles is constrained so that the centre of mass remains fixed at the origin of the XYZ axes.

The general definition, allowing for constraints, of the momentum P_s *conjugate* to the coordinate Q_s is

$$P_s = \partial(T - V)/\partial \dot{Q}_s. \tag{2.74}$$

Since the potential energy is independent of the velocities, we obtain $P_{Xr} = \partial T_{rve}/\partial \dot{X}_r$, $P_{Yr} = \partial T_{rve}/\partial \dot{Y}_r$ and $P_{Zr} = \partial T_{rve}/\partial \dot{Z}_r$, where T_{rve} is given in equation (2.56); we do not obtain simple relations such as $P_{Xr} = M_r \dot{X}_r$. Inverting the equations obtained for the generalized momenta as functions of the velocities, one obtains the velocities as functions of the momenta. Substituting for the velocities in equation (2.73) leads to the classical Hamiltonian H_{rve} as a function of the coordinates and momenta. Replacing P_{Xr} by $\hat{P}_{Xr} = -i\hbar\partial/\partial X_r$, P_{Yr} by $\hat{P}_{Yr} = -i\hbar\partial/\partial Y_r$ and P_{Zr} by $\hat{P}_{Zr} = -i\hbar\partial/\partial Z_r$,

the quantum mechanical rovibronic Hamiltonian for an l-particle molecule is obtained as

$$\hat{H}_{rve} = - (\hbar^2/2) \sum_{r=2}^{l} (\partial^2/\partial X_r^2 + \partial^2/\partial Y_r^2 + \partial^2/\partial Z_r^2)/m_r$$

$$+ (\hbar^2/2M) \sum_{r,s=2}^{l} (\partial^2/\partial X_r\partial X_s + \partial^2/\partial Y_r\partial Y_s + \partial^2/\partial Z_r\partial Z_s)$$

$$+ \sum_{r<s=1}^{l} \frac{C_r C_s e^2}{4\pi \epsilon_0 R_{rs}} \tag{2.75}$$

where $M = \sum m_r$ is the mass of the molecule.

The rovibronic Schrödinger equation is given by

$$\hat{H}_{rve} \Phi_{rve}(X_2, Y_2, Z_2, \ldots, Z_l) = E_{rve} \Phi_{rve}(X_2, Y_2, Z_2, \ldots, Z_l). \tag{2.76}$$

In section 3.2, we introduce *spin* and it is explained there that because of the presence of spin, electrons and many nuclei have a magnetic dipole moment and that some nuclei have an electric quadrupole moment as well. The internal energy of a molecule is affected by the presence of these moments. The term in the Hamiltonian that arises from the fact that each electron has a spin magnetic dipole moment is called $\hat{H}_{es}$ and the term arising from the presence of the nuclear spin moments is called $\hat{H}_{hfs}$. The principal terms in $\hat{H}_{es}$ arise from the interaction of the electron spin magnetic moments with each other (the electron *spin–spin* interaction) and from their interaction with the magnetic moments generated by the orbital motion of the electrons (the electron *spin–orbit* interaction). $\hat{H}_{hfs}$

contains similar magnetic terms involving the nuclear spin–spin and nuclear spin–orbit interactions, as well as terms for nuclear spin–electron orbit and nuclear spin–electron spin interactions. For nuclei that have an electric quadrupole moment there is an additional term involving its interaction with the electronic charge gradient at the nucleus. $\hat{H}_{es}$ and $\hat{H}_{hfs}$ can give rise to splittings of the energy levels called electronic fine structure splittings and nuclear hyperfine structure splittings, respectively.

Adding the sum of $\hat{H}_{es}$ and $\hat{H}_{hfs}$ to the electrostatic potential energy gives the complete electromagnetic interaction energy between the particles, so that the complete quantum mechanical Hamiltonian for the internal dynamics of a molecule (that is, everything except translation) is

$$\hat{H}_{int} = \hat{H}_{rve} + \hat{H}_{es} + \hat{H}_{hfs}. \tag{2.77}$$

2.7   The angular momentum operator

The classical observable of orbital angular momentum J for a system of l particles, in the centre-of-mass axis system, has an X component given by

$$J_X = \sum_{r=1}^{l} (Y_r P_{Zr} - Z_r P_{Yr}) \tag{2.78}$$

and so the operator for it is given by

$$\hat{J}_X = -i\hbar \sum_{r=1}^{l} \left(Y_r \frac{\partial}{\partial Z_r} - Z_r \frac{\partial}{\partial Y_r} \right). \tag{2.79}$$

The operators representing $\hat{J}_Y$ and $\hat{J}_Z$ are obtained from equation (2.79) by cyclic permutation of X, Y and Z. The square of the orbital angular momentum operator is given by

$$\hat{J}^2 = \hat{J}_X^2 + \hat{J}_Y^2 + \hat{J}_Z^2. \tag{2.80}$$

The commutators $[X_r, P_{Xr}]$, $[Y_r, P_{Yr}]$ and $[Z_r, P_{Zr}]$ are each $i\hbar$, from equation (2.17) but all other 'cross commutators' $[X_r, P_{Yr}]$, $[P_{Xr}, P_{Yr}]$, $[X_r, P_{Xs}]$ etc are zero. Using these results, it can be shown that the operator $\hat{J}^2$ commutes with $\hat{J}_X$, $\hat{J}_Y$ or $\hat{J}_Z$, and that these four operators each commute with the rovibronic Hamiltonian $\hat{H}_{rve}$ given in equation (2.75). However, the three angular momentum component operators do not commute with each other and we have the commutation relation

$$[\hat{J}_X, \hat{J}_Y] = i\hbar \hat{J}_Z \tag{2.81}$$

with two others obtained by cyclically permuting X, Y and Z. Referring $\hat{J}$ to molecule-fixed axes x, y and z, as we will do when we derive the rotational Hamiltonian (see section 5.5), we obtain the commutation relation

$$[\hat{J}_x, \hat{J}_y] = -i\hbar \hat{J}_z \tag{2.82}$$

with two others being obtained by cyclic permutation of x, y and z. Note the opposite signs in these two equations.

The simultaneous eigenfunctions of $\hat{H}_{\mathrm{rve}}$, $\hat{\boldsymbol{J}}^2$ and $\hat{J}_Z$ are such that

$$\hat{\boldsymbol{J}}^2 \Phi_{\mathrm{rve}} = J(J+1)\hbar^2 \Phi_{\mathrm{rve}} \qquad (2.83)$$

and

$$\hat{J}_Z \Phi_{\mathrm{rve}} = m\hbar \Phi_{\mathrm{rve}} \qquad (2.84)$$

where the total orbital angular momentum quantum number $J = 0, 1, 2, \ldots$ and the projection quantum number m has one of the $2J+1$ values $0, \pm 1, \pm 2, \ldots, \pm J$. Stationary state eigenfunctions of $\hat{H}_{\mathrm{rve}}$ can be labelled using J and m and, for a given value of J, the state has a $2J+1$ fold m-degeneracy.

If it is necessary to consider the magnetic interactions of the electron spin, we must use $\hat{H}_{\mathrm{rve}} + \hat{H}_{\mathrm{es}}$, which commutes with the square of the sum of the total orbital angular momentum (now called $\hat{\boldsymbol{N}}$) and the total electron-spin angular momentum $\hat{\boldsymbol{S}}$; this sum is called $\hat{\boldsymbol{J}}$, i.e.

$$\hat{\boldsymbol{J}}^2 = (\hat{\boldsymbol{N}} + \hat{\boldsymbol{S}})^2 \qquad (2.85)$$

and the good quantum numbers J and m now refer to the sum of the orbital and electron spin angular momenta. To include the effect of the nuclear hyperfine Hamiltonian, we use $\hat{H}_{\mathrm{int}}$; see equation (2.77). This Hamiltonian commutes with the square of the total angular momentum $\hat{\boldsymbol{F}}^2$, which is the square of the sum of $\hat{\boldsymbol{N}}$, $\hat{\boldsymbol{S}}$ and $\hat{\boldsymbol{I}}$, where the latter is the total nuclear spin angular momentum. Thus,

$$\hat{\boldsymbol{F}}^2 = (\hat{\boldsymbol{J}} + \hat{\boldsymbol{I}})^2 = (\hat{\boldsymbol{N}} + \hat{\boldsymbol{S}} + \hat{\boldsymbol{I}})^2. \qquad (2.86)$$

The eigenstates of $\hat{H}_{\mathrm{int}}$ can be labelled using the total angular momentum quantum number F and the projection quantum number $m_F = 0, \pm 1, \pm 2, \ldots, \pm F$. Angular momentum is discussed further in sections 5.5.2 and 14.5, and in problems 5.7–5.11.

2.8   The dipole moment operator and line strengths

To calculate the intensities of the lines in an absorption spectrum, we need the line strengths. Having accurate line strengths is important if one wants to use a measured spectrum to determine the concentration of the species being observed or if one wants to know the predicted intensities in a theoretically simulated spectrum. To calculate line strengths, we use the rovibronic wavefunctions in integrals that are called *transition moments*.

For a gas phase sample illuminated by a weak electromagnetic radiation field the line strength of an electric dipole transition between all possible states Φ''_{rve} having energy E''_{rve}, and all possible states Φ'_{rve} having energy E'_{rve}, is

$$S(f \leftarrow i) = \sum_{\Phi'_{\text{rve}}, \Phi''_{\text{rve}}} \sum_{A=X,Y,Z} \left| \int \Phi'^*_{\text{rve}} \mu_A \Phi''_{\text{rve}} \, d\tau \right|^2 \qquad (2.87)$$

where $d\tau = dX_2 \, dY_2 \, dZ_2 \ldots dX_l \, dY_l \, dZ_l$ is the volume element for integration over the internal coordinate space of the l particles.

In equation (2.87), μ_A is the component of the molecular electric dipole moment along the A axis and it is given by

$$\mu_A = \sum_r C_r e A_r \qquad (2.88)$$

where $C_r e$ and A_r are the charge and A coordinate of the rth particle (nuclei or electron) in the molecule, with $A = X, Y$ or Z.

The integral that is squared and summed over in equation (2.87) is a component of the electric dipole transition moment; its square is a component of the electric dipole *transition probability*. In the case of degeneracies, that is if there is more than one eigenfunction Φ''_{rve} (or Φ'_{rve}) corresponding to the eigenvalue E'_{rve}(or E''_{rve}), we obtain the line strength by adding the individual transition probabilities for all transitions between the degenerate states; this is why there is the sum over Φ''_{rve} and Φ'_{rve} in equation (2.87). Notice that translation is completely removed here. That is because the translational energy of a molecule is unaffected by a weak radiation field; a weak radiation field can only change the internal state of a molecule.

Electromagnetic radiation consists of oscillating electric and magnetic fields, both of which contribute to its energy. Above we have discussed the intensity of resonantly absorbed electric field energy, and we have expressed this in terms of the electric dipole transition moment integral in equation (2.87). A molecule can also resonantly absorb magnetic field energy and this can be expressed in terms of a magnetic dipole transition moment integral. However, the line strength of a typical magnetic dipole transition is about 10^{-5} of a typical electric dipole transition and so we usually ignore it, just as we usually ignore the extremely weak contribution to the electric field absorption line strength from electric quadrupole absorption but if the electric dipole intensity is low for some reason these weak contributions to the line strength may have to be considered. In electron spin resonance spectroscopy and nuclear magnetic resonance spectroscopy, the absorption process involves changing electron or nuclear spin states for which the electric dipole transition moment is zero; they are magnetic dipole transitions. Similarly, the electric field energy absorbed in the infrared region of the spectrum by low density molecular hydrogen gas results from electric quadrupole absorption since there is no electric dipole absorption.

2.9 Matrices and matrix algebra

In this section we give a brief review of the most important definitions required when using matrices. It can be looked over cursorily on a first reading. A *matrix* is an array of numbers (called elements) arranged in rows and columns; for example

$$G = \begin{bmatrix} 2 & 4 \\ 3 & 5 \end{bmatrix} \tag{2.89}$$

is a matrix. The matrix in equation (2.89) has two rows and two columns; it is a *square matrix* but matrices are not necessarily square. An $n \times n$ square matrix (having n rows and n columns) is said to be n-dimensional. In a general matrix, A say, the element occurring at the intersection of the ith row and jth column is called A_{ij}. Thus, from equation (2.89), we have $G_{11} = 2$, $G_{12} = 4$, $G_{21} = 3$ and $G_{22} = 5$.

The *transpose* of a matrix A, say, is obtained by interchanging each element A_{ij} with the element A_{ji} and the matrix is written $\tilde{A}$. Thus, from equation (2.89),

$$\tilde{G} = \begin{bmatrix} 2 & 3 \\ 4 & 5 \end{bmatrix}. \tag{2.90}$$

If a matrix is equal to its transpose, then the matrix is said to be *symmetric*.

The *Hermitian conjugate* (or *conjugate transpose*) $A^\dagger$ of a matrix A is obtained by taking the complex conjugate of the transpose of the matrix. Thus,

$$A^\dagger = (\tilde{A})^* \tag{2.91}$$

and

$$(A^\dagger)_{ij} = A_{ji}^*. \tag{2.92}$$

A matrix that is equal to its Hermitian conjugate is *Hermitian*.

The sum of the diagonal elements of a square matrix is the *trace* of the matrix; the Greek letter chi (χ) is used for it. From equations (2.89) and (2.90), we have

$$\chi(G) = \chi(\tilde{G}) = 7. \tag{2.93}$$

The *determinant* of an $n \times n$ square matrix A is written as

$$
\det A = |A| = \begin{vmatrix}
A_{11} & A_{12} & A_{13} & \cdots & A_{1n} \\
A_{21} & A_{22} & A_{23} & \cdots & A_{2n} \\
A_{31} & A_{32} & A_{33} & \cdots & A_{3n} \\
\cdots & \cdots & \cdots & \cdots & \cdots \\
A_{n1} & A_{n2} & A_{n3} & \cdots & A_{nn}
\end{vmatrix}. \tag{2.94}
$$

The value of the determinant is given by the sum

$$
|A| = \sum (-1)^h A_{1r_1} A_{2r_2} A_{3r_3} \ldots A_{nr_n} \tag{2.95}
$$

where the summation is over all $n!$ possible permutations of the order of the r_i. The $(n!)/2$ terms in the sum involving an even permutation of the order of the column labels r_i from the standard numerical order $123\ldots n$ have h even [and are, hence, multiplied by $(-1)^h = +1$ in the sum] and the $(n!)/2$ terms involving an odd permutation have h odd [and are hence multiplied by $(-1)^h = -1$ in the sum]. An even (odd) permutation involves the product of an even (odd) number of pair interchanges. We give two examples to show how one uses this equation.

For the matrix G in equation (2.89), the determinant involves $n! = 2! = 2$ terms: $G_{11}G_{22}$ having no permutation of the order of the r_i so that $h = 0$, i.e. h even, and $G_{12}G_{21}$ having a single permutation of 1 with 2 so that $h = 1$, i.e. h odd. Thus, in the determinant sum $G_{11}G_{22} = 2 \times 5$ is preceded by $+1$ (since h is even) and $G_{12}G_{21} = 4 \times 3$ is preceded by -1 (since h is odd). We can thus write

$$
|G| = \begin{vmatrix} 2 & 4 \\ 3 & 5 \end{vmatrix} = (2 \times 5) - (4 \times 3) = -2. \tag{2.96}
$$

As a further example, the determinant of the three-dimensional matrix

$$
D = \begin{bmatrix} 1 & 2 & 3 \\ 4 & 5 & 6 \\ 7 & 8 & 9 \end{bmatrix} \tag{2.97}
$$

involves $n! = 3! = 6$ terms:

- $D_{11}D_{22}D_{33}$ having $h = 0$ (the r_i are in the standard order),
- $D_{11}D_{23}D_{32}$ having $h = 1$ (2 and 3 are exchanged),
- $D_{12}D_{21}D_{33}$ having $h = 1$ (1 and 2 are exchanged),
- $D_{13}D_{22}D_{31}$ having $h = 1$ (1 and 3 are exchanged),
- $D_{12}D_{23}D_{31}$ having $h = 2$ (1 and 2 are exchanged, and then 1 and 3), and
- $D_{13}D_{21}D_{32}$ having $h = 2$ (1 and 3 are exchanged, and then 1 and 2).

Three have h even (and their product is preceded by $+1$ in the determinant sum), and three have h odd (and their product is preceded by -1 in the determinant sum). Using this result, but writing out the sum in a way that shows how the determinant can be written by building it up from the determinants of 2×2 sub-matrices within the 3×3 matrix, we have

$$|D| = 1(5 \times 9 - 6 \times 8) - 2(4 \times 9 - 6 \times 7) + 3(4 \times 8 - 5 \times 7) = 0. \quad (2.98)$$

We set up electronic wavefunctions as *Slater determinants* in equation (3.28) and, in this application, the most significant property that follows from the definition of a determinant, given in equation (2.95), is that the determinant of a matrix will change sign if two rows are interchanged, or if two columns are interchanged. From this, it follows that the determinant of a matrix will vanish if two rows are identical or if two columns are identical.

The *product* of an $n \times m$ matrix A (having n rows and m columns) and an $m \times q$ matrix B in the order AB is an $n \times q$ matrix C where the ijth element of C is given by

$$C_{ij} = \sum_{k=1}^{m} A_{ik} B_{kj}. \quad (2.99)$$

For example, if the matrices A and B are

$$A = \begin{bmatrix} -\frac{1}{2} & \frac{\sqrt{3}}{2} \\ -\frac{\sqrt{3}}{2} & -\frac{1}{2} \end{bmatrix} \quad \text{and} \quad B = \begin{bmatrix} -\frac{1}{2} & -\frac{\sqrt{3}}{2} \\ \frac{\sqrt{3}}{2} & -\frac{1}{2} \end{bmatrix} \quad (2.100)$$

then equation (2.99) gives the $(1, 1)$ element of the product matrix C as

$$\begin{aligned} C_{11} &= A_{11} \times B_{11} + A_{12} \times B_{21} \\ &= (-1/2) \times (-1/2) + (\sqrt{3/2}) \times (\sqrt{3/2}) \\ &= 1. \end{aligned} \quad (2.101)$$

The $(1, 2)$ element of the product matrix C is given by

$$\begin{aligned} C_{12} &= A_{11} \times B_{12} + A_{12} \times B_{22} \\ &= (-1/2) \times (-\sqrt{3/2}) + (\sqrt{3/2}) \times (-1/2) \\ &= 0. \end{aligned} \quad (2.102)$$

We can use equation (2.99) to similarly determine that $C_{21} = 0$ and that $C_{22} = 1$. Thus, we can write out the product matrix C in full as

$$C = \begin{bmatrix} -\frac{1}{2} & \frac{\sqrt{3}}{2} \\ -\frac{\sqrt{3}}{2} & -\frac{1}{2} \end{bmatrix} \begin{bmatrix} -\frac{1}{2} & -\frac{\sqrt{3}}{2} \\ \frac{\sqrt{3}}{2} & -\frac{1}{2} \end{bmatrix} = \begin{bmatrix} 1 & 0 \\ 0 & 1 \end{bmatrix}. \quad (2.103)$$

From equation (2.99), we see that for the multiplication between A and B to be possible the matrices A and B must be *conformable*, i.e. the number of columns in A must be equal to the number of rows in B. This means that, for example, A and B can both be m-dimensional square matrices and their product C will be an

m-dimensional square matrix. A can be a square matrix with dimension m, B can be a single column matrix with length m and their product will be a column matrix with length m. A can be a single row matrix of length m, B can be a square matrix of dimension m and their product will be a single row matrix of length m. A can be a single row matrix of length m, B can be a be a single column matrix with length m and their product will be a single number. Other possibilities are obtained by choosing particular values for n and q in equation (2.99). Matrix multiplication, like quantum mechanical operator multiplication, is not necessarily commutative.

An n-dimensional square matrix E that has unity in all diagonal positions and zero in all off-diagonal positions is called an n-dimensional *unit matrix*. The matrix C in equation (2.103) is a two-dimensional unit matrix. It is customary to use the letter E for a unit matrix. If we multiply an n-dimensional square matrix A by the n-dimensional unit matrix E, the result is A. That is the matrix E plays the role in matrix multiplication that unity plays in the ordinary algebraic multiplication of numbers. Square matrices having all off-diagonal elements equal to zero are said to be *diagonal*, and the unit matrix is a special case of a diagonal matrix in which all diagonal elements are unity.

If the product of two n-dimensional square matrices A and B is the n-dimensional unit matrix E, i.e. if

$$AB = E \qquad\qquad (2.104)$$

then we say that one matrix is the inverse (or reciprocal) of the other, and we write

$$A^{-1} = B \qquad \text{or} \qquad B^{-1} = A. \qquad\qquad (2.105)$$

The matrices A and B in equation (2.100) are the inverse of each other; their product is the two-dimensional unit matrix from equation (2.103). Only square matrices can have a unique inverse and efficient computer routines exist for finding matrix inverses. However, the inverse of a matrix will not exist if the determinant of the matrix is zero; a matrix having a determinant that is zero is said to be *singular*. The matrix D in equation (2.97) is singular from equation (2.98). If the inverse of a matrix is equal to the transpose of the matrix, then the matrix is *orthogonal*; the matrices A and B in equation (2.100) are orthogonal. If the inverse of a matrix is equal to the Hermitian conjugate of the matrix, then the matrix is *unitary*.

2.10   Problems

2.1   Using equation (2.1), together with the fact that $\exp(2in\pi) = 1$ only if n is a positive or negative integer, determine the eigenvalues E_n and eigenfunctions $\psi_n(\alpha)$ of the operator $\mathrm{d}/\mathrm{d}\alpha$ for the situation where α is an angle, which means that the eigenfunctions are restricted to satisfy $\psi_n(\alpha + 2\pi) = \psi_n(\alpha)$. Determine also the eigenfunctions and eigenvalues of the operator $-\mathrm{i}\,\mathrm{d}/\mathrm{d}\alpha$. Which of the operators $\mathrm{d}/\mathrm{d}\alpha$ and $-\mathrm{i}\,\mathrm{d}/\mathrm{d}\alpha$ is Hermitian?

2.2 Are the eigenfunctions in problem 2.1 also eigenfunctions of the operator $\mathrm{d}^2/\mathrm{d}\alpha^2$?

2.3 A particle of mass M moving on a circular path of radius R has angular coordinate α. The energy of the particle is $P^2/2M$ and its angular momentum is $J = PR$. By expressing the energy in terms of the angular momentum and then substituting the quantum mechanical operator $\hat{J} = -\mathrm{i}\hbar\mathrm{d}/\mathrm{d}\alpha$, determine the Hamiltonian, the Schrödinger equation and the quantized energies. The product MR^2 that scales the energy level spacings is called the *moment of inertia* of an orbiting particle. Determine the energy level spacings for orbiting atoms having various values of M and R.

2.4 Show how equations (2.38)–(2.40) are obtained from equations (2.34) and (2.37).

2.5 Use equations (2.34)–(2.37) for the 2×2 perturbation problem to determine S, c^+ and c^- for the situation with a constant off-diagonal matrix element H_{12} of $10\,\mathrm{cm}^{-1}$ but with the zero-order level separation Δ being 0, 1, 10, 100 and $1000\,\mathrm{cm}^{-1}$, respectively. Compare the results with the approximate values obtained using equations (2.38)–(2.40).

2.6 Prove that the ij element of the n-dimensional square matrix D that is the product of three n-dimensional square matrices A, B and C in the order ABC is given by

$$D_{ij} = \sum_{k=1}^{n}\sum_{l=1}^{n} A_{ik} B_{kl} C_{lj}. \qquad (2.106)$$

2.7* Prove that if any n-dimensional square matrix R say, is premultiplied by the non-singular n-dimensional square matrix Q, and postmultiplied by the inverse matrix Q^{-1}, then the character of the resultant matrix $S = QRQ^{-1}$ is the same as that of the matrix R, i.e. prove

$$\chi(S) = \chi(QRQ^{-1}) = \chi(R) = \sum_{i=1}^{n} R_{ii}. \qquad (2.107)$$

2.8 Evaluate the determinants of the matrices A and B given in equation (2.100).

Chapter 3

Electronic states

From equation (2.75) for $\hat{H}_{rve}$, we see that the rovibronic Schrödinger equation (2.76) does not involve molecular parameters, such as bond lengths and angles, and that the only quantities occurring are the masses and charges of the l particles (nuclei and electrons) that make up the molecule. Thus, we can easily set up the Schrödinger equation for any molecule. One might think that we could then simply use numerical methods to solve it. However, even using the most efficient numerical methods, current computers do not have enough power for this to be possible with the required precision except for three- and four-particle systems such as H_2^+ and H_2. This will change as computer power increases.

> For most molecules, to solve the rovibronic Schrödinger equation accurately, we are forced to make approximations and then to correct for them as best we can. The approximations introduce concepts that allow us to *understand* molecules.

Such concepts as electronic state, molecular orbital, electronic configuration, potential energy surface, equilibrium structure, force constant, electronic and vibrational angular momentum and Coriolis coupling constant come about because of approximations that are introduced. However, these concepts are only satisfactory and useful if the approximations that lead to their introduction are reasonably valid.

3.1 The Born–Oppenheimer approximation

To solve the rovibronic Schrödinger equation, we change coordinates so that it separates into simpler Schrödinger equations. Approximations have to be made but by choosing appropriate coordinates the approximations are minimized. The first separation we make is that of the electronic motion from the nuclear motion. To do this we change coordinates in $\hat{H}_{rve}$, as given in equation (2.75), from $(X_2, Y_2, Z_2, \ldots, Z_l)$ to $(\xi_2, \eta_2, \zeta_2, \ldots, \zeta_l)$, where the $\xi\eta\zeta$ axis system is parallel to the XYZ axis system but has origin at the nuclear centre of mass rather than the molecular centre of mass. This choice of origin allows us to refer the motion of the electrons to the positions of the nuclei and it gives a kinetic energy expression that is completely separable into an electronic kinetic energy $\hat{T}_e$ and a nuclear kinetic energy $\hat{T}_N$. In these new coordinates, the rovibronic Hamiltonian can be written as

$$\hat{H}_{rve} = \hat{T}_e + \hat{T}_N + V_{ee} + V_{NN} + V_{Ne} \tag{3.1}$$

and the rovibronic Schrödinger equation is

$$[\hat{T}_e + \hat{T}_N + V_{ee} + V_{NN} + V_{Ne}]\Phi_{rve} = E_{rve}\Phi_{rve} \tag{3.2}$$

where the electrostatic potential energy given in equation (2.75) has been written as $(V_{ee} + V_{NN} + V_{Ne})$. V_{ee} is the sum of all the electron–electron electrostatic repulsions and it only involves the coordinates of the electrons. V_{NN} is the sum of all the nuclear–nuclear electrostatic repulsions and it only involves the coordinates of the nuclei. V_{Ne} is the sum of all the electron–nuclear electrostatic attractions and it involves the coordinates of the nuclei and electrons.

Although the kinetic energy is completely separable into electronic and nuclear parts in these coordinates, the potential energy is not because of the presence of the electron–nuclear attraction term V_{Ne}. This part of V is the glue that holds the molecule together and we cannot just neglect it. We cannot chose coordinates in such a way that the potential function separates into two non-interacting parts where one part just involves the nuclear coordinates and the other the electron coordinates and we cannot follow the simple separation of variables procedure that we used to separate translational and the internal degrees of freedom in section 2.5.

All is not lost, however. We saw at the end of section 2.5.1 that

because electrons are so much lighter than nuclei, the first excited electronic state is at a much higher energy than the first excited vibrational state. Knowledge of this leads us to treat the rovibronic Schrödinger equation in a special way, by using the Born–Oppenheimer approximation, to separate it into nuclear and electronic parts.

To understand this approximation, it is helpful to view a molecule as having a nuclear framework that rotates and vibrates, while at the same time the electron cloud is continually modifying its shape so as to conform to the instantaneous nuclear geometry. It would be more appropriate to think of the electronic wavefunction as varying with the nuclear coordinates and, from it, one can calculate the electronic probability distribution. Using this idea, the electronic wavefunction is obtained in the Born–Oppenheimer approximation by solving the rovibronic Schrödinger equation with the nuclei fixed at an appropriate geometry and with only the electronic coordinates as variables. This means that in equation (3.2), we put $T_N = 0$ (the nuclear kinetic energy is zero because the nuclei are held fixed), and neglect V_{NN} (since we are only concerned with the electron dynamics) to give

$$[\hat{T}_e + V_{ee} + V_{Ne}]\Phi_{elec,n} = \hat{H}_{elec}\Phi_{elec,n} = V_{elec,n}\Phi_{elec,n} \qquad (3.3)$$

where a particular nuclear geometry is chosen in V_{Ne}, and n labels the successive *electronic states* $(n = 1, 2, ...)$. This equation is the electronic Schrödinger equation, and it is solved at many different nuclear geometries to yield the electronic wavefunctions $\Phi_{elec,n}$ (which are functions of the electronic coordinates) and energies $V_{elec,n}$; $\Phi_{elec,n}$ and $V_{elec,n}$ are each a parametric function of nuclear geometry.

The calculation of the electronic wavefunction as described above has been achieved by holding the nuclei fixed. To calculate the energies for the nuclear motion, we must allow them to move under the constraint of the electrostatic nuclear–nuclear repulsion potential energy term V_{NN} in equation (3.1) but we must also include the constraint imposed by the fact that the electronic energy V_{elec} depends on the nuclear geometry. The need to include this extra constraint is easy to appreciate. Suppose the nuclei move from one geometry to another in which V_{NN} is higher and that simultaneously V_{elec} is also higher; in this case, the nuclei have to work against the combined energy of $(V_{NN} + V_{elec})$ and this function (which depends on the nuclear geometry) provides the *potential energy surface* for the nuclear motion. For each electronic state n, there will be a different potential energy surface $(V_{NN} + V_{elec,n})$ and a different nuclear motion (rotation–vibration) Schrödinger equation given by

$$[\hat{T}_N + V_{NN} + V_{elec,n}]\Phi_{rv,nj} = E^0_{rve,nj}\Phi_{rv,nj} \qquad (3.4)$$

where $j(= 1, 2, ...)$ labels the rotation–vibration states in the same way that n labels the electronic states from equation (3.3).

In equation (3.4) $E^0_{rve,nj}$ is the rovibronic energy for the jth rotation–vibration level in the nth electronic state, within the Born–Oppenheimer approximation. We rewrite the equation so that the zero of energy in each electronic state is the minimum value of $(V_{NN} + V_{elec,n})$, which we call the *electronic energy* $E_{elec,n}$ of the electronic state n[1] and we obtain the rotation–vibration Schrödinger equation as

$$[\hat{T}_N + V_{N,n}]\Phi_{rv,nj} = \hat{H}_{rv}\Phi_{rv,nj} = E_{rv,nj}\Phi_{rv,nj} \qquad (3.5)$$

---

[1] The electronic term value in cm^{-1} is called $T_e(n)$.

where

$$V_{N,n} = V_{NN} + V_{elec,n} - E_{elec,n} \qquad (3.6)$$

and

$$E_{rv,nj} = E^0_{rve,nj} - E_{elec,n}. \qquad (3.7)$$

To solve the electronic and rotation–vibration Schrödinger equations, we refer the electrons and nuclei to molecule-fixed xyz axes in order to separate rotation. These axes have origin at the nuclear centre of mass, like the $\xi\eta\zeta$ axes, but they are attached to the molecule so that they rotate with it (see section 4.1).

In making a summary of this, it is helpful to represent the nuclear coordinates as $\boldsymbol{R}_N$ and the electronic coordinates as $\boldsymbol{r}_{elec}$. In the Born–Oppenheimer approximation, the rovibronic eigenfunctions are the products

$$\Phi^0_{rve,nj}(\boldsymbol{R}_N, \boldsymbol{r}_{elec}) = \Phi_{elec,n}(\boldsymbol{R}_N, \boldsymbol{r}_{elec})\Phi_{rv,nj}(\boldsymbol{R}_N) \qquad (3.8)$$

and the rovibronic eigenvalues are the sum

$$E^0_{rve,nj} = E_{elec,n} + E_{rv,nj}. \qquad (3.9)$$

We have reduced the problem of solving the $(3l - 3)$-dimensional rovibronic Schrödinger equation (3.2), to one of solving two differential equations: Equation (3.3), which is the $3(l-N)$-dimensional electronic Schrödinger equation for $\Phi_{elec,n}(\boldsymbol{R}_N, \boldsymbol{r}_{elec})$ and $V_{elec,n}(\boldsymbol{R}_N)$ (and $E_{elec,n}$), and equation (3.5), which is the $(3N-3)$-dimensional rotation–vibration Schrödinger equation for $\Phi_{rv,nj}(\boldsymbol{R}_N)$ and $E_{rv,nj}$. Fortunately, it is usually the case that only the very lowest eigenstates of equation (3.3) are of interest and efficient variational methods have been developed to obtain them even for molecules having many electrons (see section 3.3).

The Born–Oppenheimer approximation introduces the concepts of electronic state and electronic potential energy surface $V_{N,n}(\boldsymbol{R}_N)$. Potential energy surfaces are independent of isotopic substitution because the nuclear masses do not enter[2] equation (3.3). The nuclear geometry at the minimum of the potential energy surface of an electronic state is the *equilibrium geometry* of that state; it is the geometry at which the nuclei would naturally come to rest if they moved classically on the surface and it is the geometry at which $V_{N,n}(\boldsymbol{R}_N) = 0$. The *structure* of a molecule is its structure at the equilibrium configuration of its ground electronic state within the Born–Oppenheimer approximation.

The exact rovibronic wavefunctions Φ_{rve} are (by definition) eigenfunctions of the rovibronic Hamiltonian $\hat{H}_{rve}$ given in equation (3.2) but the functions Φ^0_{rve} are not since it can be shown that

$$\hat{H}_{rve}\Phi^0_{rve,nj} = E^0_{rve,n}\Phi^0_{rve,nj} + H' \qquad (3.10)$$

[2] Within the Born–Oppenheimer approximation, a small nuclear-mass dependent electron kinetic energy term $\hat{T}'_e$ is neglected.

where $H' = 0$ only if

$$\hat{T}_N \Phi_{\text{elec},n}(\boldsymbol{R}_N, \boldsymbol{r}_{\text{elec}}) \Phi_{\text{rv},nj}(\boldsymbol{R}_N) = \Phi_{\text{elec},n}(\boldsymbol{R}_N, \boldsymbol{r}_{\text{elec}}) \hat{T}_N \Phi_{\text{rv},nj}(\boldsymbol{R}_N) \quad (3.11)$$

that is only if $\hat{T}_N$ and $\Phi_{\text{elec},n}$ commute. Put another way, $H' = 0$ only if the effect of $\hat{T}_N$ acting on all the $\Phi_{\text{elec},n}$ is neglected. However, this is not the whole story and the energy separation between electronic states also enters.

The separation in energy between the zero-order Born–Oppenheimer states $\Phi_{\text{elec},n} \Phi_{\text{rv},nj}$ and $\Phi_{\text{elec},m} \Phi_{\text{rv},mk}$ is

$$\Delta(nj; mk) = (E_{\text{elec},m} + E_{\text{rv},mk}) - (E_{\text{elec},n} + E_{\text{rv},nj}). \quad (3.12)$$

The off-diagonal matrix element of the rovibronic Hamiltonian between these states is

$$H(nj; mk) = \int \Phi^*_{\text{elec},n} \Phi^*_{\text{rv},nj} \hat{T}_N \Phi_{\text{elec},m} \Phi_{\text{rv},mk} \, d\tau \quad (3.13)$$

since the electronic off-diagonal matrix elements of $(T_e + V_{ee} + V_{Ne})$ and of V_{NN} vanish at all nuclear geometries. From equations (2.31)–(2.40), we see that the extent to which these two levels perturb each other depends on the ratio $H(nj; mk)/\Delta(nj; mk)$. If this ratio is small, there will be little mixing of the states, i.e. little breakdown of the Born–Oppenheimer approximation.

For almost all molecules, the excited electronic states are at energies well above the ground electronic state. Thus, the Born–Oppenheimer approximation is almost invariably a good approximation for the levels of the ground electronic state. A significant breakdown of the Born–Oppenheimer approximation often occurs in excited electronic states if there are other excited electronic states nearby in energy.

3.2   Spin and the Pauli exclusion principle

Electrons and most nuclei have an intrinsic angular momentum called *spin angular momentum*. The electron spin angular momentum operator for one electron is written $\hat{s}$ and the nuclear spin angular momentum operator for one nucleus is written $\hat{i}$; these operators have space-fixed components $\hat{s}_Z$ and $\hat{i}_Z$. Expressions for these operators are not obtained from any classical angular momentum expressions and, in non-relativistic quantum theory, the existence of spin would be another postulate. The theory of relativity is essential for understanding the origin of spin and we will not go into that. Suffice it to say that an electron has a spin angular momentum and that the eigenvalue of its square $\hat{s}^2$ is $s(s + 1)\hbar^2$, where the spin angular momentum quantum number (or spin, for short) $s = 1/2$. For a nucleus, the eigenvalue of the square of the spin angular momentum $\hat{i}^2$ is $I(I + 1)\hbar$ where I is the spin of the nucleus. The eigenvalue of $\hat{s}_Z$ is $s_Z\hbar$, where $s_Z = \pm 1/2$, and the eigenvalue of $\hat{i}_Z$ is $I_Z\hbar$, where $I_Z = -I, -I + 1, \ldots, I$. For the elements from H through Ne in the periodic table, their nuclei have spin I as follows:

$I = 0$ for ^{4}He, ^{12}C, ^{16}O, ^{18}O, ^{20}Ne and ^{22}Ne nuclei
$I = 1/2$ for ^{1}H, ^{3}He, ^{13}C, ^{15}N and ^{19}F nuclei
$I = 1$ for ^{2}H ($= $D), ^{6}Li and ^{14}N nuclei
$I = 3/2$ for ^{7}Li, ^{9}Be, ^{11}B and ^{21}Ne nuclei
$I = 5/2$ for ^{17}O nuclei and
$I = 3$ for ^{10}B nuclei.

There are no stable nuclei having a spin I of 2.

Electrons and nuclei with non-zero spin have an associated magnetic dipole moment; each nucleus having spin $I > 1/2$ also has an associated electric quadrupole moment. These moments have effects on atomic and molecular energies and it was the observation of these effects that first lead to the experimental inference of the presence of 'spin' by Goudsmit and Uhlenbeck 1925[3]. The interactions that arise from these moments contribute with the electrostatic interaction V of equation (2.45) to give the full electromagnetic interaction energy between the particles in a molecule. We have written these spin terms as $\hat{H}_{es} + \hat{H}_{hfs}$ in equation (2.77). The expressions for these two spin terms in the Hamiltonian involve the values of the spin magnetic dipole moments of the particles and the values of any nuclear electric quadrupole moments, together with the spin angular momentum operators.

The fundamental fact is that the spin of an electron or nucleus is not something that can be understood classically and there is no spin coordinate in the classical sense. One must not be mislead by the name 'spin' in thinking of an electron or nucleus as spinning to generate a magnetic dipole moment.

Spin labels are introduced by writing the complete internal wavefunction (i.e. for everything except translation) as $\Phi_{int}(\mathbf{R}_N, \mathbf{r}_{elec}, \sigma_1, \ldots, \sigma_l)$, where each σ_i can take one of the $2s + 1$, or $2I + 1$, discrete values of the projection of the spin of the particle onto the Z direction. Electrons have a spin s of $1/2$, and protons have a spin $I = 1/2$. Thus, for electrons or protons, the projection σ can be $+1/2$ (called the α spin state) or $-1/2$ (called the β spin state). In general, for a nucleus with spin I the value of σ can be any one of the $2I + 1$ values $-I$, $-I + 1, \ldots, +I$. What we actually have here is a set of functions corresponding to all possible values of $\sigma_1, \ldots, \sigma_l$; this constitutes a set of *spin components* of the wavefunction. For example, for a one-electron system [where $\mathbf{r}$ gives the xyz coordinates of the electron] the spin component electronic wavefunctions would be $\Phi(\mathbf{r}, \alpha)$ and $\Phi(\mathbf{r}, \beta)$ as σ is α or β, respectively; the complete electronic wavefunction would be written $\Phi(\mathbf{r}, \sigma)$.

---

[3] Goudsmit G and Uhlenbeck S 1925 *Naturwissenschaften* **13** 953.

The fifth postulate of quantum mechanics (stated fully in section 9.1) requires the introduction of spin; when applied to electrons, it gives rise to the *Pauli exclusion principle*.

The Pauli exclusion principle states that the complete wavefunction Φ for a molecule (*including spin*) is changed in sign if the space coordinates and spin of any two electrons in it are interchanged.

So if we only exchange the pair of electrons labelled i and j in a molecule we must have

$$\Phi(\ldots, x_i, y_i, z_i, \sigma_i, \ldots, x_j, y_j, z_j, \sigma_j, \ldots)$$
$$= -\Phi(\ldots, x_j, y_j, z_j, \sigma_j, \ldots, x_i, y_i, z_i, \sigma_i, \ldots). \tag{3.14}$$

The electronic wavefunction (with the inclusion of spin) is *antisymmetric* (i.e. multiplied by -1) with respect to the exchange of a pair of electrons.

3.3   Electronic wavefunctions and energies

From equation (3.3), after changing to molecule-fixed xyz coordinates, the electronic Schrödinger equation is

$$\hat{H}_{\text{elec}} \Phi_{\text{elec}} = V_{\text{elec}} \Phi_{\text{elec}} \tag{3.15}$$

where

$$\hat{H}_{\text{elec}} = -\frac{\hbar^2}{2m_e} \sum_i \nabla_i^2 + \sum_{i<j} \frac{e^2}{4\pi\epsilon_0 R_{ij}} - \sum_{\alpha,i} \frac{C_\alpha e^2}{4\pi\epsilon_0 R_{i\alpha}} \tag{3.16}$$

and

$$\nabla_i^2 = \frac{\partial^2}{\partial x_i^2} + \frac{\partial^2}{\partial y_i^2} + \frac{\partial^2}{\partial z_i^2}. \tag{3.17}$$

In equation (3.16), i and j run over the $n = (l - N)$ electrons and α runs over the N nuclei. We initially make the drastic approximation of neglecting the term that describes the electrostatic repulsions between the electrons:

$$V_{\text{ee}} = \sum_{i<j} \frac{e^2}{4\pi\epsilon_0 R_{ij}}. \tag{3.18}$$

This reduces $\hat{H}_{\text{elec}}$ to the very approximate (va) electronic Hamiltonian:

$$\hat{H}_{\text{elec}}^{\text{va}} = \sum_i \left\{ -\frac{\hbar^2}{2m_e} \nabla_i^2 - \sum_\alpha \frac{C_\alpha e^2}{4\pi\epsilon_0 R_{i\alpha}} \right\} = \sum_i \hat{h}_i. \tag{3.19}$$

With this approximation, $\hat{H}_{\text{elec}}^{\text{va}}$ is separable into the sum of n identical Hamiltonians, each of which is the Hamiltonian for one electron (having coordinates $r = xyz$) moving in the field of the N nuclei, with the nuclei held fixed at a particular molecular geometry. The one-electron Schrödinger equation is

$$\hat{h}\phi_k^0(r) = \varepsilon_k^0\phi_k^0(r) \tag{3.20}$$

where, for notational convenience, we omit the dependence on nuclear coordinates in the expressions for $\phi_k^0(r)$ and ε_k^0. The eigenfunctions and eigenvalues of $\hat{H}_{\text{elec}}^{\text{va}}$ are

$$\Phi_{\text{elec}}^{\text{va}} = \phi_a^0(r_1)\phi_b^0(r_2)\ldots\phi_\lambda^0(r_n) \tag{3.21}$$

and

$$V_{\text{elec}}^{\text{va}} = \varepsilon_a^0 + \varepsilon_b^0 + \cdots + \varepsilon_\lambda^0. \tag{3.22}$$

From each of the normalized spatial orbitals $\phi_k^0(r)$, we can form two normalized spin-orbitals

$$\chi_k^0(r, \sigma) = \phi_k^0(r)\,|s, \sigma\rangle \tag{3.23}$$

where $|s, \sigma\rangle$ is an electrons spin function, $s = 1/2$ and $\sigma = \alpha$ or β.

3.3.1 The Slater determinant

For a two-electron molecule, such as H_2, HeH^+ or H_3^+, we could write an electronic wavefunction as the *Hartree product*

$$\Phi_1^{\text{HP}} = \chi_i^0(r_1, \sigma_1)\chi_j^0(r_2, \sigma_2). \tag{3.24}$$

But, Φ_1^{HP} is not an acceptable wavefunction since if we exchange the electron coordinates in it we do not obtain $-\Phi_1^{\text{HP}}$, as we should if it satisfied the Pauli exclusion principle and were antisymmetric; instead we get

$$\Phi_2^{\text{HP}} = \chi_i^0(r_2, \sigma_2)\chi_j^0(r_1, \sigma_1). \tag{3.25}$$

However, from Φ_1^{HP} and Φ_2^{HP}, we can construct a properly antisymmetric two-electron function as (where the $2^{-1/2}$ assures normalization):

$$\Phi^{(2)} = 2^{-1/2}[\chi_i^0(r_1, \sigma_1)\chi_j^0(r_2, \sigma_2) - \chi_i^0(r_2, \sigma_2)\chi_j^0(r_1, \sigma_1)]. \tag{3.26}$$

We can use determinant notation [see equation (2.95)] to rewrite it as

$$\Phi^{(2)} = \frac{1}{\sqrt{2}}\begin{vmatrix} \chi_i^0(r_1, \sigma_1) & \chi_j^0(r_1, \sigma_1) \\ \chi_i^0(r_2, \sigma_2) & \chi_j^0(r_2, \sigma_2) \end{vmatrix}. \tag{3.27}$$

The ground electronic state of the H_2 molecule is described by such a determinant as this where the two spin-orbitals have the same spatial orbitals and one has an

α spin component while the other has a β spin component, to give a closed-shell state with paired electron spins. Both electrons could not have the same spin in such a situation since then the wavefunction would be symmetric with respect to their exchange and this is not allowed by the Pauli exclusion principle. Such a state is 'excluded' from existence and another way of stating the Pauli exclusion principle for electrons is to say that no two can be in the same state.

Equation (3.27) can be generalized to produce an n-electron wavefunction as the following *Slater determinant* sum of $n!$ terms:

$$\Phi^{(n)} = \frac{1}{\sqrt{n!}} \begin{vmatrix} \chi_i^0(r_1,\sigma_1) & \chi_j^0(r_1,\sigma_1) & \cdots & \chi_k^0(r_1,\sigma_1) \\ \chi_i^0(r_2,\sigma_2) & \chi_j^0(r_2,\sigma_2) & \cdots & \chi_k^0(r_2,\sigma_2) \\ \vdots & \vdots & \vdots & \vdots \\ \chi_i^0(r_n,\sigma_n) & \chi_j^0(r_n,\sigma_n) & \cdots & \chi_k^0(r_n,\sigma_n) \end{vmatrix}. \tag{3.28}$$

The interchange of two rows of the determinant in equation (3.28) is equivalent to the exchange of the spatial coordinates and spin of two of the electrons. From the properties of determinants [see equation (2.95)], this will cause the function to change its sign. Also, if two electrons occupied the same spin-orbital, it would mean that two columns of the determinant would be equal and the determinant would vanish.

> The Slater determinant expression for an n-electron wavefunction treats all the electrons equivalently and it guarantees that the Pauli exclusion principle is satisfied. It is the sum of $n!$ terms.

Equation (3.28) is a Slater determinant wavefunction for n electrons occupying the n spin-orbitals $(\chi_i^0, \chi_j^0, \ldots, \chi_k^0)$. In a shorthand notation, we write it as

$$\Phi^{(n)} = |\chi_i^0, \chi_j^0, \ldots, \chi_k^0\rangle. \tag{3.29}$$

3.3.2 The Hartree–Fock approximation and molecular orbitals

In the previous section, the electron–electron repulsion term V_{ee} in $\hat{H}_{elec}$ is completely neglected. This term causes *electron correlation* since it prevents the electrons from moving independently of each other; as a result, the motion of each electron depends on (or is *correlated* with) the motions of all the other electrons in the molecule. There is a way of partly including V_{ee} while still maintaining an orbital-product description of the wavefunction. To explain how this is achieved, we focus on the determination of the electronic ground state wavefunction and energy. In this approach, we partly include V_{ee} by making the *Hartree–Fock* approximation. The ground-state wavefunction is expanded as a Slater determinant in Hartree–Fock spin-orbitals χ, where these are eigenfunctions of the effective one-electron Hamiltonian (called the Fock operator):

$$\hat{h}_i^F = -\frac{\hbar^2}{2m_e}\nabla_i^2 - \sum_\alpha \frac{C_\alpha e^2}{4\pi\epsilon_0 R_{i\alpha}} + V_i^{HF}. \tag{3.30}$$

The function V_i^{HF} is the average potential experienced by electron i in the field of the other $(n-1)$ electrons and this potential depends on the spin-orbitals of the other electrons, i.e. on the eigenfunctions of $\hat{h}_i^F$.

Thus, the Hartree–Fock equation

$$\hat{h}^F \chi_\lambda(r,\sigma) = \varepsilon_\lambda \chi_\lambda(r,\sigma) \tag{3.31}$$

must be solved iteratively. This procedure is called the *self-consistent field approximation* (SCF) method. The $\chi_\lambda(r,\sigma)$ are *molecular orbitals* (MOs).

The SCF procedure involves making an initial guess for K orthogonal spin-orbitals (where $K > n$) and calculating V_i^{HF} for each electron. The Hartree–Fock equation is then solved to obtain new spin-orbitals. The new spin-orbitals are used to calculate new V_i^{HF} and the Hartree–Fock equation is solved again. This procedure is repeated until the V_i^{HF} do not change significantly; this indicates that self-consistency has been achieved.

The n spin-orbitals with the lowest energies obtained in this procedure are called *occupied* spin-orbitals and the Slater determinant formed from them is the Hartree–Fock ground-state wavefunction $\Psi_{elec,gs}^{HF}$. The ground-state electronic energy is obtained as the expectation value

$$V_{elec,gs}^{HF} = \langle \Psi_{elec,gs}^{HF} | \hat{H}_{elec} | \Psi_{elec,gs}^{HF} \rangle \tag{3.32}$$

and it is not given as the sum of the orbital energies ε_λ from the Hartree–Fock equation (3.31). The Hartree–Fock procedure gives K spin-orbitals, where the upper $(K - n)$ orbitals are called *unoccupied*, or *virtual*, orbitals. The larger K is the more complete is the set of basis functions and the lower and better will be the expectation value $V_{elec,gs}^{HF}$ by the variational theorem. Thus, calculations are made with the largest practicable value for K.

Most electronic ground states are closed-shell with no unpaired spin and a restricted Hartree–Fock calculation is made in which the α and β spin states are constrained to have the same spatial orbitals. In this very common situation, the Slater determinant (in the shorthand notation) can be written as

$$\Phi_{elec,gs}^{RHF} = |\chi_1, \bar{\chi}_1, \chi_2, \bar{\chi}_2, \ldots, \chi_{n/2}, \bar{\chi}_{n/2}\rangle \tag{3.33}$$

where χ_i has α spin and $\bar{\chi}_i$ has β spin.

3.3.3 MOs as linear combinations of atomic orbitals

In a Hartree–Fock calculation, the usual MO trial functions are linear combinations of atomic orbitals (LCAO), where the atomic orbitals are centred on the nuclei in the molecule. At each nucleus the AOs could be taken as the wavefunctions of the one-electron atom, i.e. the eigenfunctions of the Hamiltonian

$$\hat{h}^{(Z)} = -\frac{\hbar^2}{2m_e}\left(\frac{\partial^2}{\partial x^2} + \frac{\partial^2}{\partial y^2} + \frac{\partial^2}{\partial z^2}\right) - \frac{Z\,e^2}{4\pi\epsilon_0\sqrt{x^2+y^2+z^2}} \tag{3.34}$$

which describes an electron at the position $r = xyz$ interacting with a nucleus of charge Ze at the origin of the xyz axis system. The eigenvalues of $\hat{h}^{(Z)}$ are

$$E_n = -\frac{Z^2 m_e e^4}{2n^2(4\pi\epsilon_0)^2\hbar^2} \qquad n = 1, 2, 3, \ldots \tag{3.35}$$

Each eigenfunction of $\hat{h}^{(Z)}$ is specified by the value of three quantum numbers n (which determines the energy E_n), l and m, where l takes the n values $0, 1, \ldots (n-1)$, and m takes the $2l + 1$ values $l, l - 1, \ldots, -l$. As a result, the level E_n has a degeneracy given by $\sum_{l=0}^{(n-1)}(2l + 1) = n^2$. States with $l = 0, 1, 2$ and 3 are called s, p, d and f states, respectively. The single state with energy E_1 is called the 1s state and the four states with energy E_2 are called the 2s, $2p_x$, $2p_y$ and $2p_z$ states.

The eigenfunctions of $\hat{h}^{(Z)}$, called *Slater type orbitals* (STOs), are normally not used as the AOs. Instead, for reasons of computational convenience, it is customary to use *Gaussian type orbitals* (GTOs) which have the same angular dependence and the same rotational symmetry properties. For example, the form of a Gaussian 1s orbital is

$$\phi(r) = (2\zeta/\pi)^{3/4}e^{-\zeta r^2} \tag{3.36}$$

where ζ is adjustable. The exponent in a GTO involves r^2 whereas the exponent in each eigenfunction of $\hat{h}^{(Z)}$ involves r. Using the GTOs $\phi_\mu(r)$, we write trial spatial MO functions as the linear combinations

$$\psi_i = \sum_{\mu=1}^{K} C_{\mu i}\phi_\mu(r) \tag{3.37}$$

and the trial spin-orbitals would be $\psi_1, \bar{\psi}_1, \psi_2, \bar{\psi}_2, \ldots, \psi_K, \bar{\psi}_K$. The $C_{\mu i}$ are adjusted in the SCF procedure to minimize (and improve) $V_{\text{elec,gs}}^{\text{HF}}$ and this gives the optimum MOs.

3.3.4 Configuration interaction

Electron correlation, i.e. the effect of the electron–electron repulsion term V_{ee}, can be more completely allowed for than it is in a Hartree–Fock calculation. The

Hartree–Fock ground-state single-determinant wavefunction is written in its most general form in terms of the n occupied spin-orbitals as

$$\Phi_{\text{elec,gs}}^{\text{HF}} = |\chi_1, \chi_2, \ldots, \chi_a, \chi_b, \ldots, \chi_n\rangle. \tag{3.38}$$

In the calculation, $2K$ spin-orbitals are used and we form a singly excited determinant by promoting an electron from the occupied orbital χ_a to the unoccupied orbital χ_r:

$$\Phi_a^r = |\chi_1, \chi_2, \ldots, \chi_r, \chi_b, \ldots, \chi_n\rangle. \tag{3.39}$$

A doubly excited determinant would be

$$\Phi_{ab}^{rs} = |\chi_1, \chi_2, \ldots, \chi_r, \chi_s, \ldots, \chi_n\rangle \tag{3.40}$$

and we can define triply, quadruply, etc excited determinants. Each of these determinants Ψ is defined by specifying the n orbitals from which it is formed and such a set of MOs is called an *electron configuration*. A better approximation to the ground-state energy than $V_{\text{elec,gs}}^{\text{HF}}$ is obtained by diagonalizing $\hat{H}_{\text{elec}}$ in the functions $\Phi_{\text{elec,gs}}^{\text{HF}}, \Phi_a^r, \Phi_{ab}^{rs}$, which we truncate at some systematically chosen point, such as after all singly and doubly excited determinants. It is the electron correlation term V_{ee} that gives rise to the non-vanishing off-diagonal matrix elements of $\hat{H}_{\text{elec}}$ in this calculation, which is called a *configuration interaction* (CI) calculation. It leads to a ground-state wavefunction $\Phi_{\text{elec,gs}}^{\text{CI}}$ that is a linear combination of $\Phi_{\text{elec,gs}}^{\text{HF}}$ and the excited functions Φ_a^r, Φ_{ab}^{rs}, etc used in the calculation. The ground-state potential energy surface, from equation (3.6), is given by

$$V_{\text{N,gs}}^{\text{CI}} = V_{\text{NN}} + V_{\text{elec,gs}}^{\text{CI}} - E_{\text{elec,gs}}^{\text{CI}} \tag{3.41}$$

where $V_{\text{elec,gs}}^{\text{CI}}$ is the CI electronic energy and $E_{\text{elec,gs}}^{\text{CI}}$ is the minimum value of $V_{\text{NN}} + V_{\text{elec,gs}}^{\text{CI}}$. In giving the result of such a so-called *ab initio* calculation of a potential energy surface, it is necessary to quote the nature and number of the AO basis set functions used and to describe the configurations included in the CI calculation.

3.4 Molecular orbital theory

We can get a qualitative understanding of the electronic structure of a molecule by considering the trial LCAO-MOs in equation (3.37). The AOs are shown in figure 3.1. The 2s orbital has spherical symmetry and the $2p_x$ orbital has cylindrical symmetry with a *node* at its centre; at a node, an orbital has the value zero and it changes sign. The sign of the $2p_x$ orbital is chosen so that it has positive values for $x > 0$ and negative values for $x < 0$; the zy plane is a *nodal plane*. The $2p_y$ and $2p_z$ orbitals are rotated versions of the $2p_x$ orbital.

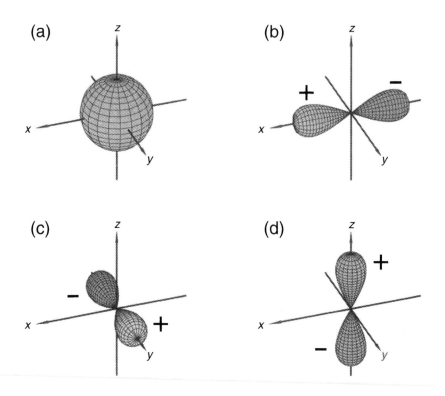

Figure 3.1. The $n = 2$ atomic orbitals: (a) the 2s orbital, (b) the $2p_x$ orbital, (c) the $2p_y$ orbital and (d) the $2p_z$ orbital.

3.4.1 Bonding and antibonding orbitals

To introduce the central concept of *bonding* and *antibonding* MOs, we consider the hydrogen molecule and, in the basis of the AOs $1s(H_1)$ and $1s(H_2)$, we diagonalize the one-electron Hamiltonian [see equation (3.19)]

$$\hat{h} = -\frac{\hbar^2}{2m_e}\left(\frac{\partial^2}{\partial x^2} + \frac{\partial^2}{\partial y^2} + \frac{\partial^2}{\partial z^2}\right) - \frac{e^2}{4\pi\epsilon_0 R_{e1}} - \frac{e^2}{4\pi\epsilon_0 R_{e2}} \qquad (3.42)$$

where xyz are the coordinates of the electron and R_{e1} and R_{e2} are the distances of the electron from the protons 1 and 2, respectively. In figure 3.2, we show two hydrogen atoms, at a separation of R_{HH}, with their 1s AOs. The coordinates of an electron in an axis system with axes parallel to the xyz axis but with origin at proton 1 are given by

$$(x', y', z') = \left(x - \frac{R_{HH}}{2}, y, z\right). \qquad (3.43)$$

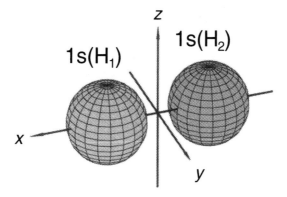

Figure 3.2. Two hydrogen atoms, H_1 and H_2, separated by R_{HH} and located symmetrically about the origin on the x axis; each has a 1s orbital centred on it.

The chain rule gives

$$\frac{\partial}{\partial x'} = \frac{\partial x}{\partial x'}\frac{\partial}{\partial x} = \frac{\partial}{\partial x} \tag{3.44}$$

so we can write

$$\hat{h} = -\frac{\hbar^2}{2m_e}\left(\frac{\partial^2}{\partial(x')^2} + \frac{\partial^2}{\partial(y')^2} + \frac{\partial^2}{\partial(z')^2}\right) - \frac{e^2}{4\pi\epsilon_0 R_{e1}} - \frac{e^2}{4\pi\epsilon_0 R_{e2}}. \tag{3.45}$$

The first two terms here are the Hamiltonian of a hydrogen atom consisting of proton 1 and an electron and, since $1s(H_1)$ is a hydrogen atom eigenfunction with energy

$$E_1 = -\frac{m_e e^4}{2(4\pi\epsilon_0)^2\hbar^2} < 0 \tag{3.46}$$

from equation (3.35), we have

$$\hat{h}[1s(H_1)] = E_1[1s(H_1)] - \frac{e^2}{4\pi\epsilon_0 R_{e2}}[1s(H_1)]. \tag{3.47}$$

Consequently, the off-diagonal matrix element

$$\langle 1s(H_2)|\hat{h}|1s(H_1)\rangle = E_1 \int 1s(H_2)^*1s(H_1)\,dx'\,dy'\,dz'$$

$$- \frac{e^2}{4\pi\epsilon_0}\int 1s(H_2)^*\frac{1}{R_{e2}}1s(H_1)\,dx'\,dy'\,dz'. \tag{3.48}$$

Since the AOs $1s(H_1)$ and $1s(H_2)$ are real and positive, the off-diagonal matrix element is real and negative and we write

$$Q = -\langle 1s(H_2)|\hat{h}|1s(H_1)\rangle > 0. \tag{3.49}$$

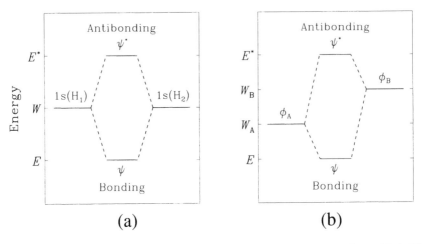

Figure 3.3. (a) The energies of the bonding orbital ψ and the antibonding orbital ψ^* relative to the energy of the two AOs $1s(H_1)$ and $1s(H_2)$ and (b) the formation of a bonding and an antibonding orbital by interaction of two non-equivalent AOs ϕ_A and ϕ_B.

The two functions $1s(H_1)$ and $1s(H_2)$ are equivalent, and we define

$$W = \langle 1s(H_1)|\hat{h}|1s(H_1)\rangle = \langle 1s(H_2)|\hat{h}|1s(H_2)\rangle. \tag{3.50}$$

Thus, the Hermitian matrix of $\hat{h}$ in the chosen basis is

$$\begin{bmatrix} W & -Q \\ -Q & W \end{bmatrix}. \tag{3.51}$$

Using equations (2.30)–(2.37) the MOs are

$$\psi = \frac{1}{\sqrt{2}}[1s(H_1) + 1s(H_2)] \quad \text{and} \quad \psi^* = \frac{1}{\sqrt{2}}[1s(H_1) - 1s(H_2)] \tag{3.52}$$

with energies

$$E = W - Q \quad \text{and} \quad E^* = W + Q \tag{3.53}$$

respectively, where $E < E^*$ since $Q > 0$.

With both electrons in the orbital ψ (one with α spin and the other with β spin), the energy is $2(W - Q)$ which is lower than the electronic energy $2W$ of two non-interacting hydrogen atoms. In the orbital ψ, the two $1s$ orbitals add together in the region of space between the two nuclei and form a bond. The MO ψ is a *bonding orbital*. In contrast, the MO ψ^* has energy $2(W + Q)$ and it has a node midway between the nuclei. Electrons occupying this orbital have a lower probability of being found in the region of space between the two nuclei and there is no bonding. The MO ψ^* is an *antibonding orbital*.

In figure 3.3(a), we represent the energetics of the isolated AOs, the bonding MO and the antibonding MO. Bonding and antibonding orbitals can also form by interaction of AOs that are not equivalent. Two such orbitals ϕ_A and ϕ_B give rise to a matrix

$$\begin{bmatrix} W_A & -Q \\ -Q & W_B \end{bmatrix} \tag{3.54}$$

analogous to that in equation (3.51) but now $W_A \neq W_B$. The two AOs ϕ_A and ϕ_B give rise to a bonding and an antibonding orbital; the energies of these orbitals are

$$E = \tfrac{1}{2}(W_A + W_B) - \tfrac{1}{2}\sqrt{4Q^2 + (W_A - W_B)^2} \tag{3.55}$$

and

$$E^* = \tfrac{1}{2}(W_A + W_B) + \tfrac{1}{2}\sqrt{4Q^2 + (W_A - W_B)^2} \tag{3.56}$$

from equations (2.31)–(2.34). As shown in figure 3.3(b), $E < W_A (< W_B$ in our example), so also in the case of non-equivalent AOs interacting, the formation of the bonding orbital leads to an energy lowering.

3.4.2 Hybridization

Rather than using simple AOs, a better understanding of the electronic structure of a molecule is often obtained if we use *hybrid orbitals* and we illustrate this using hybrid orbitals for the carbon atom. Carbon 2s, $2p_x$, $2p_y$ and $2p_z$ AOs are degenerate, from equation (3.35), and any linear combination has the same energy. Using matrix multiplication notation, we form four orthonormal symmetrical linear combinations

$$\begin{bmatrix} \phi_1^{(sp^3)} \\ \phi_2^{(sp^3)} \\ \phi_3^{(sp^3)} \\ \phi_4^{(sp^3)} \end{bmatrix} = \frac{1}{2} \begin{bmatrix} 1 & 1 & 1 & 1 \\ 1 & -1 & 1 & -1 \\ 1 & 1 & -1 & -1 \\ 1 & -1 & -1 & 1 \end{bmatrix} \begin{bmatrix} 2s \\ 2p_x \\ 2p_y \\ 2p_z \end{bmatrix} \tag{3.57}$$

which are known as sp^3 *hybrid orbitals*; all four are represented in figure 3.4(a). Each has a large lobe with a positive value and a much smaller lobe with a negative value; these smaller lobes are just visible for $\phi_3^{(sp^3)}$ and $\phi_4^{(sp^3)}$ in figure 3.4(a). Each hybrid orbital has cylindrical symmetry, and the entire structure of orbitals shown in figure 3.4(a) has tetrahedral symmetry; the 'outward tips' of the four positive lobes are at the vertices of a tetrahedron (see figure 6.7).

The sp^3 hybrid orbitals form bonding and antibonding orbitals with AOs centred on other atoms. For example, a bond can be formed by interaction with the 1s orbital of a hydrogen atom as indicated in figure 3.4(b). Thus, the electronic structure of the CH_4 molecule is described as having two electrons in the C(1s) AO and two electrons in each of the four bonding orbitals formed from a carbon atom sp^3 hybrid orbital and a hydrogen 1s AO.

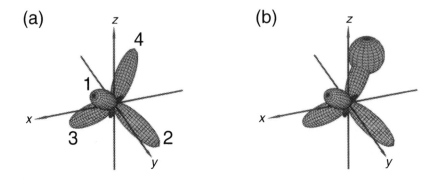

Figure 3.4. (a) sp^3 hybrid orbitals $\phi_j^{(\mathrm{sp}^3)}$ ($j = 1, 2, 3, 4$) from equation (3.57) centred on a carbon nucleus. (b) The sp^3 hybrid orbital $\phi_4^{(\mathrm{sp}^3)}$ interacting with the 1s AO on a hydrogen atom.

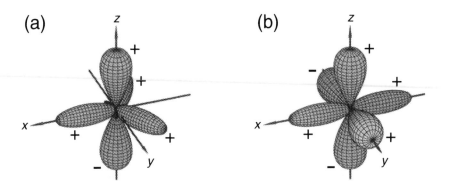

Figure 3.5. (a) The $2p_z$ AO and the three sp^2 hybrid orbitals centred on a carbon nucleus. (b) The $2p_y$ and $2p_z$ AOs and the two sp$_x$ hybrid orbitals centred on a carbon nucleus.

The water molecule can also be described using the concept of sp^3 hybrid orbitals. The H$_2$O molecule has 10 electrons and a bond angle of about 105°. This is close to the tetrahedral angle of 109.5°. The electronic structure is described by saying that there are two electrons in the O(1s) AO and two in each of two bonding orbitals formed from an oxygen atom sp^3 hybrid orbital and a hydrogen 1s AO. The other two oxygen atom sp^3 orbitals are each then said to contain *lone pairs* of electrons that do not participate in the bonding.

Other useful basis sets for carbon atoms are sp^2 and sp hybrid orbitals. sp^2

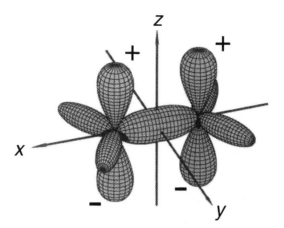

Figure 3.6. The electronic orbital structure of the two carbon nuclei in an ethylene molecule C_2H_4. The signs of the p_z AOs are indicated.

hybrids are formed as linear combinations of 2s, $2p_x$ and $2p_y$ AOs using

$$\begin{bmatrix} \phi_1^{(sp^2)} \\ \phi_2^{(sp^2)} \\ \phi_3^{(sp^2)} \end{bmatrix} = \begin{bmatrix} \frac{1}{\sqrt{3}} & \frac{2}{\sqrt{6}} & 0 \\ \frac{1}{\sqrt{3}} & -\frac{1}{\sqrt{6}} & \frac{1}{\sqrt{2}} \\ \frac{1}{\sqrt{3}} & -\frac{1}{\sqrt{6}} & -\frac{1}{\sqrt{2}} \end{bmatrix} \begin{bmatrix} 2s \\ 2p_x \\ 2p_y \end{bmatrix}. \tag{3.58}$$

The sp^2 basis functions are pictured in figure 3.5(a), where the unhybridized $2p_z$ AO is as given in figure 3.1(d). Two sp hybrid orbitals are formed as the linear combinations

$$\begin{bmatrix} \phi_1^{(sp)} \\ \phi_2^{(sp)} \end{bmatrix} = \frac{1}{\sqrt{2}} \begin{bmatrix} 1 & 1 \\ 1 & -1 \end{bmatrix} \begin{bmatrix} 2s \\ 2p_x \end{bmatrix}. \tag{3.59}$$

The sp basis functions are pictured in figure 3.5(b), where the unhybridized $2p_y$ and $2p_z$ AOs are as given in figure 3.1(c)–(d).

To understand the electronic structure of the ethylene molecule, we visualize two sp^2-hybridized carbon atoms as shown in figure 3.6. The sp^2-hybrid orbitals with lobes along the x axis interact to form bonding and antibonding orbitals; two electrons occupy the bonding orbital to form a 'σ' bond. The situation in figure 3.6, where the two p_z orbitals are parallel to each other allows the p_z orbitals to form a bonding 'π' orbital and an antibonding π^* orbital:

$$\pi = \frac{1}{\sqrt{2}}[2p_z(1) + 2p_z(2)] \quad \text{and} \quad \pi^* = \frac{1}{\sqrt{2}}[2p_z(1) - 2p_z(2)]. \tag{3.60}$$

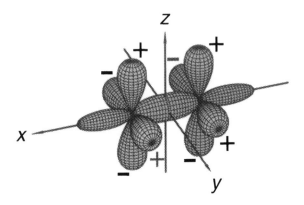

Figure 3.7. The electronic orbital structure of the two carbon nuclei in an acetylene molecule C_2H_2. The signs of the p_y and p_z AOs are indicated.

In the ground electronic state, two electrons occupy the bonding π orbital and there is a σ and π *double bond*. The sp^2 hybrid orbitals all lie in the xy plane and each of them form bonding and antibonding orbitals with the 1s orbital on a hydrogen atom. This rationalizes why ethylene is planar in its electronic ground state. In the planar geometry, the π bond can form and so this geometry is energetically favourable.

For an acetylene molecule, two sp-hybridized carbon atoms form a σ bond as shown in figure 3.7. Lining up the AOs as shown in the figure, the two p_y orbitals form a bonding MO π_y and an antibonding MO π_y^* and the two p_z orbitals form analogous MOs π_z and π_z^*. With two electrons in each of the two bonding orbitals π_y and π_z, two π bonds are formed and since the σ bond is also present, there is a *triple bond*. The carbon atoms each have two 'unused' hybrid orbitals, one directed in the positive x direction and one directed in the negative x direction, and by letting each of these orbitals bond with the 1s orbital on a hydrogen atom, we obtain the acetylene molecule HCCH. This rationalizes why acetylene is linear.

As discussed in section 3.4.1, bond formation lowers the energy. The energy lowering from the formation of a σ bond is larger than that from the formation of a π bond because the AOs that give rise to a σ bond interact more than those that give rise to a π bond. The π MOs are the highest occupied MOs (HOMOs) and the electrons in these orbitals are the ones that predominantly determine the properties of the molecule. These are the 'outer' electrons by means of which the molecule interacts with the external world. More scientifically speaking, the lowest excited electronic states of the molecule are obtained by promoting electrons in the HOMOs to the lowest unoccupied MOs (LUMOs), and the energetically most favourable reactions involving the molecule (i.e. the reactions with the lowest activation energies) are those brought about by a rearrangement of the π orbitals.

3.4.3   The Hückel approximation and benzene

So far, we have considered the situation in which two AOs, centred on two neighbouring nuclei in a molecule, interact to form bonding and antibonding MOs. Sometimes it is necessary to consider the simultaneous interaction between more than two AOs centred on more than two nuclei; in this case, we speak about the *delocalization* of the electrons. An approximate method used in this situation is the Hückel approximation which we describe for the benzene and buta-1,3-diene molecules.

If we take six sp^2-hybridized carbon atoms as shown in figure 3.5(a) and arrange them in a regular hexagon so that, for each carbon atom, the xy plane in figure 3.5(a) lies in the plane of the hexagon, then for each carbon atom, two of the sp^2 hybrid orbitals interact with the analogous hybrid orbitals on the neighbouring carbon atoms to form σ bonds. Each carbon atom has an 'unused' hybrid orbital with the positive lobe pointing away from the centre of the hexagon and this orbital can form a σ bond with the 1s orbital of a hydrogen atom. As a result, we obtain the σ-*bond skeleton* of benzene C_6H_6 as shown in figure 3.8. Figure 3.8 also shows the molecule-fixed axis system xyz employed for benzene.

In addition to the sp^2 hybrid orbitals in the xy plane, each carbon atom in the benzene molecule also has centred on it the $2p_z$ orbital shown in figure 3.5(a). We denote the $2p_z$ orbital centred on carbon atom $\mu(= 1, 2, \ldots, 6)$ as ϕ_μ and these π AOs are shown in figure 3.9. Each of the ϕ_μ orbitals in the hexagon must interact to the same extent with each of its two neighbours and we have to consider all six ϕ_μ simultaneously when we construct MOs.

In the Hückel approximation for benzene, we construct the 6×6 matrix of the Hamiltonian $\hat{h}$, just like the 2×2 matrix in equation (3.51), in the basis of the AOs ϕ_μ, where $\mu = 1, 2, \ldots, 6$. The following approximations are made:

- We consider only diagonal matrix elements $\langle\phi_\mu|\hat{h}|\phi_\mu\rangle$ and nearest-neighbour off-diagonal matrix elements $\langle\phi_\mu|\hat{h}|\phi_\nu\rangle$, where ϕ_μ and ϕ_ν are centred on neighbouring nuclei.
- The overlap integrals between the AOs are neglected and we set

$$\langle\phi_\mu|\phi_\nu\rangle = \delta_{\mu\nu} \tag{3.61}$$

where $\delta_{\mu\nu}$ is a Kronecker delta.

In the case of benzene, all carbon atoms are equivalent and all interactions between neighbouring carbon atoms are equivalent, so with the customary definitions

$$\alpha = \langle\phi_1|\hat{h}|\phi_1\rangle \quad \text{and} \quad \beta = \langle\phi_1|\hat{h}|\phi_2\rangle \tag{3.62}$$

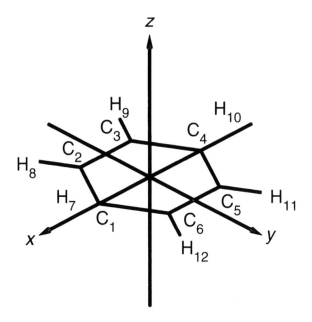

Figure 3.8. The σ-bond skeleton of the benzene molecule C_6H_6 and the molecule-fixed *xyz* axis system.

we obtain for the matrix of $\hat{h}$, the Hückel matrix

$$\mathbf{H}_\pi = \begin{bmatrix} \alpha & \beta & 0 & 0 & 0 & \beta \\ \beta & \alpha & \beta & 0 & 0 & 0 \\ 0 & \beta & \alpha & \beta & 0 & 0 \\ 0 & 0 & \beta & \alpha & \beta & 0 \\ 0 & 0 & 0 & \beta & \alpha & \beta \\ \beta & 0 & 0 & 0 & \beta & \alpha \end{bmatrix}. \tag{3.63}$$

It can be shown that when we choose the ϕ_μ orbitals such that they are all positive above the plane of the benzene molecule and all negative below this plane, then we have $\beta < 0$, in the same way that $-Q < 0$ in equation (3.51). The diagonalization of the Hückel matrix is greatly facilitated if we use symmetry and this is done in section 10.2.

3.4.4 Polyene chain molecules

For the polyene chain molecule $H_2C(CH)_{n-2}CH_2$, there is also electron delocalization and we can use the Hückel approximation to obtain general expressions for the π MOs and the associated orbital energies. An example of such a molecule is the buta-1,3-diene molecule $CH_2(CH)_2CH_2$ in its *cis*-planar

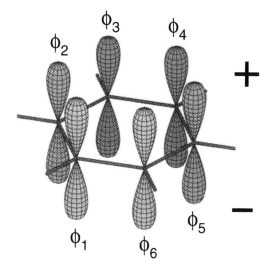

Figure 3.9. The six π orbitals ϕ_μ ($\mu = 1, 2, \ldots, 6$) centred on the carbon nuclei in benzene C_6H_6; each is a carbon atom $2p_z$ AO. All ϕ_μ have positive values above the molecular plane and negative values below this plane.

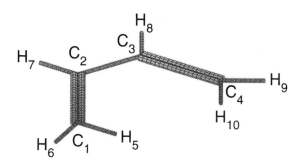

Figure 3.10. The buta-1,3-diene molecule $CH_2(CH)_2CH_2$ in its *cis*-planar configuration. Thin cylinders represent single bonds and double cylinders represent localized double bonds.

configuration, which we show with localized double bonds in figure 3.10. As in the case of benzene, to allow for π electron delocalization, we use as basis functions for the π MOs of a general n-carbon-atom polyene chain molecule one p AO ϕ_μ($\mu = 1, 2, 3, 4, \ldots, n$) centred on each carbon atom. We see from figure 3.10 that the four carbon atoms in buta-1,3-diene are not all equivalent. If, in $H_2C(CH)_{n-2}CH_2$, we label the carbon atom at one end as 1, the next one as 2 and so on until we label the last one as n, then carbon atoms 1 and n are equivalent, 2 and $n-1$ are equivalent, 3 and $n-2$ are equivalent and so on. Even though the carbon atoms are not all equivalent, we make approximations analogous to those made for benzene: All diagonal matrix elements $\langle \phi_\mu | \hat{h} | \phi_\mu \rangle$ are given a common value α and all off-diagonal matrix elements $\langle \phi_\mu | \hat{h} | \phi_{\mu+1} \rangle$ or $\langle \phi_{\mu+1} | \hat{h} | \phi_\mu \rangle$ connecting π orbitals centred on neighbouring nuclei are given a common value β. Thus, the $n \times n$ Hückel matrix for $H_2C(CH)_{n-2}CH_2$ is

$$\begin{bmatrix} \alpha & \beta & 0 & 0 & 0 & \ldots & 0 & 0 \\ \beta & \alpha & \beta & 0 & 0 & \ldots & 0 & 0 \\ 0 & \beta & \alpha & \beta & 0 & \ldots & 0 & 0 \\ \vdots & \vdots & \vdots & \vdots & \vdots & \ldots & \alpha & \beta \\ 0 & 0 & 0 & 0 & 0 & \ldots & \beta & \alpha \end{bmatrix}. \tag{3.64}$$

The 6×6 version of this matrix differs from that in equation (3.63). The eigenvalues of this $n \times n$ matrix can be determined analytically[4] as

$$E_j = \alpha + 2\cos\left(\frac{j\pi}{n+1}\right)\beta \tag{3.65}$$

where $j = 1, 2, 3, 4, \ldots, n$. The MO eigenfunction associated with E_j is

$$\psi_j = \sqrt{\frac{2}{n+1}} \sum_{\mu=1}^{n} \sin\left(\frac{\mu j \pi}{n+1}\right) \phi_\mu. \tag{3.66}$$

The four p AOs ϕ_μ ($\mu = 1, 2, 3, 4$) considered for buta-1,3-diene are shown in figure 3.11. For $n = 4$, we obtain from equation (3.65) the energies

$$(E_1, E_2, E_3, E_4) = (\alpha + 1.618\beta, \alpha + 0.618\beta, \alpha - 0.618\beta, \alpha - 1.618\beta) \tag{3.67}$$

where, since $\beta < 0$, $E_1 < E_2 < E_3 < E_4$, with the associated orbitals from equation (3.66)

$$\psi_1 = 0.37\phi_1 + 0.60\phi_2 + 0.60\phi_3 + 0.37\phi_4 \tag{3.68}$$
$$\psi_2 = 0.60\phi_1 + 0.37\phi_2 - 0.37\phi_3 - 0.60\phi_4 \tag{3.69}$$
$$\psi_3 = 0.60\phi_1 - 0.37\phi_2 - 0.37\phi_3 + 0.60\phi_4 \tag{3.70}$$

and

$$\psi_4 = 0.37\phi_1 - 0.60\phi_2 + 0.60\phi_3 - 0.37\phi_4. \tag{3.71}$$

We will discuss these orbitals further in section 10.3, where we analyse their symmetry properties and, in section 10.4, where we employ them for determining the path of the butadiene-cyclobutene conversion reaction.

[4] Coulson C A and Longuet-Higgins H C 1947 *Proc. R. Soc.* A **192** 16.

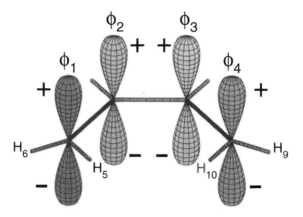

Figure 3.11. The four p AOs ϕ_μ centred on the carbon nuclei C_μ ($\mu = 1, 2, 3, 4$) in the buta-1,3-diene molecule in its *cis*-planar configuration. For clarity only, the protons 5, 6, 9, and 10 are labelled; the other labels can be inferred from figure 3.10.

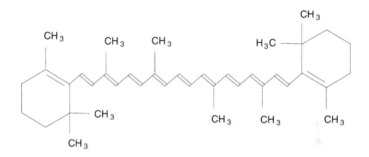

Figure 3.12. The molecule β-carotene $C_{40}H_{56}$.

3.5 Problems

3.1 In *ab initio* calculations, atomic units are used. The atomic unit of length is the *Bohr radius* $a_0 = (4\pi\epsilon_0\hbar^2)/(m_e e^2)$ and the atomic unit of energy is the *Hartree energy* $E_h = e^2/(4\pi\epsilon_0 a_0)$. Calculate a_0 in units of m, E_h in units of J, $E_h/(hc)$ in units of cm^{-1} and E_h/h in units of Hz; $m_e = 9.109\,382\,6 \times 10^{-31}$ kg.

3.2 Express the energies of the one-electron atom [equation (3.35)] in units of E_h.

3.3 Transform the electronic Hamiltonian in equation (3.16) to a form where the energy is expressed in units of E_h and all lengths in units of a_0. Show that this form of the Hamiltonian does not formally depend on the values of the fundamental constants e, m_e, $\hbar$, etc.

3.4   Verify that the four Hückel orbitals ψ_1, ψ_2, ψ_3 and ψ_4 given for the buta-1,3-diene molecule in equations (3.68)–(3.71) are normalized and orthogonal. Show that these functions are MO eigenfunctions in the Hückel approximation by deriving the matrix elements $\langle \psi_j | \hat{h} | \psi_k \rangle$ ($j, k = 1, 2, 3, 4$), in terms of the Hückel matrix elements $\langle \phi_\mu | \hat{h} | \phi_\nu \rangle$ ($\mu, \nu = 1, 2, 3, 4$) in the 4×4 version of equation (3.64).

3.5   Calculate the Hückel energies and orbitals for prop-2-enyl (allyl), i.e. the polyene chain molecule $H_2C(CH)CH_2$. How many nodes does each of the orbitals have? Indicate the positions of the nodes. Which orbital is bonding and which is antibonding?

3.6   In a crude model for the π-electron system of β-carotene (figure 3.12), we assume that the 22 π electrons can move freely along the molecular chain, taken to have length L. In this model the orbital energies are those of a particle in a one-dimensional box [the corresponding expression for a three-dimensional box is given in equation (2.72)] $E_n = h^2 n^2 / (8 m_e L^2)$, $n = 1, 2, 3, \ldots$. The 22 π electrons are distributed in the lowest 11 energy levels according to the Pauli exclusion principle. The electronic transition from the 11th to the 12th level is observed to occur at the wavelength 451 nm. Calculate the value of L.

3.7   The eigenfunction $\psi_{100}(r, \theta, \varphi)$ for the one-electron atom [i.e. the wavefunction corresponding to the energy E_1 given in equation (3.35) for $n = 1$] with nuclear charge number Z is given by

$$\psi_{100}(r, \theta, \varphi) = \frac{1}{\sqrt{\pi}} \left(\frac{Z}{a_0} \right)^{3/2} \exp\left(-\frac{Zr}{a_0} \right).$$

The polar coordinates r, θ and φ define the position of the electron in an xyz axis system with the nucleus at the origin: $(x, y, z) = (r \sin \theta \cos \varphi, r \sin \theta \sin \varphi, r \cos \theta)$ and the volume element $d\tau = dx\, dy\, dz = r^2\, dr\, \sin \theta\, d\theta\, d\varphi$.

(a) Show that $\psi_{100}(r, \theta, \varphi)$ is normalized, i.e. that

$$\int_0^\infty r^2\, dr \int_0^{2\pi} d\varphi \int_0^\pi \sin \theta\, d\theta\, |\psi_{100}(r, \theta, \varphi)|^2 = 1.$$

(b) Calculate the average value of r,

$$\langle r \rangle = \int_0^\infty r^3\, dr \int_0^{2\pi} d\varphi \int_0^\pi \sin \theta\, d\theta\, |\psi_{100}(r, \theta, \varphi)|^2$$

for the hydrogen atom and for He^+.

(c) Calculate the most probable value of r, i.e. the position of the maximum of the radial probability density function

$$P(r) = r^2 \int_0^{2\pi} d\varphi \int_0^\pi \sin \theta\, d\theta\, |\psi_{100}(r, \theta, \varphi)|^2$$

for the hydrogen atom and for He$^+$.

(d) Calculate the probability of finding the electron at a distance from the nucleus larger than the Bohr radius a_0.

(e) Sketch $P(r)$ as a function of r and interpret the previous results in terms of the sketch.

Use the following expressions:

$$\int u^2 e^{bu}\, du = \frac{u^2 e^{bu}}{b} - \frac{2e^{bu}}{b^3}(bu - 1)$$

$$\int_0^\infty u^n e^{-cu}\, du = \frac{n!}{c^{n+1}}.$$

Chapter 4

Vibrational states

4.1 Space-fixed and molecule-fixed axes

In section 2.4, the classical expression for the energy of a molecule consisting of l particles, nuclei and electrons, was written down using XYZ axes. The XYZ axes have both origin and orientation fixed in space. In section 2.5, the translational energy was separated by introducing XYZ axes parallel to the XYZ axes but with their origin at the molecular centre of mass. The XYZ axes thus have space-fixed orientation but a molecule-fixed origin. After separating out the translation, the internal (rovibronic) energy is expressed using the (X_r, Y_r, Z_r) coordinates of the particles but now there are three *constraints* on these coordinates [see equation (2.51)]:

$$\sum_{r=1}^{l} m_r X_r = \sum_{r=1}^{l} m_r Y_r = \sum_{r=1}^{l} m_r Z_r = 0. \tag{4.1}$$

These constraints arise because the molecular centre of mass is fixed within the XYZ axis system to be at its origin.

To separate the electronic energy within the Born–Oppenheimer approximation, we introduced new axes, the $\xi\eta\zeta$ axes, before equation (3.1). The $\xi\eta\zeta$ axes have the same space-fixed orientation as the XYZ and XYZ axes, but their origin is at the centre of mass of the N nuclei in the molecule. From equation (3.5), the classical expression for the rotation–vibration energy in the $\xi\eta\zeta$ axis system is

$$E_{\mathrm{rv}} = \tfrac{1}{2} \sum_{i=1}^{N} m_i (\dot{\xi}_i^2 + \dot{\eta}_i^2 + \dot{\zeta}_i^2) + V_{\mathrm{N},n}(\xi_i, \eta_i, \zeta_i) \tag{4.2}$$

where the potential energy $V_{\mathrm{N},n}$ is defined in equation (3.6)[1]. Since the $\xi\eta\zeta$ axes have their origin at the nuclear centre of mass, the nuclear coordinates in

---

[1] From here on in this chapter, we drop the extra subscript n on $V_{\mathrm{N},n}$ that specifies the electronic state.

equation (4.2) are subject to the three translational constraints

$$\sum_{i=1}^{N} m_i \xi_i = \sum_{i=1}^{N} m_i \eta_i = \sum_{i=1}^{N} m_i \zeta_i = 0. \tag{4.3}$$

It is possible to separate the translational energy completely and exactly by referring the nuclei and electrons to the translating XYZ axes.

It is not possible to separate completely the electronic energy by introducing the $\xi \eta \zeta$ axes and we cannot separate completely the rotational energy either. We do the best we can by referring the nuclei (and electrons) to rotating *molecule-fixed xyz* axes, with their origin at the nuclear centre of mass like the $\xi \eta \zeta$ axes but that are fixed to the molecule and rotate with it.

When the molecule is in its equilibrium configuration, the coordinates of nucleus i in the xyz axis system are written (x_i^e, y_i^e, z_i^e) and, at a displaced configuration, the Cartesian *vibrational displacement coordinates* $\Delta \alpha_i$ are given by

$$\Delta x_i = (x_i - x_i^e) \qquad \Delta y_i = (y_i - y_i^e) \qquad \text{and} \qquad \Delta z_i = (z_i - z_i^e). \tag{4.4}$$

As a result of the fact that the $\xi \eta \zeta$ axes translate with the molecule, there are three translational constraint equations for the $3N$ coordinates ξ_i, η_i and ζ_i and these are given in equation (4.3). Within the xyz axis system, there are three similar translational constraint equations that tie the origin to the nuclear centre of mass. However, there are now also three rotational constraint equations that tie the xyz axes so that they rotate with the molecule. The translational and rotational constraint equations are formulated in terms of the $3N$ vibrational displacement coordinates $\Delta \alpha_i$ by requiring that the translational normal coordinates T_α, given in equation (4.11), and rotational normal coordinates, where R_x is given in equation (4.12), vanish. It can be shown that the explicit form of the rotational constraint equations obtained by setting $R_x = R_y = R_z = 0$, where the R_α are as defined here, minimize the terms in the rotation–vibration Hamiltonian that spoil the separation of rotation.

From here on we will use atomic masses rather than nuclear masses in the equations used to set up the rotation–vibration Hamiltonian in order to allow for the fact that the electrons move with the nuclei. This is an effect of the breakdown of the Born–Oppenheimer approximation.

4.2   The vibrational Hamiltonian

The classical expression for the vibrational energy, using molecule-fixed xyz coordinates, is

$$E_{vib} = \tfrac{1}{2} \sum_{i=1}^{3N} m_i \dot{u}_i^2 + V_N(u_i) \tag{4.5}$$

where the $3N$ u_i are defined as $(u_1, \ldots, u_{3N}) = (\Delta x_1, \ldots, \Delta z_N)$ and the six constraints $T_\alpha = R_\alpha = 0$ apply with $\alpha = x$, y or z [see equations (4.11) and (4.12)]. Since V_N, and its first derivative with respect to any u_i, are zero at equilibrium, the Taylor's series expansion about equilibrium is

$$V_N = \frac{1}{2} \sum_{i,j=1}^{3N} k_{ij} u_i u_j + \frac{1}{6} \sum_{i,j,k=1}^{3N} k_{ijk} u_i u_j u_k + \frac{1}{24} \sum_{i,j,k,l=1}^{3N} k_{ijkl} u_i u_j u_k u_l + \cdots \tag{4.6}$$

where the k_{ij}, k_{ijk} and k_{ijkl} are constants (the *force* constants). The lowest order terms in the expansion are quadratic and, for small displacements, it is a satisfactory approximation to express the potential by the quadratic terms alone, V_N^0; this is the *harmonic-oscillator approximation*. The higher-order terms are cubic, quartic, etc *anharmonicity* terms. In the harmonic-oscillator approximation, the vibrational energy is

$$E_{vib}^0 = \tfrac{1}{2} \sum_{i=1}^{3N} m_i \dot{u}_i^2 + \tfrac{1}{2} \sum_{i,j=1}^{3N} k_{ij} u_i u_j \tag{4.7}$$

where the k_{ij} are the harmonic (quadratic) force constants.

A standard result from classical mechanics is that the vibrational energy of a nonlinear N-body harmonic oscillator can be written in terms of $(3N - 6)$ mass-weighted linear combinations of the u_i, called vibrational *normal coordinates* Q_r, in such a way that the vibrational energy becomes

$$E_{vib}^0 = \tfrac{1}{2} \sum_{r=1}^{3N-6} [\dot{Q}_r^2 + \lambda_r Q_r^2] \tag{4.8}$$

with no cross terms in the kinetic or potential energy expressions[2]. The constants λ_r depend on the masses and the harmonic force constants. In writing this equation, we have removed the six linear combinations of the u_i that correspond to the three translational and the three rotational normal coordinate T_α and R_α that are constrained to be zero in the molecule-fixed xyz axis system. Each of the $(3N - 6)$ vibrational normal coordinates Q_r describes a collective *normal mode*

---

[2] Linear molecules, treated in section 5.3.2, have $(3N - 5)$ normal coordinates since they have only two rotational degrees of freedom.

of vibration. Because there are no cross terms between the normal modes for a harmonic oscillator, there is no coupling between them and, if displaced according to a normal mode coordinate, an N-body harmonic oscillator would then vibrate in that mode with all the nuclei moving with the same frequency according to simple harmonic motion. The normal modes of the water molecule are shown in figure 11.6.

The Q_r are a linear function of the u_i and, in terms of the $3N\Delta\alpha_i$, we have

$$m_i^{1/2}\Delta\alpha_i = \sum_{r=1}^{3N} l_{\alpha i,r} Q_r. \tag{4.9}$$

The l matrix is orthogonal, that is

$$\sum_{\alpha}\sum_{i=1}^{N} l_{\alpha i,r} l_{\alpha i,s} = \delta_{rs} \tag{4.10}$$

and the elements involving $r = 1$ to $(3N - 6)$ relate the mass-weighted Cartesian displacement coordinates to the $(3N - 6)$ vibrational normal coordinates. These elements of the l matrix determine the form of the vibrational normal coordinates and they depend on the atomic masses and on the harmonic force constants k_{ij}: they are determined in a so-called FG calculation[3], as are the λ_r. Different electronic states will have different harmonic force constants k_{ij} and, hence, the elements of the l matrix, the λ_r, and the form of the normal coordinates, will differ.

The elements of the l matrix involving $r = (3N - 5)$ to $3N$ relate the mass-weighted Cartesian displacement coordinates to the three translational normal coordinates Q_{3N-5}, Q_{3N-4} and $Q_{3N-3}(= T_x, T_y$ and $T_z)$, and to the three rotational normal coordinates Q_{3N-2}, Q_{3N-1} and $Q_{3N}(= R_x, R_y$ and $R_z)$, that are constrained to be zero by the choice of the xyz axes. The translational normal coordinate $T_\alpha(\alpha = x, y, z)$ is given by

$$T_\alpha = M^{-1/2}\sum_{i=1}^{N} m_i^{1/2}(m_i^{1/2}\Delta\alpha_i) \tag{4.11}$$

where m_i is the mass of atom i and M is the total mass of all the atoms in the molecule. The rotational normal coordinate R_x is given by

$$R_x = (I_{xx}^e)^{-1/2}\sum_{i=1}^{N} m_i^{1/2}[y_i^e(m_i^{1/2}\Delta z_i) - z_i^e(m_i^{1/2}\Delta y_i)] \tag{4.12}$$

where I_{xx}^e is defined in equation (5.4). The coordinates R_y and R_z are obtained by cyclically permuting xyz in equation (4.12). The expressions for I_{yy}^e and

[3] See Bunker P R and Jensen P 1998 *Molecular Symmetry and Spectroscopy* 2nd edn (Ottawa: NRC Research Press)

I_{zz}^e are given in equations (5.5) and (5.6), respectively. The factors $M^{-1/2}$ in equation (4.11) and $(I_{xx}^e)^{-1/2}$ in (4.12) are required in order to make the l matrix [equation (4.9)] orthogonal so that equation (4.10) is satisfied.

To obtain the quantum mechanical vibrational Hamiltonian from the classical expression for the vibrational energy given in equation (4.8), we first express it in terms of the Q_r and conjugate momenta P_r and then replace the P_r by $\hat{P}_r = -i\hbar\partial/\partial Q_r$. From equation (2.74), we have (since the potential energy does not depend on the velocities):

$$P_r = \partial T_{\text{vib}}/\partial\dot{Q}_r, = \dot{Q}_r. \tag{4.13}$$

The quantum mechanical harmonic-oscillator Hamiltonian is thus obtained from equation (4.8) as

$$\hat{H}_{\text{vib}}^0 = \tfrac{1}{2}\sum_{r=1}^{3N-6}(\hat{P}_r^2 + \lambda_r Q_r^2). \tag{4.14}$$

Substituting equation (4.9) for the u_i in terms of the Q_r into equation (4.6) for V_N, we obtain the anharmonic correction in terms of the normal coordinates as

$$V_N^{\text{anh}} = \frac{1}{6}\sum_{r,s,t}\Phi_{rst}Q_rQ_sQ_t + \frac{1}{24}\sum_{r,s,t,u}\Phi_{rstu}Q_rQ_sQ_tQ_u + \cdots \tag{4.15}$$

and the complete vibrational Hamiltonian is

$$\hat{H}_{\text{vib}} = \hat{H}_{\text{vib}}^0 + V_N^{\text{anh}}. \tag{4.16}$$

4.3   Vibrational wavefunctions and energies

The zero-order (harmonic-oscillator) vibrational Schrödinger equation is obtained from equation (4.14) as

$$\left[\tfrac{1}{2}\sum_{r=1}^{3N-6}(\hat{P}_r^2 + \lambda_r Q_r^2)\right]\Phi_{\text{vib}} = E_{\text{vib}}\Phi_{\text{vib}} \tag{4.17}$$

which separates into $3N - 6$ normal mode wave equations. We can, therefore, write the eigenfunctions and eigenvalues as

$$\Phi_{\text{vib}} = \Phi_{v_1}(Q_1)\Phi_{v_2}(Q_2)\ldots\Phi_{v_{3N-6}}(Q_{3N-6}) \tag{4.18}$$

and

$$E_{\text{vib}} = E_{v_1} + E_{v_2} + \cdots + E_{v_{3N-6}} \tag{4.19}$$

where $\Phi_{v_r}(Q_r)$ and E_{v_r} are from the one-dimensional harmonic-oscillator equation:

$$\tfrac{1}{2}(\hat{P}_r^2 + \lambda_r Q_r^2)\Phi_{v_r}(Q_r) = E_{v_r}\Phi_{v_r}(Q_r). \tag{4.20}$$

Equation (4.20) is a standard differential equation for which the eigenfunctions are known analytical functions of $\gamma_r (= \lambda_r^{1/2}/\hbar)$ and Q_r:

$$\Phi_{v_r} = N_{v_r} H_{v_r}(\gamma_r^{1/2} Q_r) \exp(-\gamma_r Q_r^2/2) \tag{4.21}$$

where the normalizing constant is

$$N_{v_r} = \gamma_r^{1/4}/(\pi^{1/2} 2^{v_r} v_r!)^{1/2} \tag{4.22}$$

and $H_{v_r}(\gamma_r^{1/2} Q_r)$ is a Hermite polynomial for which the first four values are

$$H_0(\gamma_r^{1/2} Q_r) = 1 \tag{4.23}$$
$$H_1(\gamma_r^{1/2} Q_r) = 2\gamma_r^{1/2} Q_r \tag{4.24}$$
$$H_2(\gamma_r^{1/2} Q_r) = 4\gamma_r Q_r^2 - 2 \tag{4.25}$$

and

$$H_3(\gamma_r^{1/2} Q_r) = 8\gamma_r^{3/2} Q_r^3 - 12\gamma_r^{1/2} Q_r. \tag{4.26}$$

In general $H_{v_r}(\gamma_r^{1/2} Q_r)$ contains $(\gamma_r^{1/2} Q_r)$ to the powers $v_r, v_r - 2, v_r - 4, \ldots, 1$ or 0, i.e. either all even powers or all odd powers up to v_r as v_r is even or odd, respectively. This means that the eigenfunction $\Phi_{v_r}(Q_r)$ is an even(odd) function of Q_r as v_r is even(odd), i.e.

$$\Phi_{v_r}(Q_r) = (-1)^{v_r} \Phi_{v_r}(-Q_r). \tag{4.27}$$

From the form of the wavefunctions, it can be seen that the number of nodes in Φ_{v_r} (i.e. the number of places where Φ_{v_r} changes sign) equals v_r and that the more nodes there are, the greater the value of the related eigenvalue will be. The eigenvalues of the harmonic-oscillator wave equation (4.20) are given by

$$E_{v_r} = (v_r + \tfrac{1}{2})\hbar^2 \gamma_r. \tag{4.28}$$

The harmonic vibrational energy quantum $\hbar^2 \gamma$ is usually expressed as a frequency ν in Hz (cycles s^{-1}) or as a wavenumber ω_e in cm^{-1}, where

$$\hbar^2 \gamma = \hbar \lambda^{1/2} = h\nu = hc\omega_e. \tag{4.29}$$

In figures 4.1 and 4.2, we show the wavefunctions and term values for two different harmonic-oscillator potentials: In one, the force constant λ is 3.19×10^{29} s^{-2} ($\omega_e = 3000$ cm^{-1}) and, in the other, it is 3.19×10^{27} s^{-2} ($\omega_e = 300$ cm^{-1}). These figures show that the vibrational term values are further apart and the vibrational displacements smaller, if the force constant is larger.

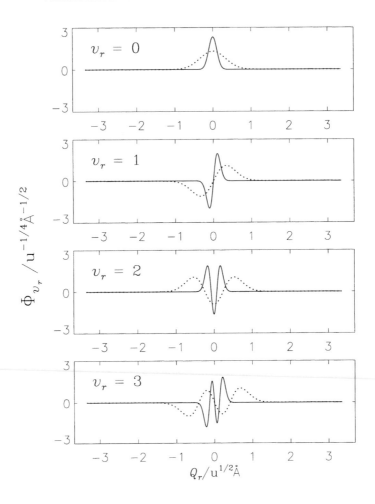

Figure 4.1.   Harmonic-oscillator wavefunctions Φ_{v_r} [equation (4.21)] for $\omega_e = 3000 \ cm^{-1}$ (full curves) and $300 \ cm^{-1}$ (dotted curves).

The measurement of the vibrational term value separations leads to a determination of the force constants and, hence, to a determination of the rigidity (or strength) of the bonds and bond angles.

For the $^{12}C^{16}O$ molecule, the single stretching normal vibration has a harmonic wavenumber of $2169.8 \ cm^{-1}$ and, thus, the vibrational term values in

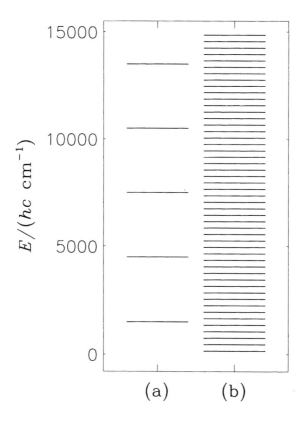

Figure 4.2. The term values for two different harmonic-oscillator potentials: In (a) the force constant λ is 3.19×10^{29} s^{-2} ($\omega_e = 3000$ cm^{-1}), and in (b) $\lambda = 3.19 \times 10^{27}$ s^{-2} ($\omega_e = 300$ cm^{-1}).

the harmonic-oscillator approximation are given by

$$G_{\text{vib}} = \omega_e(v + 1/2) = 2169.8(v + 1/2) \tag{4.30}$$

where $G_{\text{vib}} = (E_{\text{vib}}/hc)$ cm^{-1} and $v = 0, 1, \ldots$. These energies are plotted in figure 4.3. The *zero-point energy* is the energy of the $v = 0$ level; from equation (4.30) the zero-point term value for ^{12}C^{16}O is 1084.9 cm^{-1}.

For a diatomic molecule, if the bond length r is stretched from its equilibrium value by an amount Δr, then, in the harmonic-oscillator approximation, the restoring force F trying to re-establish the equilibrium value for r is a linear function of the displacement and has the magnitude

$$F = k_{rr} \Delta r \tag{4.31}$$

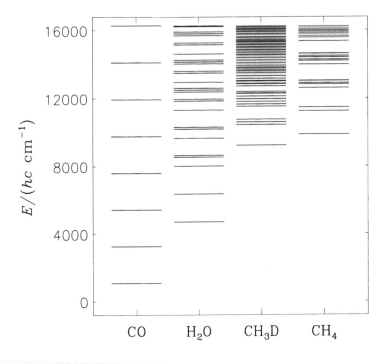

Figure 4.3. The lowest vibrational energies for several simple molecules.

where k_{rr} is the force constant for the bond. It can be shown that

$$k_{rr} = (2\pi c \omega_e)^2 \mu \qquad (4.32)$$

where $\mu = m_A m_B / (m_A + m_B)$ is the reduced mass of the diatomic molecule AB. From these equations, we calculate, using $\omega_e = 2169.8$ cm^{-1}, that for the $^{12}C^{16}O$ molecule $k_{rr} = 1902$ kg s^{-1}, so that if the CO bond is stretched[4] 0.1 Å, the restoring force will be 1.902×10^{-8} N. For the Na$_2$ molecule, $\omega_e = 159$ cm^{-1} which gives $k_{rr} = 17.1$ kg s^{-1} and a restoring force of only 1.71×10^{-10} N when the Na–Na bond is stretched by 0.1 Å. The Na$_2$ bond is thus considerably weaker (in that it resists being stretched much less) than the CO bond.

For the water molecule, there are three normal modes of vibration: Q_1 with harmonic vibrational wavenumber $\omega_1 = 3832.0$ cm^{-1}, Q_2 with harmonic vibrational wavenumber $\omega_2 = 1648.9$ cm^{-1} and Q_3 with harmonic vibrational wavenumber $\omega_3 = 3942.5$ cm^{-1}. Thus, the vibrational term values for H$_2$O in the harmonic-oscillator approximation are given by

$$G_{vib} = 3832.0(v_1 + 1/2) + 1648.9(v_2 + 1/2) + 3942.5(v_3 + 1/2) \qquad (4.33)$$

[4] 1 Å = 1 ångström = 10^{-10} m.

where $v_r = 0, 1, \ldots$. We use this expression to plot the vibrational energies of the water molecule in figure 4.3. The $v_1 = v_2 = v_3 = 0$ zero-point term value for the H_2O molecule is 4711.7 cm^{-1} from equation (4.33).

If two normal vibrations Q_a and Q_b, say, with conjugate momenta $\hat{P}_a$ and $\hat{P}_b$, have identically the same values of $\lambda_a = \lambda_b (= \lambda)$, then we treat the two modes together as a two-dimensional harmonic oscillator (tdho):

$$\hat{H}_{tdho} = \tfrac{1}{2}[\hat{P}_a^2 + \hat{P}_b^2 + \lambda(Q_a^2 + Q_b^2)]. \tag{4.34}$$

The energy is given by

$$E(v_a, v_b) = [(v_a + 1/2) + (v_b + 1/2)]\hbar^2\gamma \tag{4.35}$$

where $\gamma = \lambda^{1/2}/\hbar$. The energy depends only on the sum $(v_a + v_b)$ so that, for example, the four levels with $(v_a, v_b) = (3, 0), (2, 1), (1, 2)$ and $(0, 3)$ are degenerate. Each level has a degeneracy of $(v_a + v_b + 1)$. It is useful to introduce the coordinates Q and α according to

$$Q_a = Q \cos \alpha \quad \text{and} \quad Q_b = Q \sin \alpha. \tag{4.36}$$

This leads to the introduction of the *vibrational angular momentum* operator $\hat{M}$:

$$\hat{M} = (Q_a \hat{P}_b - Q_b \hat{P}_a) = -i\hbar \frac{\partial}{\partial \alpha} \tag{4.37}$$

which commutes with $\hat{H}_{tdho}$. The eigenfunctions of $\hat{H}_{tdho}$ could be written as the product of one-dimensional harmonic oscillator functions in Q_a and Q_b but instead we use the (Q, α) coordinates and introduce the quantum numbers v and l, to give eigenfunctions that are written as

$$\Psi_{v,l} = F_{v,l}(Q)e^{il\alpha} \tag{4.38}$$

where

$$\hat{M}\Psi_{v,l} = l\hbar\Psi_{v,l} \tag{4.39}$$

and $v = (v_a + v_b)$. The energy is given by

$$E_v = (v + 1)\hbar^2\gamma \tag{4.40}$$

which is $(v + 1)$-fold degenerate as the vibrational angular momentum quantum number l assumes one of the $(v + 1)$ values $-v, -v + 2, \ldots, +v$. $F_{v,l}(Q)$ is a function of Q that involves even, or odd, powers of $(\gamma^{1/2}Q)$, as v is even, or odd, respectively.

The CH_3D molecule has nine normal modes: Three (Q_r with $r = 1, 2$ and 3) are one-dimensional modes, and six (Q_{ra} and Q_{rb} with $r = 4, 5$ and 6) form three two-dimensional modes. The vibrational term values for CH_3D, in the harmonic-oscillator approximation, are given by

$$
\begin{aligned}
G_{vib} = {} & 3064.1(v_1 + 1/2) + 2282.1(v_2 + 1/2) \\
& + 1362.0(v_3 + 1/2) + 3156.5(v_4 + 1) \\
& + 1521.7(v_5 + 1) + 1206.2(v_6 + 1)
\end{aligned}
\tag{4.41}
$$

where $v_r = 0, 1, \ldots$. We use this expression to plot the vibrational energies of the CH_3D molecule in figure 4.3. The zero-point term value for the CH_3D molecule is 9238.5 cm^{-1}.

Some molecules have three-dimensional harmonic oscillator modes, for which the energy is given by

$$
E_v = (v + 3/2)\hbar^2\gamma
\tag{4.42}
$$

where $v = 0, 1, \ldots$. The harmonic-oscillator eigenfunctions for this case are written $\Psi_{v,l,n}(Q, \alpha, \beta)$, where α and β are vibrational angular coordinates, and l and n are vibrational angular momentum quantum numbers restricted to the values $l = v, v - 2, v - 4, \ldots, 1$ or 0, and $n = -l, -l + 1, \ldots, l - 1, l$.

The methane molecule CH_4 has nine normal modes of vibration: One (Q_1) is a one-dimensional mode, two $(Q_{2a}$ and $Q_{2b})$ form a two-dimensional mode, and six $(Q_{ra}, Q_{rb}$ and Q_{rc}, with $r = 3$ and 4) form two three-dimensional modes. The vibrational term values for CH_4, in the harmonic-oscillator approximation, are given by

$$
\begin{aligned}
G_{vib} = {} & 3025.5(v_1 + 1/2) + 1582.7(v_2 + 1) \\
& + 3156.8(v_3 + 3/2) + 1367.4(v_4 + 3/2)
\end{aligned}
\tag{4.43}
$$

where $v_r = 0, 1, \ldots$. We use this expression to plot the vibrational energies of the CH_4 molecule in figure 4.3. The zero-point term value for the CH_4 molecule is 9881.75 cm^{-1}. As we show in section 11.3, symmetry can be used to determine the dimensionality of the normal modes.

To calculate the effects of anharmonicity and rotation–vibration coupling (see sections 4.4 and 11.5) and vibrational selection rules (see section 12.3.1), one needs the matrix elements of the normal coordinates Q and conjugate momenta $\hat{P}$ in the harmonic-oscillator basis functions. These matrix elements are given in table 4.1.

4.4 Anharmonicity

If a diatomic molecule existed with a purely harmonic potential function, it would be a very strange molecule indeed. It would mean that as one stretched it more and more the restoring force, from equation (4.31), would continue to increase linearly with stretching, even if the nuclei were kilometres apart. For real molecules, the

Table 4.1. Non-vanishing matrix elements of normal coordinate Q and momentum $\hat{P}$ for the one-dimensional harmonic-oscillator[a].

$\langle v+1 \lvert Q \rvert v \rangle = \sqrt{(v+1)/(2\gamma)}$	$\langle v+1 \lvert \hat{P} \rvert v \rangle = i\hbar\sqrt{(v+1)\gamma/2}$
$\langle v-1 \lvert Q \rvert v \rangle = \sqrt{v/(2\gamma)}$	$\langle v-1 \lvert \hat{P} \rvert v \rangle = -i\hbar\sqrt{v\gamma/2}$

[a] $\hat{P} = -i\hbar\partial/\partial Q$ and the one-dimensional harmonic-oscillator functions $\Phi_v = \lvert v \rangle$. We use $\gamma = \lambda^{1/2}/\hbar = 4\pi^2 c\omega_e/h$, where the one-dimensional harmonic-oscillator Hamiltonian is $\hat{H}_{ho} = \frac{1}{2}(\hat{P}^2 + \lambda Q^2)$.

restoring force eventually goes to zero as the molecule is stretched to dissociation. The potential function, of necessity, must have some anharmonicity, although it may be slight in the region around equilibrium.

For a diatomic molecule, the lowest-order anharmonic correction is the cubic term

$$V_N^{\text{anh},3} = \tfrac{1}{6}\Phi_{111}Q^3 \tag{4.44}$$

and we determine its effect on the vibrational energy levels by setting up the matrix of $\hat{H}^0_{\text{vib}} + V_N^{\text{anh},3}$ in the harmonic-oscillator basis functions $\lvert v \rangle$ of equation (4.21). The diagonal matrix elements of $\hat{H}^0_{\text{vib}}$ are the term values as given for CO, for example, in equation (4.30); $\hat{H}^0_{\text{vib}}$ has no off-diagonal matrix elements.

The matrix elements of the cubic term are determined by making the expansion

$$\langle v \lvert Q^3 \rvert v' \rangle = \sum_{v'',v'''} \langle v \lvert Q \rvert v'' \rangle \langle v'' \lvert Q \rvert v''' \rangle \langle v''' \lvert Q \rvert v' \rangle. \tag{4.45}$$

This equations is an implementation of the rule for multiplying matrices given in equation (2.99); perhaps this can be better appreciated if $\langle v \lvert Q \rvert v'' \rangle$ is written as $Q_{v,v''}$ etc. This is a very commonly used method for evaluating the matrix elements of a product; one can think of it as the insertion of $\sum \lvert v'' \rangle \langle v'' \rvert = 1$ and $\sum \lvert v''' \rangle \langle v''' \rvert = 1$, between members of the product. From table 4.1, the non-vanishing terms in this sum are those for which $v'' = v \pm 1$, $v''' = v'' \pm 1$, and $v' = v''' \pm 1$. Thus, the only non-vanishing matrix elements overall are those for which $v' = v \pm 1$ or $v \pm 3$. These can be written out in full as

$$\langle v \lvert Q^3 \rvert v+3 \rangle = \langle v \lvert Q \rvert v+1 \rangle \langle v+1 \lvert Q \rvert v+2 \rangle \langle v+2 \lvert Q \rvert v+3 \rangle \tag{4.46}$$

$$
\begin{aligned}
\langle v \lvert Q^3 \rvert v+1 \rangle = {} & \langle v \lvert Q \rvert v+1 \rangle \langle v+1 \lvert Q \rvert v+2 \rangle \langle v+2 \lvert Q \rvert v+1 \rangle \\
& + \langle v \lvert Q \rvert v+1 \rangle \langle v+1 \lvert Q \rvert v \rangle \langle v \lvert Q \rvert v+1 \rangle \\
& + \langle v \lvert Q \rvert v-1 \rangle \langle v-1 \lvert Q \rvert v \rangle \langle v \lvert Q \rvert v+1 \rangle
\end{aligned} \tag{4.47}
$$

$$\langle v \lvert Q^3 \rvert v-1 \rangle = \langle v \lvert Q \rvert v+1 \rangle \langle v+1 \lvert Q \rvert v \rangle \langle v \lvert Q \rvert v-1 \rangle$$

$$+ \langle v|Q|v-1\rangle\langle v-1|Q|v\rangle\langle v|Q|v-1\rangle$$
$$+ \langle v|Q|v-1\rangle\langle v-1|Q|v-2\rangle\langle v-2|Q|v-1\rangle \quad (4.48)$$
$$\langle v|Q^3|v-3\rangle = \langle v|Q|v-1\rangle\langle v-1|Q|v-2\rangle\langle v-2|Q|v-3\rangle. \quad (4.49)$$

Inserting the values from table 4.1 of the individual matrix elements of Q, one obtains

$$\langle v|Q^3|v+3\rangle = \left(1/\sqrt{8\gamma^3}\right)\sqrt{(v+1)(v+2)(v+3)} \quad (4.50)$$

$$\langle v|Q^3|v+1\rangle = \left(3/\sqrt{8\gamma^3}\right)(v+1)\sqrt{v+1} \quad (4.51)$$

$$\langle v|Q^3|v-1\rangle = \left(3/\sqrt{8\gamma^3}\right)v\sqrt{v} \quad (4.52)$$

$$\langle v|Q^3|v-3\rangle = \left(1/\sqrt{8\gamma^3}\right)\sqrt{(v-1)(v-2)(v-3)}. \quad (4.53)$$

Multiplying by $\Phi_{111}/6$ gives the matrix elements of the cubic anharmonicity term. This is much smaller than the difference in the diagonal matrix elements (which is ω_e or $3\omega_e$) and we can use the approximate expression given in equation (2.38) to evaluate the vibrational energy shift. In this way, we determine that the $v = 0$ level is pushed down by the $v = 1$ level by an amount

$$S_1 = \Phi_{111}^2/32\gamma^3\omega_e \quad (4.54)$$

and, by the $v = 3$ level, an amount

$$S_3 = \Phi_{111}^2/144\gamma^3\omega_e \quad (4.55)$$

so that the total shift down of the $v = 0$ level by the cubic anharmonicity term is $11\Phi_{111}^2/(288\gamma^3\omega_e)$. The $v = 1$ level is shifted up by cubic anharmonicity interaction with $v = 0$ and down by interactions with the $v = 2$ and 4 levels; it suffers a total shift down of $71\Phi_{111}^2/(288\gamma^3\omega_e)$. Thus, cubic anharmonicity reduces the separation between the $v = 1$ and 0 levels by $5\Phi_{111}^2/(24\gamma^3\omega_e)$.

Anharmonicity reduces the vibrational energy level separations in such a way that the levels become closer and closer together the higher up in energy they are, eventually converging to a *continuum* above dissociation. As a result, vibrational energy levels do not have the even spacing that is characteristic of the harmonic potential.

For a polyatomic molecule, anharmonicity terms in the potential function can couple vibrational states that are close in energy and for which it is necessary to diagonalize the Hamiltonian matrix since the approximation of equation (2.38) is unsatisfactory. An important example is the effect, for the water molecule, of the cubic anharmonicity term

$$V' = \tfrac{1}{2}\Phi_{122}Q_1Q_2^2. \quad (4.56)$$

V' connects the states $(v_1, v_2, v_3) = (1, 0, 0)$ and $(0, 2, 0)$ with term values, calculated using equation (4.33), that differ by 534.2 cm^{-1}. For the water molecule, vibrational states having a common value of $2v_1 + v_2$, and the same value of v_3, are relatively close in energy and are connected by V'. This is a *Fermi* resonance and the harmonic-oscillator energies and wavefunctions are a poor approximation for such levels.

Many such *vibrational resonances* occur involving other cubic or quartic anharmonicity terms. A resonance caused by the term $(1/4)\Phi_{1133}Q_1^2 Q_3^2$, such as that between the states $(2, 0, 0)$ and $(0, 0, 2)$ in the water molecule, is named a *Darling–Dennison* resonance. As we show in section 11.5, symmetry can be used to show that some of the normal coordinate force constants Φ_{rst} and Φ_{rstu} must vanish. For example, for the water molecule, Φ_{223} vanishes and there is no Fermi resonance between the states $(0, 0, 1)$ and $(0, 2, 0)$.

4.5 Tunnelling

The quantum mechanical phenomenon of *tunnelling* is explained here and we base this explanation on the one-dimensional harmonic-oscillator wave equation in cm^{-1} units:

$$\frac{1}{2hc}(\hat{P}^2 + \lambda Q^2)\Phi_v(Q) = G_v \Phi_v(Q) \tag{4.57}$$

which describes motion within a harmonic potential energy curve given by $V_{\text{harm}}/\text{cm}^{-1} = (\lambda/2hc)Q^2$. We rewrite this equation in terms of the *dimensionless normal coordinate*

$$q = \gamma^{1/2} Q. \tag{4.58}$$

Using equation (4.29), the potential energy expressed in terms of q is

$$V_{\text{harm}}/\text{cm}^{-1} = \frac{\lambda}{2hc}Q^2 = \frac{\lambda}{2hc\gamma}q^2 = \frac{\hbar\lambda^{1/2}}{2hc}q^2 = \frac{1}{2}\omega_e q^2. \tag{4.59}$$

From the chain rule,

$$\frac{d}{dQ} = \frac{dq}{dQ}\frac{d}{dq} = \sqrt{\gamma}\frac{d}{dq} \tag{4.60}$$

and the kinetic energy operator is

$$\frac{1}{2hc}\hat{P}^2 = -\frac{\hbar^2}{2hc}\frac{d^2}{dQ^2} = -\frac{\hbar^2\gamma}{2hc}\frac{d^2}{dq^2} = -\frac{1}{2}\omega_e\frac{d^2}{dq^2}. \tag{4.61}$$

From equations (4.59) and (4.61), the harmonic-oscillator equation (4.57), in terms of q, is

$$\left(-\frac{1}{2}\omega_e\frac{d^2}{dq^2} + \frac{1}{2}\omega_e q^2\right)\Phi_v(q) = G_v \Phi_v(q). \tag{4.62}$$

The potential $\omega_e q^2/2$ in equation (4.62) has a single harmonically shaped minimum centred at the $q = 0$ configuration. To introduce the concept of tunnelling, we use instead the potential energy function

$$V_t(q)/\text{cm}^{-1} = H\left[1 - \left(\frac{q}{q_e}\right)^2\right]^2 = H + H\left(\frac{q}{q_e}\right)^4 - 2H\left(\frac{q}{q_e}\right)^2 \quad (4.63)$$

where 't' is for 'tunnelling'. $V_t(q)$ has the value zero at its minima, when $q = \pm q_e$, and the value H when $q = 0$; it is a symmetrical *double minimum potential* with a *barrier* of H (expressed in cm^{-1} units) between the minima. In figure 4.4, $V_t(q)$ is drawn for $H = 10\,000$ cm^{-1} and $q_e = 5$. The Schrödinger equation is

$$\left(-\frac{1}{2}\omega_e\frac{d^2}{dq^2} + V_t(q)\right)\Phi_{t,j}(q) = G_{t,j}\Phi_{t,j}(q) \quad (4.64)$$

where the index $j = 0, 1, 2, 3, \ldots$ labels the solutions in order of ascending energy.

For any set of values of $\omega_e > 0$, $H > 0$ and $q_e > 0$, equation (4.64) can be solved numerically using the method of matrix diagonalization described in section 2.3. The complete set of orthonormal functions used as the basis function is taken to be harmonic-oscillator eigenfunctions $\Phi_v(q)$ from equation (4.62). The unknown functions $\Phi_{t,j}(q)$ in equation (4.64) are expanded in terms of the functions $\Phi_v(q)$ as described by equation (2.23):

$$\Phi_{t,j}(q) = \sum_{v=0}^{200} C_{jv}\Phi_v(q). \quad (4.65)$$

Satisfactory accuracy for the states of interest here is achieved if the expansion is truncated at $v = 200$.

The matrix elements of the tunnelling Hamiltonian

$$\hat{H}_t = -\frac{1}{2}\omega_e\frac{d^2}{dq^2} + H + H\left(\frac{q}{q_e}\right)^4 - 2H\left(\frac{q}{q_e}\right)^2 \quad (4.66)$$

are given by

$$\langle\Phi_{v''}|\hat{H}_t|\Phi_{v'}\rangle = \frac{1}{2}\omega_e\left\langle\Phi_{v''}\left|-\frac{d^2}{dq^2}\right|\Phi_{v'}\right\rangle + H\langle\Phi_{v''}|\Phi_{v'}\rangle$$
$$+ \frac{H}{q_e^4}\langle\Phi_{v''}|q^4|\Phi_{v'}\rangle - 2\frac{H}{q_e^2}\langle\Phi_{v''}|q^2|\Phi_{v'}\rangle. \quad (4.67)$$

The overlap integral $\langle\Phi_{v''}|\Phi_{v'}\rangle = \delta_{v''v'}$ (where $\delta_{v''v'}$ is the Kronecker delta) because the $\Phi_v(q)$ functions form an orthonormal basis. The other matrix elements can be evaluated using expansions like those in equations (4.45)–(4.49).

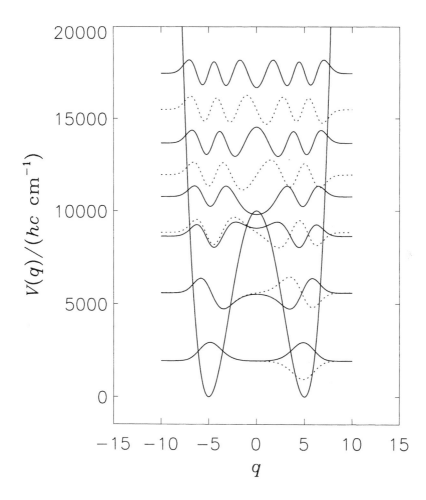

Figure 4.4. The result of calculation (a). The solution of equation (4.64) with $H = 10\,000$ cm^{-1}, $q_e = 5$ and $\omega_e = 5000$ cm^{-1}. All wavefunctions $\Phi_{t,j}(q)$ are drawn at the position of their energies with the same arbitrary ordinate scale. Wavefunctions with even (odd) j are drawn as full (dotted) curves.

The matrix elements $\langle \Phi_{v''}| - \mathrm{i}\mathrm{d}/\mathrm{d}q|\Phi_{v'}\rangle$ and $\langle \Phi_{v''}|q|\Phi_{v'}\rangle$, which are required to start the process of forming the matrix elements in equation (4.67), are obtained by setting $\gamma = 1$ and $\hbar = 1$ in the expressions of table 4.1. Diagonalizing the Hamiltonian matrix gives the term values $G_{t,j}$ and the eigenfunctions $\Phi_{t,j}(q)$.

Taking $q_e = 5$ and $\omega_e = 5000$ cm^{-1}, the eigenfunctions and eigenvalues of equation (4.64) are calculated for three different cases:

(a) $H = 10\,000$ cm^{-1}. The eigenvalues are given in table 4.2 under the heading

Table 4.2. Term values $G_{t,j}$ (in cm^{-1}) for equation (4.64).

j	(a)[a]	(b)[a]	(c)[a]
0	1 946.7	113.8	1 946.8
1	1 946.8	271.3	5 598.4
2	5 591.8	578.2	8 842.9
3	5 598.4	933.3	11 949.3
4	8 631.0	1338.5	15 487.5
5	8 842.9	1783.5	
6	10 765.6	2263.0	
7	11 949.3	2773.0	
8	13 666.3	3310.4	
9	15 487.5	3873.0	
10	17 473.7		

[a] See the text for the explanation of the calculations
(a), (b) and (c).

'(a)' and figure 4.4 shows the potential energy function and the associated
eigenfunctions.

(b) $H = 100$ cm^{-1}. The eigenvalues are given in table 4.2 under the heading
'(b)' and figure 4.5 shows the potential energy function and the associated
eigenfunctions.

(c) Here, the same Hamiltonian as in calculation (a) is used but the wavefunction
is constrained to be zero for $q < q_{min} = -10$ and $q > q_{max} = 0$ by doing the
matrix diagonalization in a basis of normalized eigenfunctions for a square
potential well [see equation (2.71) and the discussion of it]. That is, we
express each wavefunction as

$$\Phi_{t,j}(q) = \begin{cases} \sum_{n=1}^{200} C_{jn} \sqrt{\dfrac{2}{L}} \sin\left(\dfrac{n\pi(q-q_{min})}{L}\right) & \text{for } q_{min} < q < q_{max} \\ 0 & \text{otherwise} \end{cases}$$

(4.68)

where $L = q_{max} - q_{min} = 10$. The eigenvalues and eigenfunctions
are obtained by matrix diagonalization analogous to that carried out in
calculations (a) and (b). The kinetic energy operator $-(1/2)\omega_e d^2/dq^2$
is diagonal in the chosen basis and its matrix elements can be derived
analytically; the matrix elements of the potential energy operator can be
calculated by numerical integration. The eigenvalues obtained are given in
table 4.2 under the heading '(c)' and figure 4.6 shows the potential energy
function and the associated eigenfunctions.

Calculation (a) illustrates tunnelling. The states $j = 0$–5, with energy below
the barrier, form three pairs of states and the two components of each pair are

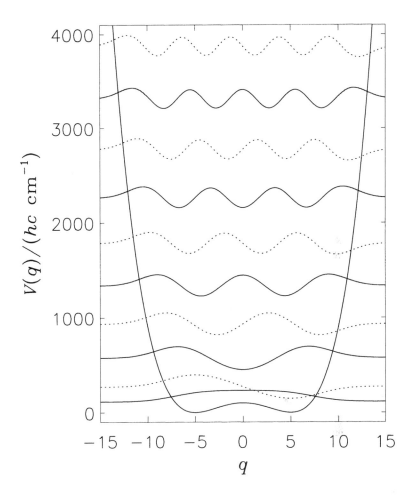

Figure 4.5. The result of calculation (b). The solution of equation (4.64) with $H = 100$ cm^{-1}, $q_e = 5$ and $\omega_e = 5000$ cm^{-1}. All wavefunctions $\Phi_{t,j}(q)$ are drawn at the position of their energies with the same arbitrary ordinate scale. Wavefunctions with j even (odd) are drawn as full (dotted) curves.

close in energy. These wavefunctions have amplitude in both wells and these states are said to 'tunnel through the barrier' since although they have energies that are too low to allow them to pass over the barrier in the classical sense each of them has a non-zero probability of being in either well. The two states with $j = 0$ and 1 are near-degenerate and the energy splitting between the pairs of components increases as the energy increases.

For a one-dimensional problem of this type, where the potential energy

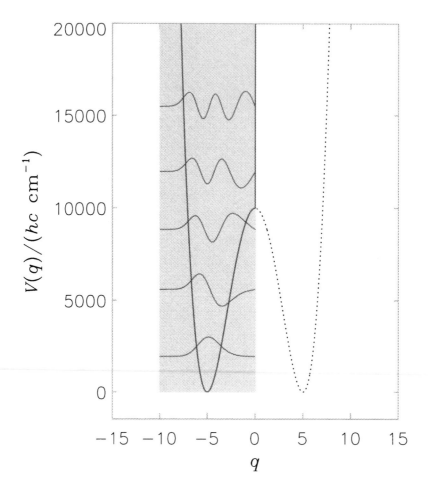

Figure 4.6. The result of calculation (c). The solution of equation (4.64) with $H = 10\,000$ cm^{-1}, $q_e = 5$ and $\omega_e = 5000$ cm^{-1} as in calculation (a) but with all wavefunctions constrained to be zero outside the shaded area for which $-10 \le q \le 0$. The potential energy function in figure 4.4 is drawn outside the shaded region as a dotted curve. All wavefunctions $\Phi_{t,j}(q)$ are drawn at the position of their energies with the same arbitrary ordinate scale.

function is an even function of q, i.e. $V_t(q) = V_t(-q)$, it can be shown rigorously that the wavefunctions with j even will be even functions of q and that the wavefunctions with j odd will be odd functions of q, i.e. for j odd $\Phi_{t,j}(q) = -\Phi_{t,j}(-q)$. Thus, the three state pairs below the barrier each comprise an even wavefunction $\Phi_{t,j}(q)$ and an odd wavefunction $\Phi_{t,j+1}(q)$, where $j = 0, 2$, or

4. We see in figure 4.4 that these two wavefunctions bear some relation to each other in that we can choose the signs of the wavefunctions so that, for $q < 0$ (and $j = 0, 2$, or 4),

$$\Phi_{t,j}(q) \approx \Phi_{t,j+1}(q) \tag{4.69}$$

whereas, for $q > 0$,

$$\Phi_{t,j}(q) \approx -\Phi_{t,j+1}(q). \tag{4.70}$$

These relations hold best for the lowest state pair with $j = 0$ and deteriorate to an increasing extent for $j = 2$ and 4. Above the barrier, there are no similar effects; the states do not form pairs. The calculated pattern of energies is a typical result of tunnelling motion. If the energy level is far below the barrier, the energies form a pair of almost degenerate states, exemplified by the $j = 0$ and 1 states here, with wavefunctions related to each other. As the energy increases towards the top of the barrier, the degeneracy is removed more and more since tunnelling through the barrier can take place to an increasing extent. This effect is seen in the pairs with $j = 2$–3 and 4–5.

Calculation (b) shows the low-barrier case. The lowest allowed state is above the barrier and there are no approximate degeneracies.

In calculation (c), we modify the potential relative to calculation (a) by making it infinite for $q < q_{min} = -10$ and for $q > q_{max} = 0$. Now no tunnelling is possible. Both the wavefunctions and energies for the three lowest states are similar to those obtained for the three state pairs in calculation (a). However, by removing access to the potential minimum at $q = 5.0$, we have removed the double degeneracy. Notice that the (non-tunnelling) energies obtained in calculation (c) are the same as the odd-j tunnelling energies from calculation (a). This is because for all states in calculation (c) and for the odd-j states in calculation (a), the wavefunctions are zero at $q = 0$, i.e. for the odd-j states in calculation (a) the wavefunctions have a node at $q = 0$. Thus, tunnelling does not symmetrically split each energy level into two; for each pair of tunnelling states, the component whose wavefunction has a node at $q = 0$ is unaffected whereas the component whose wavefunction does not have a node at $q = 0$ is pushed down.

4.6 Problems

4.1   What are the appropriate SI units for the the normal coordinate Q_r and the parameter λ_r in equation (4.14)? What are the appropriate SI units for the quantities Φ_{rst} and Φ_{rstu} in equation (4.15)?

4.2   We take the vibrational energies of the HI molecule to be given by equation (4.28). The force constant $k_{rr} = 313.8$ N m^{-1}. Calculate the energy difference between two neighbouring energy levels. Calculate the wavelength of the electromagnetic radiation that would be in resonance with this energy difference.

4.3   In the harmonic approximation, the vibrational energy of a diatomic molecule is proportional to $(v + 1/2)$ as given in equation (4.28) but, because of anharmonicity, a more accurate expression involves a correction term proportional to the parameter x_e:

$$G_v = \omega_e \left[\left(v + \frac{1}{2} \right) - x_e \left(v + \frac{1}{2} \right)^2 \right] \qquad v = 0, 1, 2, 3, \ldots.$$

For $^{14}N^{16}O$ the following term value differences are derived from experimental data: $G_1 - G_0 = 1876.06 \text{ cm}^{-1}$ and $G_2 - G_0 = 3724.20 \text{ cm}^{-1}$. Calculate ω_e, x_e and the force constant k_{rr}.

4.4   Derive the matrix elements of the kinetic energy operator $-(1/2)\omega_e \times d^2/dq^2$ in a basis of normalized eigenfunctions for a square potential well [see equations (2.71) and (4.68)]; these matrix elements are required to compute the wavefunctions shown in figure 4.6.

4.5   The eigenfunctions for the one-dimensional harmonic-oscillator are given in equation (4.21). Verify by actual calculation that the three eigenfunctions of lowest energy Φ_0, Φ_1 and Φ_2 are orthogonal. *Hint:* Transform the necessary integrals to depend on the dimensionless quantity $\gamma_r Q_r$ and use the expressions $(a > 0)$

$$\int_{-\infty}^{\infty} \exp(-ax^2) \, dx = \sqrt{\frac{\pi}{a}}$$

$$\int_{-\infty}^{\infty} x^{2n} \exp(-ax^2) \, dx = \frac{1 \cdot 3 \cdot 5 \cdot \ldots \cdot (2n-1)}{2^n a^n} \sqrt{\frac{\pi}{a}}.$$

Chapter 5

Rotational states

The zero-order rotational Hamiltonian of a molecule is the *rigid-rotor* Hamiltonian and it describes the molecule as rotating in space with its geometry fixed at equilibrium. Nonlinear molecules are categorized as being either symmetric tops, spherical tops or asymmetric tops and, in section 5.3, we discuss the rigid-rotor Schrödinger equation for each of these types of molecule. The derivation of the rigid-rotor Hamiltonian and an account of the detailed form of the wavefunctions are given at the end of the chapter in section 5.5. In section 11.5, the effects of rotation–vibration coupling, which produce centrifugal distortion and Coriolis coupling corrections to the rigid-rotor Hamiltonian, will be discussed. The rigid-rotor Hamiltonian involves the principal moments of inertia of the molecule and its eigenfunctions are functions of the Euler angle; so we begin by defining the Euler angles and by explaining what the principal moments of inertia are.

5.1 The Euler angles

The rotational coordinates are the Euler angles (θ, ϕ, χ) and these three angles specify the orientation of the molecule-fixed xyz axes, introduced in section 4.1, relative to the $\xi\eta\zeta$ axes. The angles θ and ϕ are the polar coordinates that specify the orientation of the z axis within the $\xi\eta\zeta$ axis system as shown in figure 5.1. We only need define θ to have the range $0 \leq \theta \leq \pi$ in order to cover all possible orientations of the z axis; ϕ has the range $0 \leq \phi \leq 2\pi$. In figure 5.1, we also show the 'node line' ON which is needed for the definition of the third Euler angle χ. ON is perpendicular to both the z and ζ axes and directed so that a right-handed screw is driven along ON in its positive direction by twisting it from ζ to z through θ (where $0 \leq \theta \leq \pi$). The definition of χ is shown in figure 5.2; it is the angle by which the y axis is twisted away from the ON node line measured in a right-handed sense about the z axis. A three-dimensional figure showing how the Euler angles depict the rotation of the xyz axes relative to the $\xi\eta\zeta$ axes is

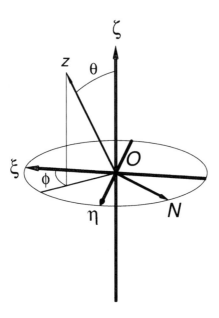

Figure 5.1. The definition of the Euler angles (θ, ϕ); these are the polar coordinates that specify the orientation of the molecule-fixed z axis in the $\xi\eta\zeta$ axis system. In this figure, we also show the 'node line' ON (see text).

shown in figure 5.3 in section 5.5.2. From this figure, one can appreciate that the node line is the line of intersection of the $\xi\eta$ and xy planes.

5.2 The principal moments of inertia

Using molecule-fixed xyz axes, the moments of inertia of a molecule are the diagonal elements of the *inertia matrix* I given by

$$I_{\alpha\alpha} = \sum_i m_i(\beta_i^2 + \gamma_i^2) \tag{5.1}$$

where $\alpha\beta\gamma$ is a permutation of xyz. The off-diagonal elements of I are

$$I_{\alpha\beta} = -\sum_i m_i\alpha_i\beta_i \tag{5.2}$$

(the $\sum_i m_i\alpha_i\beta_i$ are called *products of inertia*), where $\alpha \neq \beta$. However, it is always possible to orient the xyz axes in the molecule so that the off-diagonal elements $I_{\alpha\beta}$ vanish and, in this circumstance, the xyz axes are said to be the *principal axes* of inertia. In developing the rotational Hamiltonian of a molecule,

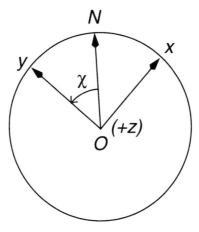

Figure 5.2. The definition of the Euler angle χ. The molecule-fixed axis system xyz is drawn with the xy plane in the plane of the page and with the z axis pointing up out of the page. The y axis forms the angle χ with the node line ON measured in a right-handed sense about the z axis.

we attach xyz molecule-fixed axes to the molecule in its equilibrium configuration so that they align with the principal axes, i.e. the xyz axes are located so that

$$\sum_i m_i x_i{}^e y_i{}^e = \sum_i m_i y_i{}^e z_i{}^e = \sum_i m_i z_i{}^e x_i{}^e = 0 \tag{5.3}$$

where the m_i are the atomic masses. Using xyz principal axes, the equilibrium *principal moments of inertia* are given by

$$I^e_{xx} = \sum_i [m_i (y^e_i)^2 + m_i (z^e_i)^2] \tag{5.4}$$

$$I^e_{yy} = \sum_i [m_i (z^e_i)^2 + m_i (x^e_i)^2] \tag{5.5}$$

$$I^e_{zz} = \sum_i [m_i (x^e_i)^2 + m_i (y^e_i)^2]. \tag{5.6}$$

The $I^e_{\alpha\alpha}$ can be calculated if we know the equilibrium bond lengths and angles and the atomic masses. The principal axes of the equilibrium configuration are labelled a, b, and c in order of increasing value of the principal moments of inertia, i.e. so that $I^e_{aa} \leq I^e_{bb} \leq I^e_{cc}$.

5.3   The rigid-rotor Hamiltonian

The rigid-rotor Hamiltonian in cm^{-1} units, written in terms of the molecule-fixed abc components of the angular momentum operator $\hat{J}$, is

$$\hat{H}_{\mathrm{rot}}^0 = \hbar^{-2}(A_e \hat{J}_a^{\,2} + B_e \hat{J}_b^{\,2} + C_e \hat{J}_c^{\,2}) \tag{5.7}$$

where the *rotational constants* (in cm^{-1}) are given by

$$A_e = \frac{\hbar^2}{2hc I_{aa}^e} \qquad B_e = \frac{\hbar^2}{2hc I_{bb}^e} \qquad \text{and} \qquad C_e = \frac{\hbar^2}{2hc I_{cc}^e}. \tag{5.8}$$

From the definition of the a, b, and c axes, i.e. with $I_{aa}^e \leq I_{bb}^e \leq I_{cc}^e$, the rotational constants must be in the order $A_e \geq B_e \geq C_e$. For any molecule, the eigenvalues of the rotational Hamiltonian depend on the values of the rotational constants and the values of the rotational constants depend on the values of the principal moments of inertia, which, in turn, depend on the bond lengths and bond angles in the molecule and on the atomic masses. For simple symmetric molecules, knowledge of the rotational constants leads to a determination of the structure, particularly if data for several isotopomers are available. If the rotational constants are in cm^{-1} and the moments of inertia are in u $\mathring{A}^2$, they are related using

$$\frac{\hbar^2}{2hc} \approx 16.858 \text{ u } \mathring{A}^2 \text{ cm}^{-1}. \tag{5.9}$$

We categorize molecules as being of four different types:

(i)   Symmetric top molecules, for which two of the $I_{\alpha\alpha}^e$ are equal
(ii) Linear molecules
(iii) Spherical top molecules, for which all three $I_{\alpha\alpha}^e$ are equal, and
(iv) Asymmetric top molecules, for which all three $I_{\alpha\alpha}^e$ are different.

5.3.1   Symmetric top molecules

Symmetric top molecules have one unique moment of inertia (about the unique principal axis) and two others that are equal to each other. The CH_3D and H_3^+ molecules are examples. For such a molecule, any axis perpendicular to the unique axis will also be a principal axis and we choose two such directions at $90°$ to each other; they have the same moments of inertia about them. The unique moment of inertia can be larger or smaller than the other two identical moments of inertia. If the unique moment is smaller than the other two, the molecule is said to be a prolate symmetric top molecule and the unique axis is the a axis; for a prolate symmetric top, $A_e > B_e = C_e$. Alternatively, if the unique moment is larger, the molecule is said to be an oblate symmetric top and the unique axis is the c axis; for an oblate symmetric top $A_e = B_e > C_e$. The CH_3D molecule is a prolate symmetric top and H_3^+ is an oblate symmetric top.

The rigid-rotor Schrödinger equation for a prolate top is

$$\hbar^{-2}[A_e\hat{J}_a^2 + B_e(\hat{J}_b^2 + \hat{J}_c^2)]\Phi_{rot}(\theta, \phi, \chi) = F_{rot}\Phi_{rot}(\theta, \phi, \chi) \quad (5.10)$$

where the rotational term value $F_{rot} = (E_{rot}/hc)$ cm^{-1}. For a prolate top the a, b and c axes are chosen as the z, x and y axes, respectively, and replacing $(\hat{J}_b^2 + \hat{J}_c^2) = (\hat{J}_x^2 + \hat{J}_y^2)$ by $(\hat{J}^2 - \hat{J}_z^2)$ [from equation (5.49)], we obtain

$$\hbar^{-2}[B_e\hat{J}^2 + (A_e - B_e)\hat{J}_z^2]\Phi_{rot}(\theta, \phi, \chi) = F_{rot}\Phi_{rot}(\theta, \phi, \chi). \quad (5.11)$$

Using the known eigenvalues of $\hat{J}^2$ and $\hat{J}_z$, see section 5.5.3, we obtain

$$F_{rot} = B_e J(J + 1) + (A_e - B_e)K^2 \quad (5.12)$$

where $J = 0, 1, 2, \ldots$, $K = |k|$ and $k = 0, \pm 1, \pm 2, \ldots \pm J$; the energies of a rigid prolate symmetric top depend on the values of the rotational constants A_e and B_e, and on the quantum numbers J and K. Equation (5.12) was used to calculate the rotational energies of the CH_3D molecule in figure 1.5 with the values of $A_e = 5.35$ cm^{-1} and $B_e = 3.95$ cm^{-1}. The wavefunctions of a symmetric top molecule are the functions

$$\Phi_{rot}(\theta, \phi, \chi) = [1/(2\pi)]^{1/2}S_{Jkm}(\theta, \phi)e^{ik\chi}. \quad (5.13)$$

These functions, and the quantum number m, are discussed in section 5.5.3.

The rotational constants depend on the molecular structure. For CH_3D, we know the tetrahedral bond angle is 109.5° and that all bond lengths are equal at equilibrium. Trigonometry gives I_{aa}^e, the moment of inertia about the CD bond (the a axis), as $3m_H[r_e \cos(109.5° - 90°)]^2$. Using this equation with $A_e = 5.35$ cm^{-1}, one determines that the CH bond length is $r_e = 1.08$ Å.

For an oblate rotor, the rotational Schrödinger equation is

$$\hbar^{-2}[B_e(\hat{J}_a^2 + \hat{J}_b^2) + C_e\hat{J}_c^2]\Phi_{rot} = F_{rot}\Phi_{rot}. \quad (5.14)$$

Choosing the c, a and b axes as the z, x and y axes, respectively, we obtain exactly the same wave equation as for the prolate rotor except that A_e is replaced by C_e. Thus, for an oblate rotor (in cm^{-1}),

$$F_{rot} = B_e J(J + 1) - (B_e - C_e)K^2 \quad (5.15)$$

and Φ_{rot} is as given in equation (5.13). For an oblate symmetric top, the rotational energies decrease with increasing K for a given value of J, whereas for a prolate rotor the rotational energies increase with increasing K for a given value of J.

The H_3^+ molecular ion is an oblate symmetric top with an equilibrium bond length r_e of 0.877 Å. The c axis is perpendicular to the molecular plane so that $C_e \approx 16.858/[3 \times 1.0078(0.5r_e/\cos 30°)^2] = 21.8$ cm^{-1}. The b axis is chosen to pass through one of the protons in the plane of the molecule, so that $B_e \approx 16.858/[2 \times 1.0078(0.5r_e)^2] = 43.5$ cm^{-1}.

5.3.2   Linear molecules

Linear molecules (i.e. molecules having a linear equilibrium structure) have only two rotational degrees of freedom and, as a consequence, an N-atomic linear molecule has $(3N - 5)$ vibrational degrees of freedom. There are $(N - 1)$ one-dimensional stretching normal modes and $(N - 2)$ two-dimensional bending normal modes. Each pair of two-dimensional normal modes (Q_{ra}, Q_{rb}) is described by the pair of coordinates (Q_r, α_r) given in equation (4.36) appropriate for a two-dimensional harmonic-oscillator and each such pair of modes has a vibrational angular momentum $l_r \hbar$ about the linear (z) axis according to equation (4.39). The total vibrational angular momentum is given by

$$l\hbar = \sum_{r=1}^{N-2} l_r \hbar. \tag{5.16}$$

At linear nuclear configurations, $\hat{H}_{\text{elec}}$ commutes with the electronic angular momentum operator $\hat{L}_z$; the electronic angular momentum about the z axis is given by $\Lambda \hbar$, where, at linearity,

$$\hat{L}_z \Phi_{\text{elec}} = \Lambda \hbar \Phi_{\text{elec}}. \tag{5.17}$$

The rigid-rotor rotational Hamiltonian[1] of a linear molecule (with z as the axis of linearity) is, in cm^{-1},

$$\hat{H}_{\text{rot}}^0 = B_e \hbar^{-2} (\hat{J}_x^2 + \hat{J}_y^2) \tag{5.18}$$

with eigenvalues

$$F_{\text{rot}} = B_e [J(J + 1) - K^2] \tag{5.19}$$

and eigenfunctions

$$\Phi_{\text{rot}} = [1/(2\pi)]^{1/2} S_{Jkm}(\theta, \phi) e^{ik\chi} \tag{5.20}$$

where k is restricted by

$$k = l + \Lambda \tag{5.21}$$

because there can be no nuclear rotational (orbital) contribution to the angular momentum about the z axis.

For a diatomic molecule, there is no vibrational angular momentum, and $l = 0$. For CO in its ground electronic state, $\Lambda = 0$ and so $k = 0$. The energy levels of $^{12}\text{C}^{16}\text{O}$ plotted in figure 1.5 were calculated using equation (5.19) with $K = 0$ and the experimentally determined value of $B_e = 1.931 \text{ cm}^{-1}$. The moment of inertia of a CO molecule is

$$I_{bb}^e = \mu r_e^2 \tag{5.22}$$

---

[1] To be technically correct, this should be called the *isomorphic* Hamiltonian. This way of treating a linear molecule leads to the introduction of χ as an extra rotational variable and this allows the use of the symmetric top wavefunctions as long as k is restricted to be the sum in equation (5.21).

where the reduced mass

$$\mu = m_C m_O / (m_C + m_O). \qquad (5.23)$$

For $^{12}C^{16}O$, using $(B_e/\text{cm}^{-1}) \approx 16.858/(I_{bb}^e/\text{u Å}^2)$, with $B_e = 1.931 \text{ cm}^{-1}$, one obtains $r_e = 1.128 \text{ Å}$.

5.3.3 Spherical top molecules

Spherical top molecules are such that the moment of inertia about any axis passing through the centre of mass is the same and the products of inertia vanish regardless of how the xyz axes are oriented in the molecule. The methane molecule CH_4 is an example of such a molecule and, for methane, we chose the xyz axes to be oriented so that the z axis is along one CH bond and the y axis in an HCH plane. For a spherical top molecule, $A_e = B_e = C_e$ and the rigid-rotor Schrödinger equation is

$$\hbar^{-2} B_e \hat{J}^2 \Phi_{\text{rot}} = F_{\text{rot}} \Phi_{\text{rot}} \qquad (5.24)$$

so that

$$F_{\text{rot}} = B_e J (J + 1) \qquad (5.25)$$

and the wavefunctions are as for a symmetric top molecule given in equation (5.13). For the methane molecule, equation (5.25) was used to calculate its rotational energies in figure 1.5 with $B_e = 5.35 \text{ cm}^{-1}$ (i.e. $r_e = 1.08 \text{ Å}$).

The rigid-rotor rotational eigenfunctions of all symmetric top, linear and spherical top molecules are the same function [given in detail in equation (5.56)] of the Euler angles θ, ϕ and χ and we see that the function does not involve the rotational constants of the molecule; we call the wavefunction the *symmetric top wavefunction* and write it $|J, k, m\rangle$. For a linear molecule, $k = l + \Lambda$.

5.3.4 Asymmetric top molecules

For an asymmetric top molecule, the three equilibrium principal moments of inertia are different from each other. Using the convention (called the I^r convention; see section 5.5.4) of identifying the a, b and c axes with the z, x and y axes, respectively, in equation (5.7), we obtain the rigid-rotor rotational Hamiltonian for an asymmetric top molecule as

$$\hat{H}_{\text{rot}}^0 = \hbar^{-2} (A_e \hat{J}_z^2 + B_e \hat{J}_x^2 + C_e \hat{J}_y^2) \qquad (5.26)$$

and, as for all molecules, it commutes with $\hat{J}^2$ and $\hat{J}_\zeta$. However, the asymmetric top rotational Hamiltonian does not commute with $\hat{J}_z$ and the symmetric top wavefunctions are, in general, not eigenfunctions of the asymmetric top Hamiltonian. In this case, the rotational eigenfunctions and eigenvalues are determined by diagonalizing the matrix of $\hat{H}_{\text{rot}}^0$ in the symmetric top basis functions $|J, k, m\rangle$. As a result, the asymmetric top rotational eigenfunctions are linear combinations of the $|J, k, m\rangle$ functions having the same value of J and m, with coefficients that depend on the rotational constants.

The detailed derivation of the asymmetric top wavefunctions and energies is given in section 5.5.4, where it is explained how each asymmetric top energy level correlates with one prolate symmetric top energy level and with one oblate symmetric top level. Because of this, the $2J + 1$ rotational levels of an asymmetric top molecule that have the same value of J are labelled $J_{K_a K_c}$, where the labels K_a and K_c indicate the K values (which are less than or equal to J) of the prolate and oblate levels to which the level correlates. In the $K_a K_c$ subscript, for a given value of J, the label K_a has the $(2J + 1)$ values $0, 1, 1, 2, 2, \ldots, J, J$ going from the bottom level to the top, whereas the K_c label has these values going from the top of the $(2J + 1)$ levels to the bottom. For example, the seven $J = 3$ levels are labelled $3_{03}, 3_{13}, 3_{12}, 3_{22}, 3_{21}, 3_{31}$ and 3_{30}, in order of increasing energy. Asymmetric rotor levels are said to be ee, eo, oe, or oo depending on whether K_a and K_c are even (e) or odd (o), respectively; e.g. the level 3_{21} is an eo level. In figure 1.5, the asymmetric top energy values for the water molecule are given using the values 27.2, 14.6 and 9.5 cm^{-1}, for A_e, B_e and C_e.

5.4 Rovibronic wavefunctions

In this chapter and in the two preceding ones, we have discussed the separation of the rovibronic Schrödinger equation into electronic, vibrational and rotational equations. Zeroth-order approximate rovibronic wavefunctions are obtained as the product of electronic, vibrational and rotational wavefunctions:

$$\Phi^0_{\text{rve},nvr} = \Phi^0_{\text{elec},n}(Q^{(n)}, r_{\text{elec}}, \sigma)\Phi^0_{\text{vib},nv}(Q^{(n)})\Phi^0_{\text{rot},nr}(\theta, \phi, \chi) \qquad (5.27)$$

where we indicate the electronic (n), vibrational (v) and rotational (r) quantum number labels and show that the normal coordinates $Q^{(n)}$ change with electronic state. In the harmonic-oscillator approximation, the vibrational wavefunction is the product of separate harmonic-oscillator wavefunctions in each of the normal coordinates.

An appropriate selection of the infinite set of product wavefunctions given in equation (5.27) can be used as a basis set in forming a truncated matrix of the complete rovibronic Hamiltonian and many different types of off-diagonal matrix element can be non-vanishing. Off-diagonal matrix elements that cause the breakdown of the Born–Oppenheimer separation of the electronic and rovibrational wavefunctions come from the nuclear kinetic energy operator $\hat{T}_N$ [see equation (3.13)]. Off-diagonal matrix elements that spoil the separation of rotation and vibration come from the rotational Hamiltonian terms given in equations (5.36) and (5.37). Off-diagonal matrix elements spoiling the normal mode separation come from the anharmonic potential energy term V_N^{anh} [see equation (4.15)].

5.5   The Hamiltonian and wavefunctions in detail

This is the final section of the chapter. In it we give many of the background details for what has been discussed earlier. It can be looked over cursorily on a first reading.

5.5.1   The derivation of the rigid-rotor Hamiltonian

The complete rotational Hamiltonian is obtained by first subtracting E_{vib} in equation (4.5) from E_{rv} in equation (4.2) to obtain E_{rot}, where this is expressed in terms of the coordinates $(\theta, \phi, \chi, Q_r)$, and their velocities, by using the inverse of equation (5.40), and by using equation (4.9). Secondly, it is necessary to convert the velocities to momenta to obtain the classical expression H_{rot}, which is then converted to the quantum mechanical rotational Hamiltonian operator. In the same way that normal coordinates are introduced in order to simplify the solution of the vibrational Schrödinger equation, so it is useful to introduce the xyz components of the angular momentum operator, in place of the momenta that are conjugate to the Euler angles, in the rotational Hamiltonian. The lengthy derivation of $\hat{H}_{\text{rot}}$ from E_{rot} is associated with the names of Podolsky, Darling and Dennison, Wilson and Howard, and Watson. The rotational Hamiltonian operator $\hat{H}_{\text{rot}}$ (in cm^{-1}) that is finally obtained is

$$\hat{H}_{\text{rot}} = \frac{1}{2hc} \sum_{\alpha,\beta} \mu_{\alpha\beta}(\hat{J}_\alpha - \hat{p}_\alpha - \hat{L}_\alpha)(\hat{J}_\beta - \hat{p}_\beta - \hat{L}_\beta) + U \tag{5.28}$$

where α and $\beta = x$, y or z; it is necessary to define $\mu_{\alpha\beta}$, $\hat{J}_\alpha$, $\hat{p}_\alpha$, $\hat{L}_\alpha$ and U.

The elements of the matrix $\boldsymbol{\mu}$, when expanded about equilibrium as a Taylor's series in the normal coordinates, are given by

$$\mu_{\alpha\beta} = \mu^e_{\alpha\beta} - \sum_r \mu^e_{\alpha\alpha} a_r^{\alpha\beta} \mu^e_{\beta\beta} Q_r + \frac{3}{4} \sum_{r,s,\gamma} \mu^e_{\alpha\alpha} a_r^{\alpha\gamma} \mu^e_{\gamma\gamma} a_s^{\gamma\beta} \mu^e_{\beta\beta} Q_r Q_s + \cdots \tag{5.29}$$

where $\mu^e_{\alpha\beta} = \{[\boldsymbol{I}^e]^{-1}\}_{\alpha\beta}$ is an element of the inverse of the moment of inertia matrix for the molecule in its equilibrium configuration (only diagonal elements $\mu^e_{\alpha\alpha}$ are non-vanishing) and the coefficients $a_r^{\alpha\beta}$ depend on the equilibrium nuclear geometry, on the atomic masses and on the elements of the l matrix, see equation (4.9), according to

$$a_r^{\alpha\beta} = 2 \sum_{i=1}^N m_i^{1/2} (\delta_{\alpha\beta} \sum_\gamma \gamma_i^e l_{\gamma i,r} - \alpha_i^e l_{\beta i,r}). \tag{5.30}$$

Symmetry can be used to determine that some of the coefficients $a_r^{\alpha\beta}$ vanish; see equation (11.44). The $\hat{J}_\alpha$ and $\hat{J}_\beta$ are xyz components of the angular momentum operator [see section 2.7 and equations (5.45)–(5.47)]. The $\hat{p}_\alpha$ and $\hat{p}_\beta$ are xyz components of the vibrational angular momentum operator; they are given by

$$\hat{p}_\alpha = \sum_{r,s} \zeta_{r,s}^\alpha Q_r \hat{P}_s \qquad (5.31)$$

where the $\zeta_{r,s}^\alpha$ (Coriolis coupling constants) depend on the l matrix according to

$$\zeta_{r,s}^x = -\zeta_{s,r}^x = \sum_{i=1}^{N} (l_{yi,r} l_{zi,s} - l_{zi,r} l_{yi,s}) \qquad (5.32)$$

and cyclically for the y and z coefficients. Symmetry can be used to determine that some of the coefficients $\zeta_{r,s}^\alpha$ vanish; see equation (11.45). The $\hat{L}_\alpha$ are the xyz components of the electronic angular momentum. The term U is given by

$$U = -\frac{\hbar^2}{8hc} \sum_\alpha \mu_{\alpha\alpha} \qquad (5.33)$$

which can be considered as a mass-dependent addition to the potential energy function.

We write $\hat{H}_{\text{rot}}$ as

$$\hat{H}_{\text{rot}} = \hat{H}_{\text{rot}}^0 + \hat{H}_{\text{rot}}' \qquad (5.34)$$

where

$$\hat{H}_{\text{rot}}^0 = \frac{1}{2hc} \sum_\alpha \mu_{\alpha\alpha}^e \hat{J}_\alpha{}^2 = \frac{1}{2hc} \sum_\alpha \frac{1}{I_{\alpha\alpha}^e} \hat{J}_\alpha{}^2 \qquad (5.35)$$

and

$$\hat{H}_{\text{rot}}' = \frac{1}{2hc} \sum_{\alpha,\beta} (\mu_{\alpha\beta} - \mu_{\alpha\beta}^e) \hat{J}_\alpha \hat{J}_\beta \qquad (5.36)$$

$$- \frac{1}{hc} \sum_{\alpha,\beta} \mu_{\alpha\beta} \hat{J}_\alpha \hat{p}_\beta \qquad (5.37)$$

$$- \frac{1}{hc} \sum_{\alpha,\beta} \mu_{\alpha\beta} \hat{J}_\alpha \hat{L}_\beta \qquad (5.38)$$

$$+ \frac{1}{2hc} \sum_{\alpha,\beta} \mu_{\alpha\beta} (\hat{p}_\alpha + \hat{L}_\alpha)(\hat{p}_\beta + \hat{L}_\beta) + U. \qquad (5.39)$$

The eigenvalues of $\hat{H}_{\text{rot}}^0$ give the energy of the molecule as it rotates with its geometry rigidly held at the equilibrium geometry; hence, it is called the rigid-rotor Hamiltonian. It does not involve the vibrational degrees of freedom and it has been obtained from $\hat{H}_{\text{rot}}$ by neglecting $\hat{H}_{\text{rot}}'$, that is by neglecting all but the leading term in equation (5.29), by neglecting the vibrational and electronic

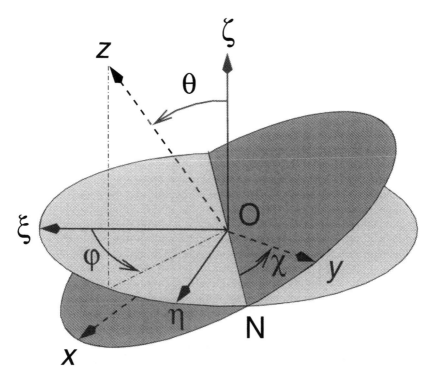

Figure 5.3. The definition of the Euler angles (θ, ϕ, χ) that relate the orientation of the molecule-fixed xyz axes to the $\xi\eta\zeta$ axes. The origin of both axis systems is at the nuclear centre of mass O and the node line ON is directed so that a right-handed screw is driven along ON in its positive direction by twisting it from ζ to z through θ where $0 \leq \theta \leq \pi$. ϕ and χ have the ranges 0 to 2π. χ is measured from the node line.

angular momenta, $\hat{p}_\alpha$ and $\hat{L}_\alpha$ and by neglecting U. Because of the presence of $\hat{H}'_{\text{rot}}$, we cannot separate rotation completely from the vibrational and electronic motion. The terms in equations (5.36) and (5.37) cause centrifugal distortion and vibrational Coriolis coupling, and the effects of these rotation–vibration coupling terms are discussed in section 11.5.

5.5.2 The Euler angles and angular momentum

The definition of the Euler angles (θ, ϕ, χ) that specify the orientation of the xyz axes within the $\xi\eta\zeta$ axes is given in the three-dimensional figure 5.3. We 'pulled this apart' in figures 5.1 and 5.2 to clarify the separate definitions of (θ, ϕ) and χ, respectively.

The coordinates (x_r, y_r, z_r) of particle r (a nucleus or an electron) in a molecule are related to its (ξ_r, η_r, ζ_r) coordinates, using the notation of matrix multiplication, by:

$$
\begin{bmatrix} x_r \\ y_r \\ z_r \end{bmatrix} = \begin{bmatrix} \lambda_{x\xi} & \lambda_{x\eta} & \lambda_{x\zeta} \\ \lambda_{y\xi} & \lambda_{y\eta} & \lambda_{y\zeta} \\ \lambda_{z\xi} & \lambda_{z\eta} & \lambda_{z\zeta} \end{bmatrix} \begin{bmatrix} \xi_r \\ \eta_r \\ \zeta_r \end{bmatrix} \tag{5.40}
$$

where

$$
\lambda_{x\xi} = \cos(x\hat{O}\xi) \qquad \text{etc} \tag{5.41}
$$

and these are elements of the direction cosine matrix. The direction cosines are functions of the Euler angles:

$$
\begin{aligned}
\lambda_{x\xi} &= \cos\theta\cos\phi\cos\chi - \sin\phi\sin\chi \\
\lambda_{x\eta} &= \cos\theta\sin\phi\cos\chi + \cos\phi\sin\chi \\
\lambda_{x\zeta} &= -\sin\theta\cos\chi \\
\lambda_{y\xi} &= -\cos\theta\cos\phi\sin\chi - \sin\phi\cos\chi \\
\lambda_{y\eta} &= -\cos\theta\sin\phi\sin\chi + \cos\phi\cos\chi \\
\lambda_{y\zeta} &= \sin\theta\sin\chi \\
\lambda_{z\xi} &= \sin\theta\cos\phi \\
\lambda_{z\eta} &= \sin\theta\sin\phi \\
\lambda_{z\zeta} &= \cos\theta.
\end{aligned} \tag{5.42}
$$

Using equation (5.40) and the elements of the direction cosine matrix, we can refer the angular momentum to molecule-fixed axes:

$$
\hat{J}_\alpha = \lambda_{\alpha\xi}\hat{J}_\xi + \lambda_{\alpha\eta}\hat{J}_\eta + \lambda_{\alpha\zeta}\hat{J}_\zeta \tag{5.43}
$$

where $\alpha = x$, y or z. From section 2.7, we have

$$
\hat{J}_\xi = -i\hbar \sum_{r=1}^{l} \left(\eta_r \frac{\partial}{\partial \zeta_r} - \zeta_r \frac{\partial}{\partial \eta_r} \right) \tag{5.44}
$$

with the expressions for $\hat{J}_\eta$ and $\hat{J}_\zeta$ being obtained by cyclic permutation of ξ, η and ζ. From these equations, we can derive

$$
\hat{J}_x = \sin\chi\,\hat{P}_\theta - \csc\theta\cos\chi\,\hat{P}_\phi + \cot\theta\cos\chi\,\hat{P}_\chi \tag{5.45}
$$

$$
\hat{J}_y = \cos\chi\,\hat{P}_\theta + \csc\theta\sin\chi\,\hat{P}_\phi - \cot\theta\sin\chi\,\hat{P}_\chi \tag{5.46}
$$

and

$$
\hat{J}_z = \hat{P}_\chi. \tag{5.47}
$$

In these equations,

$$
\hat{P}_A = -i\hbar \frac{\partial}{\partial A} \tag{5.48}
$$

where $A = \theta$, ϕ or χ. Summing the squares, we obtain

$$
\hat{J}^2 = \hat{J}_x^2 + \hat{J}_y^2 + \hat{J}_z^2. \tag{5.49}
$$

5.5.3 The symmetric top wavefunctions

The angular momentum operators $\hat{J}^2$, $\hat{J}_\zeta$ and $\hat{J}_z$ commute with each other and their (simultaneous) eigenfunctions are well known from angular momentum theory to be the so-called rotation matrices $D_{mk}^{(J)}(\phi, \theta, \chi)$. The eigenvalues of $\hat{J}^2$, $\hat{J}_z$ and $\hat{J}_\zeta$ are $J(J+1)\hbar^2$, $k\hbar$ and $m\hbar$, respectively, where the three quantum numbers J, k, and m can have the values

$$J = 0, 1, 2, \ldots \quad k = 0, \pm 1, \pm 2, \ldots, \pm J \quad \text{and} \quad m = 0, \pm 1, \pm 2, \ldots, \pm J. \tag{5.50}$$

The symmetric top Hamiltonian [see, for example, equation (5.11)] commutes with $\hat{J}^2$, $\hat{J}_\zeta$ and $\hat{J}_z$, and so its eigenfunctions are these same D functions, appropriately normalized.

Normalized symmetric top eigenfunctions are

$$\Phi_{\mathrm{rot}}(\theta, \phi, \chi) = [(2J+1)/(8\pi^2)]^{1/2}[D_{mk}^{(J)}(\phi, \theta, \chi)]^* \tag{5.51}$$

$$= [1/(2\pi)]^{1/2} S_{Jkm}(\theta, \phi) e^{ik\chi} \tag{5.52}$$

$$= |J, k, m\rangle \tag{5.53}$$

where

$$S_{J0m}(\theta, \phi) = Y_{Jm}(\theta, \phi) \tag{5.54}$$

$$= [1/(2\pi)]^{1/2} \Theta_{Jm}(\theta) e^{im\phi}, \tag{5.55}$$

$Y_{Jm}(\theta, \phi)$ is a spherical harmonic function and $\Theta_{Jm}(\theta)$ is a normalized associated Legendre polynomial. The explicit form of the wavefunction $|J, k, m\rangle$ is

$$N\left\{\sum_\sigma (-1)^\sigma \frac{(\cos \tfrac{1}{2}\theta)^{2J+k-m-2\sigma}(-\sin \tfrac{1}{2}\theta)^{m-k+2\sigma}}{\sigma!(J-m-\sigma)!(m-k+\sigma)!(J+k-\sigma)!}\right\} e^{im\phi} e^{ik\chi} \tag{5.56}$$

where

$$N = [(J+m)!(J-m)!(J+k)!(J-k)!(2J+1)/(8\pi^2)]^{1/2}.$$

The index σ in the sum runs from 0 or $(k-m)$, whichever is the larger, up to $(J-m)$ or $(J+k)$, whichever is the smaller.

5.5.4 The asymmetric top wavefunctions and energies

The three principal axes of an asymmetric top molecule are by convention labelled a, b and c, so that a is the axis about which the moment of inertia is the least, and c is the axis about which it is the greatest, i.e. so that $I_{aa} < I_{bb} < I_{cc}$. Depending on whether the z axis is identified with the a, b or c axis, we name the convention adopted as type I, II or III. We add a superscript r or l depending on whether

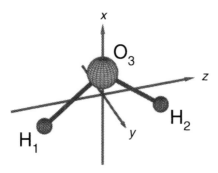

Figure 5.4. A water molecule in its equilibrium configuration, for the ground electronic state, with molecule-fixed (x, y, z) axes attached using a I^r convention.

Table 5.1. Angular momentum[a] matrix elements.

$\langle J, k, m\|\hat{J}^2\|J, k, m\rangle$	$=$	$J(J+1)\hbar^2$
$\langle J, k, m\|\hat{J}_z\|J, k, m\rangle$	$=$	$k\hbar$
$\langle J, k-1, m\|\hat{J}_m^+\|J, k, m\rangle$	$=$	$\hbar[J(J+1) - k(k-1)]^{1/2}$
$\langle J, k+1, m\|\hat{J}_m^-\|J, k, m\rangle$	$=$	$\hbar[J(J+1) - k(k+1)]^{1/2}$

[a] In a basis of symmetric top wavefunctions, where $\hat{J}_m^\pm = \hat{J}_x \pm i\hat{J}_y$.

a right- or left-handed xyz axis system is used. As an example, for the water molecule, we adopt a I^r convention (see figure 5.4) in which the z axis is located so that H_2 has a positive z coordinate, the x axis is located so that the oxygen nucleus has a positive x coordinate, and the y axis is located so that the axis system is right-handed.

If we use the I^r convention for an asymmetric top molecule, the rigid-rotor rotational Hamiltonian, from equation (5.7), is

$$\hat{H}_{\text{rot}}^0 = \hbar^{-2}(A_e\hat{J}_z^2 + B_e\hat{J}_x^2 + C_e\hat{J}_y^2) \qquad (5.57)$$

and, as we stated after equation (5.26), it commutes with $\hat{J}^2$ and $\hat{J}_\zeta$ but not with $\hat{J}_z$. This means that the asymmetric top eigenfunctions will be linear combinations of $|J, k, m\rangle$ functions that have the same values of J, m but different values of k.

The rotational eigenfunctions and eigenvalues for an asymmetric top molecule are determined by diagonalizing the matrix of the rigid-rotor

Hamiltonian using the symmetric top functions $|J, k, m\rangle$ as basis functions. To do this, we rewrite the Hamiltonian as

$$\hat{H}_{rot}^0 = \hbar^{-2}\{\tfrac{1}{2}(B_e + C_e)\hat{J}^2 + [A_e - \tfrac{1}{2}(B_e + C_e)]\hat{J}_z^2$$
$$+ \tfrac{1}{4}(B_e - C_e)[(\hat{J}_m^+)^2 + (\hat{J}_m^-)^2]\} \qquad (5.58)$$

and we use the results in table 5.1 to determine that the matrix elements of the operators occurring in this expression are:

$$\langle J, k, m|\hat{J}^2|J, k, m\rangle = J(J+1)\hbar^2 \qquad (5.59)$$
$$\langle J, k, m|\hat{J}_z^2|J, k, m\rangle = k^2\hbar^2 \qquad (5.60)$$
$$\langle J, k-2, m|(\hat{J}_m^+)^2|J, k, m\rangle = \{[J(J+1) - (k-1)(k-2)]$$
$$\times [J(J+1) - k(k-1)]\}^{1/2}\hbar^2 \qquad (5.61)$$

and

$$\langle J, k+2, m|(\hat{J}_m^-)^2|J, k, m\rangle$$
$$= \{[J(J+1) - (k+1)(k+2)][J(J+1) - k(k+1)]\}^{1/2}\hbar^2. \qquad (5.62)$$

Using a type I′ convention, the k quantum number refers to angular momentum around the a axis. It is called k_a if one needs to distinguish it from the situation that would obtain if we used a III′ convention since then the k quantum number would refer to angular momentum about the c axis and be called k_c. Using a I′ convention, we call it k.

Because there are no off-diagonal matrix elements between $|J, k, m\rangle$ basis functions having different J or m values, the Hamiltonian matrix can be organized into *blocks* so that each block is characterized by particular values of J and m; there are no off-diagonal matrix elements between the blocks. This is depicted in figure 5.5.

The Hamiltonian matrix is *block diagonal* and we can diagonalize each block separately. Each block is a $(2J + 1) \times (2J + 1)$ square matrix with rows and columns labelled by the $(2J + 1)$ values of $k = -J, -J + 1, \ldots, +J$. For a given value of J, there are $(2J + 1)$ identical blocks like this as $m = -J, -J + 1, \ldots, +J$ and this gives a $(2J + 1)$-fold m-degeneracy to every rotational level. To calculate the energies, we set up and diagonalize only the $m = 0$ blocks. The $(J, m = 0)$ blocks can be further block diagonalized, as we show later, but first we look at the $J = 0$ and 1 situations.

For $J = m = 0$, we have a single basis function $|0, 0, 0\rangle = (8\pi^2)^{-(1/2)}$ and the diagonal matrix element is zero. Thus, for the asymmetric top, $\Phi_{rot}(J = 0) = (8\pi^2)^{-(1/2)}$ and $F_{rot} = 0$.

For $J = 1$ and $m = 0$, there are three basis functions $|1, -1, 0\rangle$, $|1, 0, 0\rangle$ and $|1, +1, 0\rangle$, and the $(J = 1, m = 0)$ block is a 3×3 matrix. If we use the plus and minus combinations of the $K = 1$ functions:

$$|1, 1, 0; \pm\rangle = [|1, +1, 0\rangle \pm |1, -1, 0\rangle]/\sqrt{2} \qquad (5.63)$$

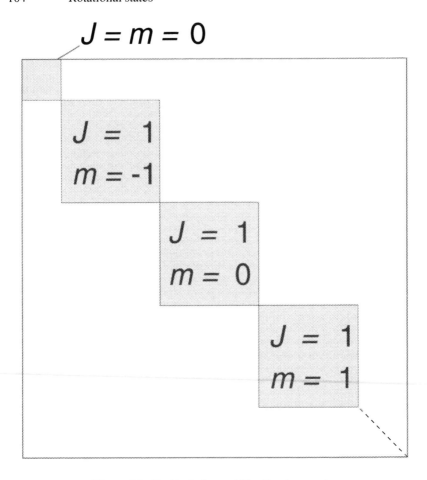

Figure 5.5. The block diagonal Hamiltonian matrix.

together with the $|1, 0, 0\rangle$ function, as the three basis functions, then the ($J = 1, m = 0$) matrix block is diagonal; all off-diagonal matrix elements are zero. That is,

$$\langle 1, 1, 0; +|\hat{H}^0_{rot}|1, 1, 0; -\rangle = \langle 1, 1, 0; \pm|\hat{H}^0_{rot}|1, 0, 0\rangle = 0. \qquad (5.64)$$

Since the matrix block is diagonal the basis functions are the eigenfunctions and the eigenvalues are the diagonal elements, which are:

$$F_{rot}(J = 1, K = 1, +) = \langle 1, 1, 0; +|\hat{H}^0_{rot}|1, 1, 0; +\rangle = A_e + B_e \quad (5.65)$$
$$F_{rot}(J = 1, K = 1, -) = \langle 1, 1, 0; -|\hat{H}^0_{rot}|1, 1, 0; -\rangle = A_e + C_e \quad (5.66)$$

and

$$F_{rot}(J = 1, K = 0) = \langle 1, 0, 0|\hat{H}^0_{rot}|1, 0, 0\rangle = B_e + C_e. \qquad (5.67)$$

The matrix elements in equations (5.64)–(5.66) are evaluated by expanding them as sums (and differences) of matrix elements involving the primitive basis functions $|1, k, 0\rangle$. The primitive matrix elements [a primitive matrix element is in equation (5.67)] are evaluated using the results in equations (5.59)–(5.62) with $\hat{H}^0_{rot}$ expressed as in equation (5.58).

For the water molecule, A_e, B_e and C_e have the values 27.2, 14.6 and 9.5 cm^{-1}. Thus the three $J = 1$ term values are 41.8, 36.7 and 24.1 cm^{-1}, and the $K_a = 1$ splitting $(B_e - C_e)$ is 5.1 cm^{-1}.

For J greater than 1, each (J, m) block of the Hamiltonian matrix factorizes into four blocks if we use sum and difference basis functions for each K value and if we separate the odd K and even K functions. The off-diagonal matrix elements between $+$ and $-$ functions, and between odd K and even K functions, vanish. The four blocks obtained are the E^+, E^-, O^+ and O^- blocks as the basis functions are $+$ or $-$ combinations of the even(E) and odd(O) K primitive basis functions $|J, k, 0\rangle$. This block diagonalization of the (J, m) blocks follows from the results in equations (5.59)–(5.62). For even J, the E^+ block has dimension $(J+2)/2$ and the other three blocks have dimension $J/2$; for odd J, the E^- block has dimension $(J - 1)/2$ and the other three blocks have dimension $(J + 1)/2$.

For $J = 2$, using $A^\pm$ to represent $E^\pm$ or $O^\pm$, the appropriate basis functions $|J, K, m; A^\pm\rangle$ are:

$$|2, 2, 0; E^+\rangle = [|2, +2, 0\rangle + |2, -2, 0\rangle]\sqrt{2} \tag{5.68}$$

$$|2, 2, 0; E^-\rangle = [|2, +2, 0\rangle - |2, -2, 0\rangle]\sqrt{2} \tag{5.69}$$

$$|2, 1, 0; O^+\rangle = [|2, +1, 0\rangle + |2, -1, 0\rangle]\sqrt{2} \tag{5.70}$$

$$|2, 1, 0; O^-\rangle = [|2, +1, 0\rangle - |2, -1, 0\rangle]\sqrt{2} \tag{5.71}$$

and

$$|2, 0, 0; E^+\rangle = |2, 0, 0\rangle. \tag{5.72}$$

The only non-vanishing off-diagonal matrix element of $\hat{H}^0_{rot}$ between the five functions is that between the two E^+ functions. From the three 1×1 blocks, we obtain energies $F_{rot}(J, A^\pm)$ given by

$$F_{rot}(2, E^-) = 4A_e + B_e + C_e \tag{5.73}$$

$$F_{rot}(2, O^+) = A_e + 4B_e + C_e \tag{5.74}$$

and

$$F_{rot}(2, O^-) = A_e + B_e + 4C_e. \tag{5.75}$$

The 2×2 block of $\hat{H}^0_{rot}$ for the $J = 2$ E^+ functions is

	$|2, 0, 0; E^+\rangle$	$|2, 2, 0; E^+\rangle$
$\langle 2, 0, 0; E^+|$	$3(B_e + C_e)$	$\sqrt{3}(B_e - C_e)$
$\langle 2, 2, 0; E^+|$	$\sqrt{3}(B_e - C_e)$	$4A_e + B_e + C_e$

From the analysis of a 2×2 matrix given in equations (2.31)–(2.40), we determine that the lower eigenvalue of the matrix is

$$F_{\text{rot}}^-(2, E^+) = 3(B_e + C_e) - S \tag{5.76}$$

and the upper eigenvalue of the matrix is

$$F_{\text{rot}}^+(2, E^+) = (4A_e + B_e + C_e) + S \tag{5.77}$$

where the energy level shift is

$$S = \sqrt{3(B_e - C_e)^2 + 4(A_e - \bar{B}_e)^2 - 2(A_e - \bar{B}_e)} \tag{5.78}$$

and $\bar{B}_e = (B_e + C_e)/2$. The eigenfunctions $\Phi_{\text{rot}}^\pm(J, E^+)$ are

$$\Phi_{\text{rot}}^-(2, E^+) = c^+|2, 0, 0; E^+\rangle - c^-|2, 2, 0; E^+\rangle \tag{5.79}$$

and

$$\Phi_{\text{rot}}^+(2, E^+) = c^+|2, 2, 0; E^+\rangle + c^-|2, 0, 0; E^+\rangle \tag{5.80}$$

where

$$c^\pm = \frac{1}{\sqrt{2}} \left\{ 1 \pm \frac{(2A_e - B_e - C_e)}{[3(B_e - C_e)^2 + (2A_e - B_e - C_e)^2]^{1/2}} \right\}^{1/2}. \tag{5.81}$$

For the water molecule, $S = 1.3$ cm^{-1}, $c^+ = 0.99$ and $c^- = 0.14$. The two E^+ functions are a mixture of $K = 0$ and $K = 2$ functions with coefficients that depend on the rotational constants. As a result, the functions are not eigenfunctions of $\hat{J}_z$ and we say that K (i.e. $K_a = |k_a|$) is not a *good* quantum number for them. However, K_a can be used as a label on the energy levels, to specify to which prolate rotor K-state the level would correlate as $(B_e \rightarrow C_e)$.

One could repeat this using a III$'$ convention. Equation (5.57) would become

$$\hat{H}_{\text{rot}}^0 = \hbar^{-2}(A_e \hat{J}_x^2 + B_e \hat{J}_y^2 + C_e \hat{J}_z^2) \tag{5.82}$$

and equation (5.58) would become

$$\hat{H}_{\text{rot}}^0 = \hbar^{-2}\{\tfrac{1}{2}(A_e + B_e)\hat{J}^2 + [C_e - \tfrac{1}{2}(A_e + B_e)]\hat{J}_z^2$$
$$+ \tfrac{1}{4}(A_e - B_e)(\hat{J}_+^2 + \hat{J}_-^2)\}. \tag{5.83}$$

We would set up the Hamiltonian matrix using sums and differences of the symmetric top functions $|J, k_c, m\rangle$ as basis functions. We would get the same energies (these are just functions of A_e, B_e and C_e) but now the eigenfunctions would be linear combinations of the functions $|J, k_c, m\rangle$ and we could label the levels using the $K_c(= |k_c|)$ label of the limiting oblate rotor as $(B_e \rightarrow A_e)$.

The degree of asymmetry in an asymmetric top is given by the value of

$$\kappa = (2B_e - A_e - C_e)/(A_e - C_e). \tag{5.84}$$

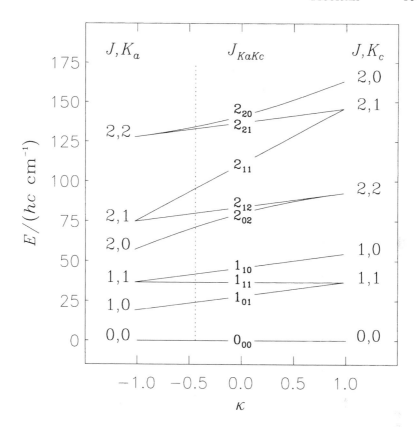

Figure 5.6. The correlation of the $J = 0, 1$ and 2 energy levels of the water molecule ($\kappa = -0.42$) with the $\kappa = \pm 1$ symmetric top limits .The value of B_e ($= 14.6$ cm^{-1}) is changed from being equal to C_e ($= 9.5$ cm^{-1}) on the left (at the prolate top limit), to being equal to A_e ($= 27.2$ cm^{-1}) on the right (at the oblate top limit).

For a prolate top, $B_e = C_e$ and $\kappa = -1$; for an oblate top, $B_e = A_e$ and $\kappa = +1$; and for an asymmetric top, $-1 < \kappa < +1$, with $\kappa = 0$ being the 'most' asymmetric when B_e is halfway between A_e and C_e.

In figure 5.6, we show the correlation of the $J = 0, 1$ and 2 energy levels of the water molecule ($\kappa = -0.42$) with the $\kappa = \pm 1$ symmetric top limits by changing the value of B_e from being equal to C_e on the left at the prolate top limit to being equal to A_e on the right at the oblate top limit.

5.6 Problems

5.1   The molecule ^{79}Br^{19}F has $B_e = 0.35717$ cm^{-1}. Calculate r_e. If the ^{79}Br^{19}F molecule were rotating classically at the constant energy given by

Table 5.2. Rotational transition wavenumbers of CH_3F (in cm^{-1}).

J'	K'	J''	K''	$\tilde{\nu}$	J'	K'	J''	K''	$\tilde{\nu}$
2	0	1	0	3.40475	4	0	3	0	6.80912
2	1	1	1	3.40470	4	1	3	1	6.80900
3	0	2	0	5.10701	4	2	3	2	6.80865
3	1	2	1	5.10692	4	3	3	3	6.80806
3	2	2	2	5.10665					

J' and K' are upper state quantum numbers, J'' and K'' are lower state quantum numbers.

equation (5.19) with $K = 0$, how many revolutions per minute would it be doing at $J = 0$, $J = 1$ and $J = 10$? *Hint*: Use the classical relation $E = I_{bb}^e \omega^2 / 2$, where ω is the angular velocity in radians per second.

5.2   For $H^{35}Cl$, $B_e = 10.5909$ cm^{-1}. Determine B_e for $H^{37}Cl$ and $D^{35}Cl$.

5.3   Determine the equilibrium bond length and bond angle in H_2O from the values of the equilibrium rotational constants: $A_e = 27.2$ cm^{-1}, $B_e = 14.6$ cm^{-1} and $C_e = 9.5$ cm^{-1}.

5.4   In a planar molecule, the equilibrium coordinates of nucleus i in the molecule-fixed axis system xyz can be written as $(x_i^e, y_i^e, 0)$. Show that, for a planar molecule,

$$I_{zz}^e = I_{xx}^e + I_{yy}^e.$$

5.5   In its equilibrium configuration CH_3F is a prolate symmetric top and the rotational term values in the rigid-rotor approximation are given by equation (5.12). In a better approximation, where we take into account rotation–vibration interaction [see section 11.5 and, in particular, equation (11.53)], the rotational term values are given by

$$F_{rot} = B_e J(J+1) + (A_e - B_e)K^2$$
$$- D_J J^2 (J+1)^2 - D_{JK} J(J+1)K^2 - D_K K^4$$

where D_J, D_{JK} and D_K are centrifugal distortion constants. In the rotational spectrum of CH_3F, the allowed transitions obey the selection rules $(J+1, K) \leftarrow (J, K)$ (see chapter 12). Table 5.2 lists the observed wavenumbers of some CH_3F rotational transitions. Which of the five parameters B_e, A_e, D_J, D_{JK} and D_K influence the measured wavenumbers? Use the wavenumbers of the two transitions $(J, K) = (2, 0) \leftarrow (1, 0)$ and $(J, K) = (3, 0) \leftarrow (2, 0)$ to determine the values of B_e and D_J. Calculate the wavenumber of the transition $(J, K) =$

$(4, 0) \leftarrow (3, 0)$ and compare with the measured value. Use the difference between the wavenumbers of the two transitions $(J, K) = (4, 2) \leftarrow (3, 2)$ and $(J, K) = (4, 3) \leftarrow (3, 3)$ to determine D_{JK}. Calculate the wavenumbers of the transitions with $K' = K'' > 0$ and compare the results with the measured values.

5.6   From equation (2.19), the commutator of the operators $\hat{A}$ and $\hat{B}$ is

$$[\hat{A}, \hat{B}] = \hat{A}\hat{B} - \hat{B}\hat{A}.$$

Show that, for four arbitrary operators $\hat{A}$, $\hat{B}$, $\hat{C}$ and $\hat{D}$,

(a) $[\hat{A} + \hat{B}, \hat{C} + \hat{D}] = [\hat{A}, \hat{C}] + [\hat{A}, \hat{D}] + [\hat{B}, \hat{C}] + [\hat{B}, \hat{D}]$
(b) $[\hat{A}, \hat{B}\hat{C}] = \hat{B}[\hat{A}, \hat{C}] + [\hat{A}, \hat{B}]\hat{C}$ and
(c) $[\hat{A}\hat{B}, \hat{C}] = \hat{A}[\hat{B}, \hat{C}] + [\hat{A}, \hat{C}]\hat{B}$.

5.7*   The angular momentum operator component $\hat{J}_\xi$ is defined in equation (5.44). By making cyclic permutations of ξ, η and ζ determine the expressions for the components $\hat{J}_\eta$ and $\hat{J}_\zeta$. Use the answers to problem 5.6 to derive the commutators $[\hat{J}_\xi, \hat{J}_\eta]$, $[\hat{J}_\zeta, \hat{J}_\xi]$ and $[\hat{J}_\eta, \hat{J}_\zeta]$. Compare these commutation relations with those given in equation (2.81) for the components of $\hat{\boldsymbol{J}}$ about the XYZ directions that also have space-fixed orientation.

5.8*   Determine $[\hat{J}^2, \hat{J}_\xi]$, $[\hat{J}^2, \hat{J}_\eta]$ and $[\hat{J}^2, \hat{J}_\zeta]$, where

$$\hat{J}^2 = \hat{J}_\xi^2 + \hat{J}_\eta^2 + \hat{J}_\zeta^2.$$

5.9*   We define the two operators

$$\hat{J}_s^+ = \hat{J}_\xi + i\hat{J}_\eta \qquad \text{and} \qquad \hat{J}_s^- = \hat{J}_\xi - i\hat{J}_\eta. \qquad (5.85)$$

Use the answers to problems 5.6 and 5.7 to derive the commutators $[\hat{J}_\zeta, \hat{J}_s^+]$ and $[\hat{J}_\zeta, \hat{J}_s^-]$.

5.10*   Assume that ψ_λ is an eigenfunction of $\hat{J}_\zeta$, i.e.

$$\hat{J}_\zeta \psi_\lambda = \hbar m \psi_\lambda. \qquad (5.86)$$

Consider the function $\hat{J}_s^+ \psi_\lambda$. Use the answers to problems 5.6, 5.7 and 5.9 to show that either this function vanishes or that it is an eigenfunction of $\hat{J}_\zeta$. Determine the corresponding eigenvalue. Repeat this determination for the function $\hat{J}_s^- \psi_\lambda$.

5.11   Repeat problems 5.7–5.10 with angular momentum components $\hat{J}_x$, $\hat{J}_y$ and $\hat{J}_z$ about the molecule-fixed directions xyz, instead of the components $\hat{J}_\xi$, $\hat{J}_\eta$ and $\hat{J}_\zeta$ about the space-fixed directions $\xi\eta\zeta$; the expressions for the operators are given in equations (5.45)–(5.47). Confirm equations (5.61) and (5.62). In solving this problem, it will be necessary to introduce $\hat{J}_m^{\pm}$ (see table 5.1) instead of $\hat{J}_s^{\pm}$ and to replace equation (5.86) by

$$\hat{J}_z \psi_\lambda = \hbar k \psi_\lambda. \qquad (5.87)$$

PART 2

SYMMETRY AND SYMMETRY GROUPS

Chapter 6

Geometrical symmetry

Some objects look exactly like their mirror image and we say that they are symmetrical or, more precisely, that they have *reflection* symmetry. Objects can also have *rotational* symmetry; for example, the letter 'T' is such that if we turn it over by rotating it through 180° about the upright axis it will look the same. Rotation and reflection symmetries are based on geometrical shape, and they are described in a precise way by introducing rotation and reflection *symmetry operations*. Complete sets of such symmetry operations form *point groups*, and the symmetry of a geometrical figure can be classified according to which point group it belongs. If we think of a molecule as being a rigid object of fixed structure, then its symmetry can be described in terms of a point group and this is useful for molecules having small amplitude vibrations in isolated electronic states.

6.1 Geometrical symmetry operations

To be precise about the geometrical symmetry of an object, it is necessary to determine the number and type of *symmetry elements* that it possesses and the symmetry operations that the symmetry elements give rise to. One type of symmetry element is a *rotational symmetry axis*. The corresponding rotational symmetry operations are rotations of the object about this axis which leave the object looking the same. Another type of symmetry element is a *reflection symmetry plane*. The corresponding reflection symmetry operation is a reflection in this plane and it leaves the object looking the same.

For the equilateral-triangle-based pyramid shown in figures 6.1, the symmetry elements are the three-fold rotational symmetry axis C_3 and the three reflection symmetry planes σ_1, σ_2 and σ_3, shown in figure 6.2. The C_3 axis generates two rotational symmetry operations, C_3 and C_3^2. They are right-handed rotations of 120° and 240°, respectively, about the axis. It is customary to use the same notation, C_3, for the rotational symmetry axis and the associated rotational symmetry operation. The positive direction of the C_3 and C_3^2 rotations is such

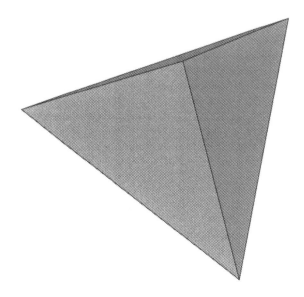

Figure 6.1. An equilateral-triangle-based pyramid. The base of the pyramid is an equilateral triangle and the pyramid height is such that the three triangular faces that form the sides are not equilateral (but isosceles).

that an observer sitting on the tip of the C_3 axis in figure 6.2 will see them as being anti-clockwise. In general, we use the notation C_n for an n-fold rotational symmetry axis. Such an axis generates $n - 1$ rotational symmetry operations $C_n, C_n^2 C_n^3, \ldots, C_n^{n-1}$, where C_n^k is a right-handed rotation of $k \times 360°/n$ about the axis (and $C_n = C_n^1$).

Each of the three reflection symmetry planes $\sigma_i (i = 1, 2, 3)$ in figure 6.2 generates one reflection symmetry operation called σ_i. The C_3 axis is at the intersection of the three σ_i planes and so each of the planes contains the axis. For a geometrical object with one C_n axis such as the pyramid in figure 6.2, a reflection symmetry plane containing the C_n axis is called a *vertical* reflection symmetry plane; σ_v (or σ_d, see below) is used as a 'generic' name for such planes. The symmetry elements of the pyramid generate the symmetry operations

$$\{C_3, C_3^2, \sigma_1, \sigma_2, \sigma_3\}. \tag{6.1}$$

As seen in figure 6.3, the symmetry elements of an equilateral triangular prism include those of the equilateral-triangle-based pyramid. The prism, however, has the additional symmetry elements shown in figure 6.4: A horizontal reflection symmetry plane σ_h, three rotational symmetry axes $C_2^{(1)}, C_2^{(2)}, C_2^{(3)}$ and a so-called *rotation–reflection axis* (or *improper axis*) S_3. The S_3 axis coincides with the C_3 axis.

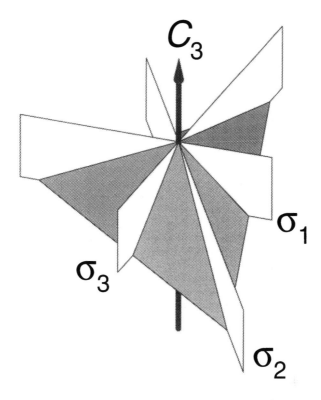

Figure 6.2. The symmetry elements of an equilateral-triangle-based pyramid.

The C_3 axis generates the two symmetry operations C_3 and C_3^2. The σ_v planes generate the three reflection operations σ_1, σ_2 and σ_3. The σ_h plane generates a reflection symmetry operation called σ_h and each of the $C_2^{(i)}$ ($i = 1, 2, 3$) axes generates a rotation $C_2^{(i)}$ of 180° about the axis in question. The rotation–reflection axis S_3 generates two *rotation–reflection* operations, S_3 and S_3^5, which we can write as

$$S_3 = \sigma_h C_3 = C_3 \sigma_h \qquad (6.2)$$

and

$$S_3^5 (= \sigma_h^5 C_3^5 = C_3^5 \sigma_h^5) = \sigma_h C_3^2 = C_3^2 \sigma_h. \qquad (6.3)$$

The S_3 operation in equation (6.2) is the *product* of σ_h and C_3. When we *multiply together* two symmetry operations R and S say, to obtain the product RS, we mean that we first carry out the operation S and then the operation R. In general, the product RS is different from the product SR but for σ_h and C_3 or C_3^2 the order does not matter as shown by equations (6.2) and (6.3). A rotation–reflection

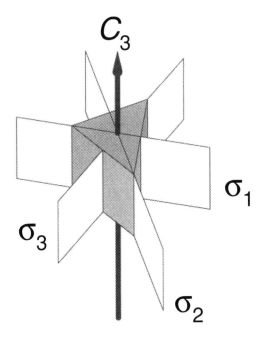

Figure 6.3. An equilateral triangular prism with the C_3 rotational symmetry axis and the three vertical reflection symmetry planes $\sigma_1, \sigma_2, \sigma_3$ shown. The prism has the additional symmetry elements shown in figure 6.4.

symmetry operation S_n is a rotation C_n about a C_n axis combined with a reflection in a reflection symmetry plane σ_h perpendicular to the C_n axis. The order in which the two operations are carried out does not matter. The symmetry elements of the equilateral triangular prism generate the symmetry operations

$$\{C_3, C_3^2, \sigma_1, \sigma_2, \sigma_3, \sigma_h, S_3, S_3^5, C_2^{(1)}, C_2^{(2)}, C_2^{(3)}\}. \qquad (6.4)$$

The rotation–reflection operation S_2 inverts every point through the intersection point of the C_2 axis and the σ_h plane. This operation is the *point group inversion i*:

$$i = S_2 = \sigma_h C_2 = C_2 \sigma_h. \qquad (6.5)$$

For the equilateral triangular pyramid, the existence of the symmetry element S_3 is a trivial consequence of the existence of the symmetry elements C_3 and σ_h. However, for n even a rotation–reflection axis S_n can exist without the C_n axis and the horizontal reflection symmetry plane σ_h being symmetry elements in their own right. We shall see an example in section 6.3 when we discuss the allene molecule.

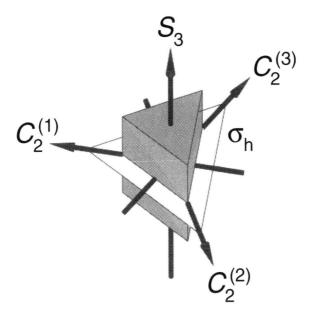

Figure 6.4. An equilateral triangular prism with the horizontal reflection symmetry plane σ_h, the three rotational symmetry axes $C_2^{(1)}$, $C_2^{(2)}$, $C_2^{(3)}$ and the rotation–reflection axis S_3 shown. The prism has these symmetry elements in addition to those shown in figure 6.3. Note that the S_3 and C_3 axes coincide.

6.2   Geometrical symmetry groups: Point groups

In figure 6.5, we illustrate the fact that C_3^2 has the same effect as the product $\sigma_2\sigma_1$; i.e. we can write

$$C_3^2 = \sigma_2\sigma_1. \tag{6.6}$$

Similarly,

$$C_3 = \sigma_3\sigma_1. \tag{6.7}$$

It is necessary to introduce the operation of 'doing nothing' for which we use the symbol E; this is called the *identity operation*. It allows us to express the effects of the following products:

$$E = \sigma_1\sigma_1 = \sigma_2\sigma_2 = \sigma_3\sigma_3 = C_3 C_3^2 = C_3^2 C_3. \tag{6.8}$$

When we add the operation E to the list of operations in equation (6.1), we obtain the set

$$\boldsymbol{C}_{3v} = \{E, C_3, C_3^2, \sigma_1, \sigma_2, \sigma_3\}. \tag{6.9}$$

We explain the $\boldsymbol{C}_{3v}$ notation later. The set of operations $\boldsymbol{C}_{3v}$ is such that the successive application of any two operations in it has the same effect as another

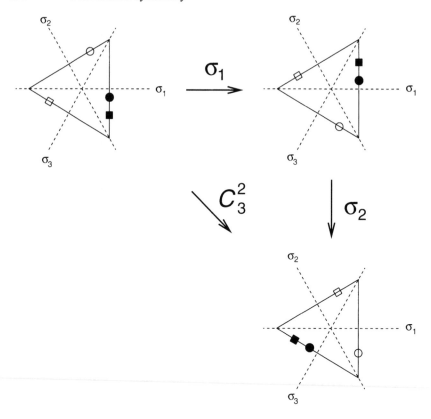

Figure 6.5. Illustrating equation (6.6) for an equilateral-triangle-based pyramid by marking four distinct points. The view is down from the tip of the C_3 axis.

operation in the set. The *multiplication table* given in table 6.1 demonstrates this. By adding the operation E to the list of symmetry operations for the equilateral triangular prism in equation (6.4), we obtain the set

$$\boldsymbol{D}_{3h} = \{E, C_3, C_3^2, \sigma_1, \sigma_2, \sigma_3, \sigma_h, S_3, S_3^5, C_2^{(1)}, C_2^{(2)}, C_2^{(3)}\}. \qquad (6.10)$$

The two sets of operations $\boldsymbol{C}_{3v}$ and $\boldsymbol{D}_{3h}$ in equations (6.9) and (6.10), respectively, are examples of a *group*. Each set is such that for any two operations in the set, R and S say, the operation T, given by $T = RS$, is present in the set. We can say that a group is 'closed' with respect to multiplication, since the product of any two of its members is present in the group. This is not the whole story and a complete definition of a group is given in section 7.4.

We can determine the symmetry elements of any geometrical object. A symmetry element can be a rotational symmetry axis, a reflection symmetry plane, a rotation–reflection symmetry axis or the *inversion centre* of a centrosymmetric object associated with the inversion symmetry operation $i = S_2$ in equation (6.5).

Table 6.1. The multiplication table of C_{3v}. Each entry is the product of first applying the operation at the top of the column and then applying the operation at the left-hand end of the row.

	E	C_3	C_3^2	σ_1	σ_2	σ_3
E:	E	C_3	C_3^2	σ_1	σ_2	σ_3
C_3:	C_3	C_3^2	E	σ_3	σ_1	σ_2
C_3^2:	C_3^2	E	C_3	σ_2	σ_3	σ_1
σ_1:	σ_1	σ_2	σ_3	E	C_3	C_3^2
σ_2:	σ_2	σ_3	σ_1	C_3^2	E	C_3
σ_3:	σ_3	σ_1	σ_2	C_3	C_3^2	E

Each of the symmetry elements generates symmetry operations as described above, and the complete set of generated symmetry operations (including E) form the *point group* of the geometrical object. Thus, C_{3v} is the point group of an equilateral-triangle-based pyramid and D_{3h} is the point group of an equilateral triangular prism. The symmetry elements of a given geometrical object will all, by necessity, intersect at one point, the centre of mass of the object, and this has given rise to the term 'point' group.

The labels customarily used for point groups (such as C_{3v} and D_{3h}) are named *Schönflies symbols*. We define some of the most important point groups here.

There are geometrical objects with no symmetry elements at all. Such an object formally has the point group

$$C_1 = \{E\}. \tag{6.11}$$

A geometrical object with one reflection symmetry plane and no other symmetry elements has the point group

$$C_s = \{E, \sigma\} \tag{6.12}$$

where σ is the reflection symmetry operation generated by the reflection symmetry plane. The group

$$C_i = \{E, i\} \tag{6.13}$$

describes the symmetry of a geometrical object with an inversion centre as the only symmetry element.

A cone and a cylinder, as shown in figure 6.6, are objects of very high symmetry; each has a C_∞ axis. Any rotation (i.e. a rotation of arbitrary rotation angle) about a C_∞ axis is a symmetry operation and there are infinitely many such rotations. In addition to the C_∞ axis, a cone has infinitely many reflection symmetry planes containing the C_∞ axis. The point group of a cone,

C_∞ C_∞

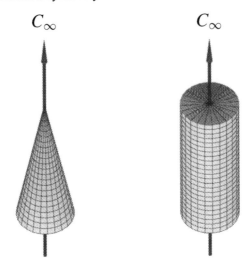

Figure 6.6. A cone and a cylinder with their C_∞ axes indicated.

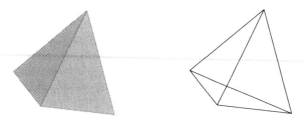

Figure 6.7. A solid tetrahedron (left) and a transparent stick model of the same tetrahedron (right).

which contains the infinitely many rotations about the C_∞ axis together with the infinitely many reflection symmetry operations generated by the reflection symmetry planes, is called $\boldsymbol{C}_{\infty v}$. In addition to the symmetry elements of a cone, a cylinder has a horizontal reflection symmetry plane σ_h (analogous to the σ_h plane of the equilateral triangular prism in figure 6.4), infinitely many C_2 axes lying in the σ_h plane and intersecting the C_∞ axis and a rotation–reflection axis S_∞ that coincides with the C_∞ axis. The point group of a cylinder is called $\boldsymbol{D}_{\infty h}$.

Another high-symmetry object is a *tetrahedron* shown in figure 6.7. A tetrahedron has four equivalent faces, each one being an equilateral triangle. Its symmetry elements are:

- Four C_3 axes (each one passing through a vertex and the centre of the opposite face)

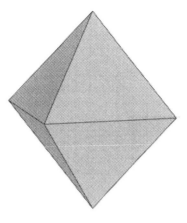

Figure 6.8. A solid octahedron.

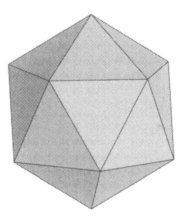

Figure 6.9. A solid icosahedron.

- Three C_2 axes, each one passing through the mid points of opposite edges
- Three S_4 axes coinciding with the C_2 axes and
- Six reflection symmetry planes, each one containing an edge and passing through the mid point of the opposite edge.

The resulting point group, with 24 elements, is called T_d.

An *octahedron*, shown in figure 6.8, has eight equivalent equilateral triangles as faces; its 48-member point group is called O_h. An *icosahedron*, shown in figure 6.9 (with 20 equivalent equilateral triangles as faces), has a 120-member point group called I_h.

Further important point groups (defined by their symmetry elements) are:

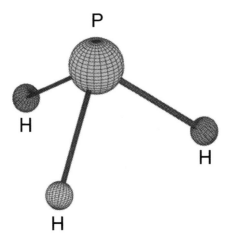

Figure 6.10. The PH_3 molecule in its equilibrium configuration.

C_n, one n-fold rotation axis;

C_{nv}, one n-fold rotation axis and n reflection planes containing this axis;

C_{nh}, one n-fold rotation axis and one reflection plane perpendicular to this axis;

D_n, one n-fold rotation axis and n C_2 axes perpendicular to it;

D_{nd}, those of D_n plus n reflection planes containing the n-fold rotation axis and bisecting the angles between the n twofold rotation axes (these vertical reflection symmetry planes are called σ_d planes);

D_{nh}, those of D_n plus a reflection plane perpendicular to the n-fold rotation axis; and

S_n, one S_n axis with n even.

6.3   The point group symmetry of molecules

The *structure* of a molecule is taken as its equilibrium structure in its ground electronic state and the symmetry of a molecule is customarily discussed using the rotation and reflection symmetry operations for this structure. In this way we determine the point group symmetry of the molecule. As examples, the ammonia NH_3 and phosphine PH_3 molecules both have the shape of an equilateral-triangle-based pyramid at equilibrium (figure 6.10) and so they both have point group symmetry C_{3v}. The three protons in the ion H_3^+ form an equilateral triangle at equilibrium (figure 6.11), and this static nuclear arrangement has D_{3h} symmetry.

The molecules HCN and CO_2 are linear at equilibrium and have the point group symmetries $C_{\infty v}$ and $D_{\infty h}$, respectively.

In its equilibrium geometry, methane CH_4 has the point group symmetry of

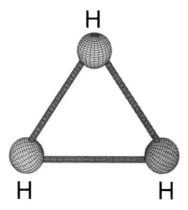

Figure 6.11. The H_3^+ ion in its equilibrium configuration.

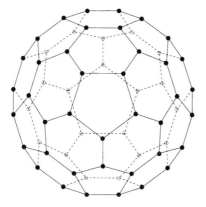

Figure 6.12. The molecule C_{60} at equilibrium. Nuclei represented as black dots are in or above the plane of the page whereas nuclei represented as grey dots are below the plane of the page.

the tetrahedron shown in figure 6.7. The four protons each occupy a vertex of the tetrahedron and the C atom is at the centre of mass. At equilibrium, sulphur hexafluoride SF_6 has octahedral symmetry. The six F nuclei form the vertices of the octahedron in figure 6.8 and the S nucleus is at the centre of mass. Less obviously, a C_{60} molecule in its equilibrium configuration (figure 6.12) has the same point group symmetry as the icosahedron in figure 6.9. The easiest way to recognize this is to notice in figure 6.12 that the 60 carbon atoms in C_{60} can be viewed as belonging to 12 disjunct five-member rings. The icosahedron must be lined up with the C_{60} molecule in such a way that each of its 12 vertices points towards the centre of one of the five-member rings. Thus, C_{60} at equilibrium has

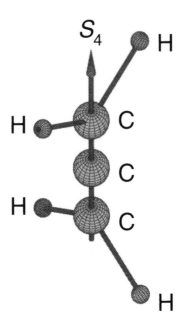

Figure 6.13. The allene molecule at equilibrium with the rotation–reflection axis (or improper axis) S_4 indicated. The two planes defined by the two CH$_2$ groups are perpendicular to each other.

point group symmetry I_h.

An allene molecule H$_2$CCCH$_2$ in its equilibrium configuration is shown in figure 6.13. This static nuclear arrangement is such that the two planes defined by the two CH$_2$ groups are perpendicular to each other. The molecule will be unchanged by a rotation of 90° about the CCC axis combined with a reflection in a plane perpendicular to this axis and containing the 'middle' C nucleus. Thus, allene has the rotation–reflection axis S_4 and the point group of allene, D_{2d}, contains the corresponding rotation–reflection operation S_4.

From the point group symmetry of a molecule's equilibrium structure, we can deduce if the molecule is a symmetric top, a linear rotor, a spherical top or an asymmetric top (section 5.3). Linear molecules are easily recognizable and have, in their equilibrium configurations, $C_{\infty v}$ or $D_{\infty h}$ symmetry. However, the following also hold:

- If the equilibrium structure of the molecule has one (and only one) C_n rotational symmetry axis or S_n rotation–reflection axis with a finite $n \geq 3$, then the molecule is a symmetric top.
- If the equilibrium structure has more than one C_n rotational symmetry axis with a finite $n \geq 3$, then the molecule is a spherical top.
- If the equilibrium structure has no rotational symmetry axes or C_2 axes only, then the molecule is an asymmetric top.

6.4 Problems

6.1 In section 6.1, we imply that an object having a rotation–reflection axis S_n with n odd must necessarily also have a C_n axis coinciding with the S_n axis and a symmetry plane σ_h perpendicular to the S_n axis. Explain this result.

6.2 Work out the multiplication table (i.e. a table analogous to table 6.1) for the set of operations D_{3h} in equation (6.10). It can be shown that each row and each column of the multiplication table contains each symmetry operation in D_{3h} once and only once.

6.3 Determine the symmetry elements for the rigid allene molecule in figure 6.13, and the symmetry operations in its point group D_{2d}.

6.4 Determine the point group symmetry of each of the letters in the phrase 'THE POINT GROUPS'.

6.5 Determine the point group symmetry of the following molecules in their electronic ground states: water H_2O, acetylene C_2H_2, ethylene C_2H_4, ethane C_2H_6, hydrogen peroxide H_2O_2, *cis* and *trans* difluoroethylene CHFCHF, boron trifluoride BF_3, methylfluoride CH_3F, and benzene C_6H_6. Determine for each molecule whether it is a symmetric top, a linear rotor, a spherical top or an asymmetric top.

Chapter 7

The symmetry of the Hamiltonian

7.1 Hamiltonian symmetry operations

We want to use the symmetry of a molecule as an aid in calculating and understanding the dynamical behaviour of the molecule. We wish to be able to do this for a molecule when it is in isolation and when it is in interaction with radiation or under the effect of an applied external static field. Point group symmetry does not allow us to do this in all circumstances. We can appreciate that this must be so because molecules are dynamical quantum mechanical objects that are not fixed at their equilibrium structure. In particular,

(i) There are molecules that can tunnel between different minima on the potential energy surface

(ii) Different electronic states of a single molecule often have different equilibrium structures; transitions or interactions between such electronic states can occur.

(iii) Atoms, nuclei and sub-nuclear particles have behaviour that can be understood using 'symmetry', yet it is clear that they cannot have the rotation and reflection symmetry of the point group. Our concept of symmetry should be general enough to include the symmetry used for such 'sub-molecular' species.

To overcome the shortcomings of geometrical symmetry operations, we define a general symmetry operation as being a transformation, such as a change in the coordinates, momenta, spin or charge of the particles in a system (the nuclei and electrons in a molecule, for example), that is such as to leave the energy of the system of particles unchanged.

> In quantum mechanics, we define a symmetry operation as a transformation that leaves the Hamiltonian for the system invariant or, equivalently, as a transformation that commutes with the Hamiltonian.

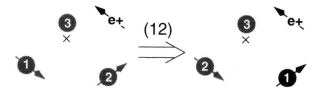

Figure 7.1. The effect of the symmetry operation (12) on a water molecule. The molecular centre of mass is indicated by a ×.

In our case the 'system' is an isolated molecule in field-free space but the definition is also applicable to atoms, nuclei and sub-nuclear particles.

This new definition of symmetry, which depends on the form of the Hamiltonian, appears to have no connection with the structure of the molecule. However, from this definition, we can recover the point group description of symmetry for the electronic and vibrational states of molecules that vibrate with small amplitude motions in isolated electronic states. By recovering point group symmetry from the more general definition, we gain a deeper understanding of what we really do to a molecule when we apply the 'rotation' and 'reflection' symmetry operations present in the molecular point group. However, with the more general definition of symmetry, we can also address problems that cannot be dealt with using point group symmetry such as conformational change, extreme non-rigidity, electronic state mixing and the effect of molecular rotation.

The more general definition of symmetry introduces symmetry operations such as the permutation of identical nuclei and the inversion of a molecule in its centre of mass (whether the molecule is centrosymmetric or not), both of which we discuss in this chapter using the H_2O and H_3^+ molecules as examples. A further symmetry operation that commutes with the molecular Hamiltonian is the overall rotation of the molecule about any axis that passes through its centre of mass; this is discussed briefly in section 7.7 but more fully in section 14.5.

7.2 Nuclear permutations and the inversion E^*

A 'snapshot' of a vibrating water molecule is shown schematically on the left-hand side of figure 7.1; it consists of two protons (labelled 1 and 2), one oxygen nucleus (labelled 3) and 10 electrons. One of the 10 electrons is shown without a label; this generic electron is indicated by **e** and the + sign next to it signifies that it is above the plane of the page. The spins of the protons and the electron are indicated using small arrows. The ^{16}O nucleus has zero spin.

On the right-hand side of figure 7.1, is shown the effect on the water molecule of the *nuclear permutation operation* (12). This is an operation in which two identical nuclei (in this case protons) are interchanged (or transposed), i.e. their positions in space and their spins are interchanged. This operation of permuting identical nuclei is to be understood just like the interchange of electrons *i* and *j* in equation (3.14). The effect of (12) can be thought of as the literal interchange of the particles 1 and 2 or, alternatively and equivalently, as the interchange of the labels 1 and 2. The nuclear permutation operation (12) does not affect the electrons in the molecule.

In the XYZ coordinate system defined in section 2.5, the particle coordinates (labelling the ten electrons 4, 5, 6, ..., 13) are written as

$$(R_1, \sigma_1, R_2, \sigma_2, R_3, \sigma_3, R_4, \sigma_4, \ldots, R_{13}, \sigma_{13}) \qquad (7.1)$$

where the spatial coordinates of the particle r are

$$R_r = (X_r, Y_r, Z_r) \qquad (7.2)$$

and the spin label is σ_r. As discussed on page 47, the spin label is not a spin 'coordinate'; however, in studying the transformation properties of wavefunctions under the effect of permutations we will for conciseness talk about 'the coordinates' of a particle when we really mean 'the space coordinates and spin labels' of the particle.

The effect of (12) on the coordinates in equation (7.1) is

$$(12)(\underbrace{R_1, \sigma_1}_{①}, \underbrace{R_2, \sigma_2}_{②}, \underbrace{R_3, \sigma_3}_{③}, \underbrace{R_4, \sigma_4}_{④}, \ldots)$$

$$= (\underbrace{R_1, \sigma_1}_{②}, \underbrace{R_2, \sigma_2}_{①}, \underbrace{R_3, \sigma_3}_{③}, \underbrace{R_4, \sigma_4}_{④}, \ldots) \qquad (7.3)$$

$$\equiv (\underbrace{R_2, \sigma_2}_{①}, \underbrace{R_1, \sigma_1}_{②}, \underbrace{R_3, \sigma_3}_{③}, \underbrace{R_4, \sigma_4}_{④}, \ldots) \qquad (7.4)$$

where the circled numbers represent the particles and the braces indicate their coordinates. The exchange of particles 1 and 2 is shown in equation (7.3) and equation (7.4) is a reordering so that the coordinates of particle 1 are given first. Omitting the circled numbers but always giving the coordinates of particle 1 first, those of 2 second, etc, gives

$$(12)(R_1, \sigma_1, R_2, \sigma_2, R_3, \sigma_3, R_4, \sigma_4, \ldots, R_{13}, \sigma_{13})$$
$$= (R_1', \sigma_1', R_2', \sigma_2', R_3', \sigma_3', R_4', \sigma_4', \ldots, R_{13}', \sigma_{13}')$$
$$\equiv (R_2, \sigma_2, R_1, \sigma_1, R_3, \sigma_3, R_4, \sigma_4, \ldots, R_{13}, \sigma_{13}). \qquad (7.5)$$

In equation (7.5) R_i, σ_i are the initial coordinates of particle i, and R_i', σ_i' are the coordinates of nucleus i after the permutation (12) has been performed.

In addition to (12), we consider also for H_2O the *inversion operation* E^* which has the effect of inverting the spatial coordinates of all the nuclei and electrons in the molecule through the molecular centre of mass as shown in

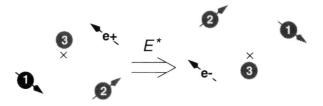

Figure 7.2. The effect of the symmetry operation E^* on a water molecule. The molecular centre of mass is indicated by a $\times$.

figure 7.2. Thus, the effect of E^* on the H$_2$O coordinates is

$$E^*(\mathbf{R}_1, \sigma_1, \mathbf{R}_2, \sigma_2, \mathbf{R}_3, \sigma_3, \mathbf{R}_4, \sigma_4, \ldots, \mathbf{R}_{13}, \sigma_{13})$$
$$= (\mathbf{R}'_1, \sigma'_1, \mathbf{R}'_2, \sigma'_2, \mathbf{R}'_3, \sigma'_3, \mathbf{R}'_4, \sigma'_4, \ldots, \mathbf{R}'_{13}, \sigma'_{13})$$
$$= (-\mathbf{R}_1, \sigma_1, -\mathbf{R}_2, \sigma_2, -\mathbf{R}_3, \sigma_3, -\mathbf{R}_4, \sigma_4, \ldots, -\mathbf{R}_{13}, \sigma_{13}) \quad (7.6)$$

where now $\mathbf{R}'_i, \sigma'_i$ are the coordinates of particle i after the inversion E^* has been performed. Since the spatial coordinates of the electrons are inverted by E^*, the electron **e** on the left-hand side of figure 7.2 is moved by E^* to be below the plane of the page and this is indicated by the $-$ sign next to it on the right-hand side of figure 7.2. The E^* operation changes the signs of all space coordinates but leaves the spin labels unchanged.

In order to study the transformations of molecular wavefunctions, we must define the effect of the operations on a function of the coordinates. The definition for the effect of a nuclear permutation operation or the inversion operation E^* on a wavefunction is that

$$R\psi(\mathbf{R}_1, \sigma_1, \mathbf{R}_2, \sigma_2, \mathbf{R}_3, \sigma_3, \mathbf{R}_4, \sigma_4, \mathbf{R}_5, \sigma_5, \ldots, \mathbf{R}_{13}, \sigma_{13})$$
$$= \psi^R(\mathbf{R}_1, \sigma_1, \mathbf{R}_2, \sigma_2, \mathbf{R}_3, \sigma_3, \mathbf{R}_4, \sigma_4, \mathbf{R}_5, \sigma_5, \ldots, \mathbf{R}_{13}, \sigma_{13})$$
$$= \psi(\mathbf{R}'_1, \sigma'_1, \mathbf{R}'_2, \sigma'_2, \mathbf{R}'_3, \sigma'_3, \mathbf{R}'_4, \sigma'_4, \mathbf{R}'_5, \sigma'_5, \ldots, \mathbf{R}'_{13}, \sigma'_{13}) \quad (7.7)$$

where ψ^R is a new function generated from ψ by applying the operation R. ψ^R is such that its value at the point $(\mathbf{R}_1, \sigma_1, \mathbf{R}_2, \sigma_2, \ldots, \mathbf{R}_{13}, \sigma_{13})$ is the same as the value of the function ψ at the point $(\mathbf{R}'_1, \sigma'_1, \mathbf{R}'_2, \sigma'_2, \ldots, \mathbf{R}'_{13}, \sigma'_{13})$ where $\mathbf{R}'_1, \sigma'_1$, etc are the coordinates of particle 1, etc after the operation R has been applied.

Thus, the effect of E^* on the function ψ is to convert it to the function ψ^{E^*} where the value of the function ψ^{E^*} at a point is given by

$$\psi^{E^*}(\mathbf{R}_1, \sigma_1, \mathbf{R}_2, \sigma_2, \mathbf{R}_3, \sigma_3, \mathbf{R}_4, \sigma_4, \ldots, \mathbf{R}_{13}, \sigma_{13})$$
$$= \psi(-\mathbf{R}_1, \sigma_1, -\mathbf{R}_2, \sigma_2, -\mathbf{R}_3, \sigma_3, -\mathbf{R}_4, \sigma_4, \ldots, -\mathbf{R}_{13}, \sigma_{13}). \quad (7.8)$$

If $f = X_1 + 3X_2 + 5X_3$, then $f^{E^*} = -X_1 - 3X_2 - 5X_3$, and $f^{(12)} = X_2 + 3X_1 + 5X_3$; each of these is a new function of the coordinates that is generated by applying the operations E^* and (12) to f.

As introduced in section 7.1, a symmetry operation is a transformation that leaves the Hamiltonian invariant. To show how this works in practice for the operations (12) and E^*, we use the elementary Hamiltonian $\hat{H}^0$, expressed in terms of (X_r, Y_r, Z_r) coordinates, that we derive from the classical energy expression without separating translation and with the neglect of spin. For the H_2O molecule

$$\hat{H}^0 = \frac{1}{2}\sum_{r=1}^{13} \frac{\hat{P}_r^{\,2}}{m_r} + \sum_{r<s=1}^{13} \frac{C_r C_s e^2}{4\pi \epsilon_0 R_{rs}} \tag{7.9}$$

where the sums run over all particles (nuclei and electrons) in the molecule, m_r is the mass and $C_r e$ is the charge of particle r (an electron has the mass m_e and the charge $-e$), R_{rs} is the interparticle distance and the quantum mechanical operator $\hat{P}_r^{\,2}$ is given by

$$\hat{P}_r^{\,2} = -\hbar^2 \left(\frac{\partial^2}{\partial X_r^2} + \frac{\partial^2}{\partial Y_r^2} + \frac{\partial^2}{\partial Z_r^2} \right). \tag{7.10}$$

In the H_2O molecule, the two protons, labelled as 1 and 2, are identical and we rewrite equation (7.9) to emphasize this:

$$\hat{H}^0 = \frac{1}{2m_H}(\hat{P}_1^{\,2} + \hat{P}_2^{\,2}) \tag{7.11}$$

$$+ \frac{1}{2m_O}\hat{P}_3^{\,2} + \frac{1}{2m_e}\sum_{r=4}^{13}\hat{P}_r^{\,2} \tag{7.12}$$

$$+ \frac{e^2}{4\pi \epsilon_0 R_{12}} + \frac{8e^2}{4\pi \epsilon_0}\left(\frac{1}{R_{13}} + \frac{1}{R_{23}} \right) \tag{7.13}$$

$$- \frac{e^2}{4\pi \epsilon_0}\sum_{r=4}^{13}\left(\frac{1}{R_{1r}} + \frac{1}{R_{2r}} \right) \tag{7.14}$$

$$- \frac{8e^2}{4\pi \epsilon_0}\sum_{r=4}^{13}\frac{1}{R_{3r}} + \sum_{r<s=4}^{13}\frac{e^2}{4\pi \epsilon_0 R_{rs}}, \tag{7.15}$$

where we have put $m_1 = m_2 = m_H$, $m_3 = m_O$, $m_4 = m_5 = \cdots = m_{13} = m_e$, $C_1 = C_2 = 1$, $C_3 = 8$, and $C_4 = C_5 = \cdots = C_{13} = -1$.
The effect of applying (12) to the Hamiltonian $\hat{H}^0$ is

- to interchange $\hat{P}_1^{\,2}$ and $\hat{P}_2^{\,2}$ in (7.11),
- to leave (7.12) unchanged,
- to leave R_{12} unchanged and to interchange R_{13} and R_{23}, in (7.13),
- to interchange R_{1r} and R_{2r} in (7.14) and
- to leave (7.15) unchanged.

The sum $\hat{P}_1^{\,2} + \hat{P}_2^{\,2}$ is unchanged, or invariant, under the operation (12) and so are all the other contributions to $\hat{H}^0$. The total Hamiltonian is invariant under (12).

In this discussion we have used a simple Hamiltonian in which we neglect the effects of the spins of the particles. However, it is clear from the definition of the word 'identical' that any permutation (i.e. relabelling) of identical nuclei in a molecule must leave the molecular Hamiltonian invariant, since identical particles have identical properties such as mass, charge and spin, that occur in the Hamiltonian.

From equation (7.6), the inversion operation E^* changes the sign of all spatial coordinates. For example, $E^*X_1 = X_1' = -X_1$, and using the chain rule, we have

$$\frac{\partial}{\partial X_1'} = \frac{\partial X_1}{\partial X_1'}\frac{\partial}{\partial X_1} = -\frac{\partial}{\partial X_1} \tag{7.16}$$

and, therefore,

$$\frac{\partial^2}{\partial (X_1')^2} = \frac{\partial^2}{\partial X_1^2}. \tag{7.17}$$

An equation like equation (7.17) can be derived for any spatial coordinate X_r, Y_r, or Z_r for $r = 1, 2, 3, \ldots, 13$, and these equations show that the operation E^* leaves the operators $\hat{P}_r^2$ [see equation (7.10)] unchanged. Therefore, the kinetic energy part of $\hat{H}^0$ is invariant under E^*. It can be seen from equation (2.46) that all inter-particle distances R_{rs} are unchanged by E^* and so the electrostatic potential energy part of $\hat{H}^0$ is invariant under this operator. Thus, the simple Hamiltonian $\hat{H}^0$ is invariant under E^*.

Unlike for a permutation of identical nuclei, it is not obvious that the complete exact Hamiltonian is invariant under the inversion operation E^*. In fact, it is not. But the term in the Hamiltonian that is not invariant to E^* is unbelievably small and its effect has been observed only in atoms but not so far in any molecule. This term arises from the so-called 'weak neutral current interaction' between nuclei and electrons and it causes 'parity violation' which we discuss in section 15.2. For all normal spectroscopic studies of molecules, we can ignore this interaction and only consider the electromagnetic interaction between the particles. When we talk about the Hamiltonian below, we will mean the electromagnetic Hamiltonian (i.e. neglecting the weak neutral current interaction) and this is invariant to E^*.

To express this using an operator equation, we consider the combined effect of $R[= (12)$ or $E^*]$ and $\hat{H}^0$ on a wavefunction ψ for the water molecule. We have shown earlier that

$$R\hat{H}^0\psi = \hat{H}^0 R\psi \tag{7.18}$$

from which is follows that[1]

$$(R\,\hat{H}^0 - \hat{H}^0\,R)\psi = [R, \hat{H}^0]\,\psi = 0 \tag{7.19}$$

that is

$$[R, \hat{H}^0] = 0 \tag{7.20}$$

[1] Where we use the commutator notation $[R, \hat{H}^0]$ as defined in equation (2.18).

and we have the definition of a symmetry operation as being a transformation operation that commutes with the Hamiltonian. Permutations of identical nuclei and the inversion E^* are symmetry operations for an isolated molecule in free space.

7.3   Symmetry labels

For the rest of the chapter, we will ignore nuclear spin because of the constraint imposed by the fifth postulate (explained in chapter 9). The rovibronic Schrödinger equation for the water molecule can be written as

$$\hat{H}_{\text{rve}}\psi_n = E_n\psi_n \tag{7.21}$$

where ψ_n is an eigenfunction of the rovibronic Hamiltonian operator $\hat{H}_{\text{rve}}$ having eigenvalue E_n. Applying the symmetry operation R [which, for H_2O, is for the moment either (12) or E^*] to the left- and right-hand sides of equation (7.21):

$$R\hat{H}_{\text{rve}}\psi_n = RE_n\psi_n. \tag{7.22}$$

Since R commutes with $\hat{H}_{\text{rve}}$ (it is a symmetry operation of $\hat{H}_{\text{rve}}$) and since R commutes with E_n (E_n is a just a constant), we can rewrite equation (7.22) as

$$\hat{H}_{\text{rve}}R\psi_n = E_nR\psi_n. \tag{7.23}$$

But we know, from equation (7.7), that $R\psi_n = \psi_n^R$ which is a new function of the coordinates. So we can write

$$\hat{H}_{\text{rve}}\psi_n^R = E_n\psi_n^R. \tag{7.24}$$

Because a symmetry operation commutes with the Hamiltonian, it generates from an eigenfunction ψ_n a 'new' eigenfunction ψ_n^R having the same eigenvalue E_n. This makes it possible to symmetry label energy levels.

To explain how energy levels are symmetry labelled, we initially focus on the labelling of non-degenerate levels. All rovibronic states of the water molecule are non-degenerate[2]. From equation (2.20), we see that, for a non-degenerate state, the 'new' eigenfunction ψ_n^R in equation (7.21) can only be a constant times the

[2] In applying permutation and inversion symmetry operations, the m-degeneracy (see section 2.7) is inconsequential in field-free space and we focus only on the $m = 0$ states here.

'original' eigenfunction ψ_n in equation (7.24), i.e. for a symmetry operation R acting on a non-degenerate wavefunction

$$R\psi_n = \psi_n^R = c^R\psi_n \qquad (7.25)$$

where c^R is a constant.

Setting $R = (12)$ in equation (7.25) and applying the (12) operation on the left- and right-hand sides of the resulting equation yields

$$(12)(12)\psi_n = (12)c^{(12)}\psi_n = c^{(12)}(12)\psi_n = (c^{(12)})^2\psi_n \qquad (7.26)$$

where we have used equation (7.25) and the fact that (12) commutes with the constant $c^{(12)}$. However, by applying (12) twice, we return to the original situation and so, by necessity,

$$(12)(12)\psi_n = \psi_n. \qquad (7.27)$$

Consequently, from equation (7.26), we must have $(c^{(12)})^2 = 1$ and, therefore,

$$c^{(12)} = \pm1. \qquad (7.28)$$

As in the case of (12), applying E^* twice takes us back to the original situation and so by setting $R = E^*$ in equation (7.25) and applying arguments analogous to those leading to equation (7.28), we find that

$$c^{E^*} = \pm1. \qquad (7.29)$$

The sign of c^{E^*} ($+$ or $-$) for a molecular state is called the *parity* of that state.

Thus, the water molecule has four kinds of state with

$$(c^{(12)}, c^{E^*}) = (+1, +1), \quad (+1, -1), \quad (-1, -1) \quad \text{or} \quad (-1, +1). \qquad (7.30)$$

Symmetry labels or *symmetry species* are introduced for the H_2O molecule by saying that wavefunctions with $(c^{(12)}, c^{E^*}) = (+1, +1)$ have A_1 symmetry, $(+1, -1)$ have A_2 symmetry, $(-1, -1)$ have B_1 symmetry and $(-1, +1)$ have B_2 symmetry. For the water molecule, each of its rovibronic states can be labelled as transforming in one of these four ways and this 'labelling' of the states plays an important role in helping us to understand the water molecule and its spectra.

7.4   Symmetry groups

Using the identity operation E introduced in equation (6.8), we can write

$$(12)(12) = E \qquad (7.31)$$

and

$$E^*E^* = E. \qquad (7.32)$$

Table 7.1. The multiplication table for H_2O. Each entry is the product of first applying the operation at the top of the column and then applying the operation at the left end of the row.

	E	(12)	E^*	$(12)^*$
E:	E	(12)	E^*	$(12)^*$
(12):	(12)	E	$(12)^*$	E^*
E^*:	E^*	$(12)^*$	E	(12)
$(12)^*$:	$(12)^*$	E^*	(12)	E

The operation $(12)^*$ is defined as the successive application of E^* and (12):

$$(12)^* = E^*(12) = (12)E^*. \qquad (7.33)$$

The operations E^* and (12) commute with each other. It can be seen that

$$(12)^*(12)^* = E. \qquad (7.34)$$

The product of two symmetry operations is itself a symmetry operation, and we now have four symmetry operations for the H_2O molecule: E, (12), E^* and $(12)^*$.

Any pair of symmetry operations can be multiplied together and, for water, the results are given in the multiplication table 7.1. An example is the result of doing first $(12)^*$ and then E^*:

$$E^*(12)^* = E^*[E^*(12)] = [E^*E^*](12) = (12). \qquad (7.35)$$

If the product of the operations R_1 and R_2 gives E (the identity), then R_1 and R_2 are the *reciprocals* of each other. The four symmetry operations for H_2O are all self-reciprocal from equations (7.31), (7.32) and (7.34).

The set of operations $\{E, (12), E^*, (12)^*\}$ forms a *symmetry group*.

A symmetry group is a set of symmetry operations that satisfy the following *group axioms*:

- The operations can be multiplied together in pairs (i.e. successively applied) and the result is a member of the group.
- One of the operations in the group is the identity operation E.
- The reciprocal of each operation is a member of the group.
- Multiplication of the operations is associative; i.e. in a multiple product the answer is independent of how the operations are associated in pairs [see the intermediate results in equation (7.35)].

Table 7.2. The symmetry species (or character table) for $C_{2v}(M)$. The protons are labelled 1 and 2. Each row of the table gives a possible combination of the constants c^R in equation (7.25).

R:	E	(12)	E^*	$(12)^*$
A_1:	1	1	1	1
A_2:	1	1	-1	-1
B_1:	1	-1	-1	1
B_2:	1	-1	1	-1

The symmetry group introduced here is the *complete nuclear permutation inversion* (CNPI) group for the water molecule.

> A CNPI group for a molecule consists of all permutations of identical nuclei, the inversion E^* and the product of E^* with all the permutations of identical nuclei; it is a symmetry group of the molecule.

The CNPI group of the water molecule is called $C_{2v}(M)$.

At the end of the previous section, the symmetry species A_1, A_2, B_1 and B_2 were introduced for the water molecule based on how the eigenfunctions transform under the operations (12) and E^*, i.e. by using the values of the transformation constants $c^{(12)}$ and c^{E^*} from equation (7.25). The transformation constants c^E and $c^{(12)^*}$ can also be determined. Since E leaves a wavefunction unchanged we must have

$$c^E = 1 \tag{7.36}$$

for all wavefunctions, and equation (7.33) implies that

$$c^{(12)^*} = c^{(12)} c^{E^*}. \tag{7.37}$$

The tabulation of all four c^R for each symmetry species produces the symmetry species table in table 7.2. In section 7.10, we use the CNPI group of the H_3^+ molecule to show how the presence of degenerate states affects the way the symmetry species table is set up and this will explain why such a table is called a *character table*. In any character table, there is always one symmetry species that has all $c^R = +1$ and it is called the *totally symmetric* symmetry species. In different groups, the totally symmetric symmetry species can have different names and we give it the general name $\Gamma^{(s)}$. For the water molecule, $\Gamma^{(s)} = A_1$.

Figure 7.3 shows the lowest rotational energy levels for water with their symmetry labels. The symmetry labels are obtained by determining how the

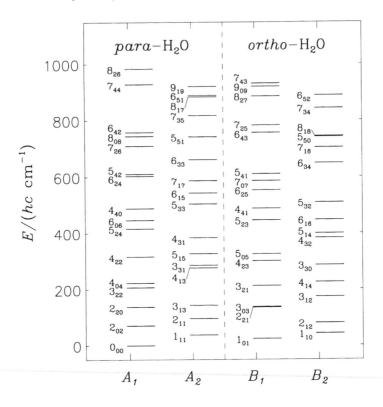

Figure 7.3. The lower rotational energies for H_2O arranged according to the symmetry labels in table 7.2. The energies are also labelled by $J_{K_a K_c}$ as explained in section 5.3.4 and the *ortho/para* designations are explained in section 9.2.

analytical expressions for the asymmetric top wavefunctions (see section 5.3.4) transform under the effect of (12) and E^*. The experimentally derived term values are shown and, because they include the small effect of $\hat{H}'_{\text{rot}}$ [see equations (5.36)–(5.38)], they are not quite equal to the rigid-rotor term values given in figure 1.5. States of A_1 or A_2 symmetry belong to *para*-H_2O, whereas states of symmetry B_1 or B_2 belong to *ortho*-H_2O; the meaning of the *ortho/para* labels will be discussed in chapter 9.

7.5   The vanishing integral rule

It is often necessary to calculate integrals of the general form

$$I = \int \psi'^* \hat{O} \psi'' \, d\tau \qquad (7.38)$$

where ψ' and ψ'' are wavefunctions and $\hat{O}$ is an operator. Without doing any numerical calculations, the symmetry labels for the states ψ' and ψ'' can be used in a simple way, in conjunction with the symmetry of $\hat{O}$, to determine if such an integral has to vanish. As a result, we can use symmetry to simplify the calculation of molecular energies and to determine selection rules for molecular transitions.

The integrand of the integral I in equation (7.38) for the water molecule can be written as

$$\psi'^* \hat{O} \psi'' = f(\boldsymbol{R}_1, \boldsymbol{R}_2, \dots, \boldsymbol{R}_{13}) = f(S) \tag{7.39}$$

where S is a general point with coordinates $(\boldsymbol{R}_1, \boldsymbol{R}_2, \dots, \boldsymbol{R}_{13})$. The symmetry species of $f(S)$ can be determined from the symmetry species of ψ'^*, $\hat{O}$ and ψ''. For the water molecule, where we do not have to consider degeneracy, the symmetry species of $f(S)$ is given by the values of the constants c_f^R determined from

$$Rf(S) = c_f^R f(S) \tag{7.40}$$

for $R = E, (12), E^*$ and $(12)^*$. If we know the symmetries of ψ'^*, $\hat{O}$ and ψ'', then it is easy to determine the c_f^R since

$$c_f^R = (c^R)'^* c_O^R (c^R)'' \tag{7.41}$$

where $(c^R)'^*$, c_O^R and $(c^R)''$ are the constants obtained by applying the symmetry operation R to ψ'^*, $\hat{O}$ and ψ'', respectively. Knowing the symmetry of ψ'^*, $\hat{O}$ and ψ'', the symmetry of the product $f(S)$ can be determined using equation (7.41).

A simple example for the water molecule is obtained by introducing the three real expressions[3]

$$\psi' = \psi'^* = X_1^2 + X_2^2 \qquad \hat{O} = X_1 + X_2 \qquad \text{and} \qquad \psi'' = X_1 - X_2. \tag{7.42}$$

By determining the effect of the symmetry operations R for the water molecule on these functions, we see that they have the symmetries A_1, A_2 and B_1 respectively. In table 7.2, the values of c^R for each of these symmetry species are given and, using these results in equation (7.41) for each R, we find that the symmetry of their product $f(S)$ is B_2. A function of symmetry A_1 multiplied by a function of symmetry A_2 multiplied by a function of symmetry B_1 gives a product of symmetry B_2 and we write

$$A_1 \otimes A_2 \otimes B_1 = B_2. \tag{7.43}$$

The *vanishing integral rule* states that, for the water molecule, the integral

$$I = \int \psi'^* \hat{O} \psi'' \, d\tau = \int f(S) \, d\tau \tag{7.44}$$

---

[3] The expressions introduced here are just simple functions of X_1 and X_2 designed to illustrate how the symmetry of a product is determined. They are not actual wavefunctions or operators.

will vanish if $f(S)$ does not have the symmetry of the totally symmetric symmetry species A_1. A general statement of the vanishing integral rule, applicable to all molecules and including the possibility of degeneracies, is given in equation (7.82).

If we use the notation that $\Gamma(A)$ is the symmetry of A, then the vanishing integral rule for the integral I in equation (7.44) can be stated by saying that I will vanish for the water molecule if

$$\Gamma(f(S)) = \Gamma(\psi'^*) \otimes \Gamma(\hat{O}) \otimes \Gamma(\psi'') \neq A_1. \tag{7.45}$$

If $\Gamma(\hat{O}) = A_1$, as is the case if $\hat{O}$ is the field-free molecular Hamiltonian $\hat{H}$, then the integral will vanish if

$$\Gamma(\psi'^*) \otimes \Gamma(\psi'') \neq A_1 \tag{7.46}$$

which is the same as saying that the integral will vanish if

$$\Gamma(\psi'') \neq \Gamma(\psi'). \tag{7.47}$$

If $f(S)$ is of A_1 symmetry, the integral I in equation (7.44) could still vanish, since the rule only states that I vanishes if $f(S)$ is *not* of A_1 symmetry. If $f(S)$ is of A_1 symmetry and if I is found by experiment to vanish, then it might indicate that there is more symmetry in the problem than one had considered; we will point out an important general example of this in section 7.7.

7.5.1   Proof of the vanishing integral rule for the water molecule

Consider the integral

$$I = \int f(R_1, R_2, R_3, R_4, \dots R_{13}) \, d\tau \tag{7.48}$$

from equation (7.44), where the volume element is

$$d\tau = dX_1 \, dY_1 \, dZ_1 \, dX_2 \, dY_2 \, dZ_2 \, dX_3 \, dY_3 \, dZ_3 \dots dX_{13} \, dY_{13} \, dZ_{13}. \tag{7.49}$$

From equation (7.5),

$$f^{(12)}(R_1, R_2, R_3, R_4, \dots R_{13}) = f(R_2, R_1, R_3, R_4, \dots R_{13}) \tag{7.50}$$

and, therefore,

$$\int f^{(12)}(R_1, R_2, R_3, R_4, \dots R_{13}) \, d\tau$$
$$= \int f(R_2, R_1, R_3, R_4, \dots R_{13}) \, d\tau. \tag{7.51}$$

We can obtain the integral on the right-hand side of equation (7.51) from that in equation (7.48) by the variable substitution $R_1, R_2 \rightarrow R_2, R_1$. Consequently, these two integrals must be equal, i.e.

$$I = \int f \, d\tau = \int f^{(12)} \, d\tau \tag{7.52}$$

where we have omitted the coordinates for brevity. Thus, if $f^{(12)} \neq f$, i.e. $c^{(12)} \neq +1$, from the transformation properties of f, then equation (7.52) can only be satisfied if $I = 0$.

For the function f^{E^*} given by

$$f^{E^*}(R_1, R_2, R_3, R_4, \ldots, R_{13})$$
$$= f(-R_1, -R_2, -R_3, -R_4, \ldots, -R_{13}) \tag{7.53}$$

we have

$$\int f^{E^*}(R_1, R_2, R_3, R_4, \ldots R_{13}) \, d\tau$$
$$= \int f(-R_1, -R_2, -R_3, -R_4, \ldots - R_{13}) \, d\tau. \tag{7.54}$$

Realizing that the limits of integration stretch symmetrically from $-\infty$ to $+\infty$ for all the spatial coordinates, it is easy to show that the integral on the right-hand side of equation (7.54), in which *all* coordinates are changed in sign, is equal to that in equation (7.48) and we have

$$I = \int f \, d\tau = \int f^{E^*} \, d\tau. \tag{7.55}$$

Thus, just as for the operation (12), if $f^{E^*} \neq f$, i.e. $c^{E^*} \neq +1$, from the transformation properties of f, then equation (7.55) can only be satisfied if $I = 0$. In summary, $I = 0$ if $c^{(12)}$ and/or $c^{E^*} \neq +1$, for f, and the vanishing integral rule for the water molecule given in equation (7.45) is proved using its CNPI group.

The general statement of the vanishing integral rule, given in equation (7.82), can be proved for any symmetry group consisting of permutations of identical nuclei and the inversion by using arguments just like those given earlier for the operations (12) and E^*.

7.6 Selection rules

The rovibronic transition moment integral

$$I_{\mathrm{TM}} = \int \Phi_{\mathrm{rve}}^{\prime *} \mu_A \Phi_{\mathrm{rve}}^{\prime\prime} \, d\tau \tag{7.56}$$

enters into the expression for the line strength of a rovibronic electric dipole transition given in equation (2.87). We continue to use the water molecule as an example. The wavefunctions Φ'_{rve} and Φ''_{rve} describe rovibronic states and, according to the discussion in section 7.3, each of them belongs to one of the symmetries in table 7.2. From the vanishing integral rule, as expressed in equation (7.45), I_{TM} will vanish if

$$\Gamma(\Phi'^*_{\text{rve}}) \otimes \Gamma(\mu_A) \otimes \Gamma(\Phi''_{\text{rve}}) \neq A_1, \tag{7.57}$$

and the rovibronic transition[4] is *forbidden* if the symmetry species of the two rovibronic states satisfies equation (7.57); to make use of this result we have to determine the symmetry of μ_A, $\Gamma(\mu_A)$.

The dipole moment component μ_A ($A = X$, Y, or Z) is given by equation (2.88). For a water molecule, the expression for μ_X is

$$\mu_X = e(X_1 + X_2) + 8e\,X_3 - e\sum_{j=4}^{13} X_j \tag{7.58}$$

and the operation (12) leaves μ_X (and μ_Y and μ_Z) unchanged, i.e. $(12)\mu_A = \mu_A$. From equation (2.88), we see that $E^*\mu_A = -\mu_A$ since $E^*A_j = -A_j$ for all A_j. Thus, each μ_A component has symmetry A_2 in table 7.2 and, from equation (7.57), a transition between a pair of levels whose symmetry product times A_2 is not A_1 will have zero transition moment and be forbidden.

Inspection of table 7.2 shows that transitions in emission or absorption that are forbidden by this symmetry rule are:

$$A_1 \leftrightarrow B_1, \; A_1 \leftrightarrow B_2, \; A_2 \leftrightarrow B_1, \; A_2 \leftrightarrow B_2. \tag{7.59}$$

Thus, *ortho*↔*para* transitions are forbidden by this symmetry rule.

7.7 The rovibronic symmetry label *J*

There are very many transitions between the levels in figure 7.3 that are not forbidden according to equation (7.59), such as $9_{19}(A_2) \leftrightarrow 0_{00}(A_1)$, but which do have zero intensity; these are examples for which the comments in the paragraph after equation (7.47) are relevant. The water molecule has more symmetry than that given in table 7.2, because, in common with all molecules, its Hamiltonian commutes with the operation of overall rotation by any amount about any axis that passes through the centre of mass of the molecule. We discuss this more fully in section 14.5 but a very brief account is appropriate here. The symmetry group consisting of E and of all overall rotations (by any amount about any axis that passes through the centre of mass of the molecule) is

---

[4] The term 'forbidden transition' can be defined in a more general way than indicated here and this is explained at the beginning of chapter 12.

called K(spatial) and it has an infinite number of elements, like the point groups $C_{\infty v}$ or $D_{\infty h}$. Using this group to symmetry label rovibronic states is equivalent to labelling the states using the rovibronic angular momentum quantum number J introduced in equation (2.83). A rovibronic wavefunction having rovibronic angular momentum quantum number J has symmetry species called $D^{(J)}$ in the group K(spatial). The dipole moment operator μ_A transforms as the symmetry species $D^{(1)}$ in the group K(spatial). Using the vanishing integral rule for the integral I_{TM} in equation (7.56) with this symmetry group and these symmetry labels leads to the rotational symmetry selection rule:

> Transitions between rovibronic states for which the angular momentum quantum number J changes by more than one unit or for which $J = 0$ in both states are forbidden.

Hence, the transition $9_{19} \leftrightarrow 0_{00}$ for the water molecule is forbidden.

Rather than formulating the selection rule on J for forbidden transitions, it is more normal to state the selection rule that applies for *allowed* transitions, i.e. for transitions that are *not* forbidden.

> Allowed rovibronic transitions $\Phi'_{\text{rve}}(J') \leftarrow \Phi''_{\text{rve}}(J'')$ satisfy the selection rule:
> $$\Delta J = 0, \pm 1 \qquad \text{but} \qquad J = 0 \leftrightarrow 0 \text{ is forbidden} \qquad (7.60)$$
> where $\Delta J = J' - J''$ is the change in J for the transition.

7.8 Diagonalizing the Hamiltonian matrix using symmetry

The vanishing integral rule can also be applied to the integrals given in equation (2.21)

$$H_{mn} = \int (\psi_m^0)^* \hat{H} \psi_n^0 \, d\tau. \qquad (7.61)$$

These integrals enter into the calculation of the molecular energies and wavefunctions described in section 2.2. The basic idea is to represent the wavefunction ψ_j of the molecule as a linear combination of basis functions ψ_n^0 as given in equation (2.23) and we choose these basis functions so that each has one of the symmetries in table 7.2. The Hamiltonian is totally symmetric in the symmetry group by definition of the symmetry operations (the Hamiltonian

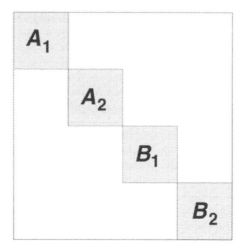

Figure 7.4. The block diagonal structure of the Hamiltonian matrix for H_2O, for a given value of J and with $m = 0$, constructed in terms of basis functions ψ_n^0 that have the symmetries in table 7.2. The elements of the matrix in the shaded areas can be non-vanishing, whereas the elements between the shaded area must vanish.

is invariant to a symmetry operation) and so, using the water molecule as an example, from equation (7.47), off-diagonal matrix elements must vanish between states of different symmetry. Consequently, the Hamiltonian matrix made up of the matrix elements in equation (7.61) becomes block diagonal. As illustrated in figure 7.4, the Hamiltonian matrix for the water molecule factorizes into four blocks, each of which corresponds to one of the four symmetries A_1, A_2, B_1 and B_2 in table 7.2. The matrix elements between basis functions of different symmetry are all zero. Because of rotational symmetry (see section 7.7), off-diagonal matrix elements also vanish if the basis functions have different values of the total angular momentum quantum number J. So each block for the water molecule, for a given value of J and with $m = 0$, factorizes into four blocks according to the symmetries in table 7.2. The block diagonalization of the Hamiltonian matrix is a general result of symmetry and it leads to great savings in computer time in actual calculations of molecular energies and wavefunctions.

7.9   The Stark effect

As long as we consider an isolated molecule in field-free space, the Hamiltonian is totally symmetric and the non-vanishing matrix elements of it can only be between basis functions of the same symmetry. However, if we subject the molecule to a constant electric field E (a *Stark field*), the Hamiltonian is modified so that it is no longer totally symmetric. As a result, the application of the vanishing integral rule leads to selection rules on the matrix elements that are less restrictive. Let us assume that the electric field E is directed along the space-fixed Z axis. In this case, the total molecular Hamiltonian of a molecule is

$$\hat{H}_{\text{Stark}} = \hat{H} - E\mu_Z \tag{7.62}$$

where $\hat{H}$ is the Hamiltonian of the isolated molecule and μ_Z is the Z-component of the electric dipole moment of the molecule [equation (2.88)]. To obtain the molecular energies, we now need the matrix elements

$$H_{mn}^{(\text{Stark})} = \int (\psi_m^0)^* \hat{H}_{\text{Stark}} \psi_n^0 \, d\tau \tag{7.63}$$

$$= \int (\psi_m^0)^* \hat{H} \psi_n^0 \, d\tau - E \int (\psi_m^0)^* \mu_Z \psi_n^0 \, d\tau. \tag{7.64}$$

As we have seen, the first integral in equation (7.64) can be non-vanishing for the water molecule when ψ_m^0 and ψ_n^0 have the same symmetry. The second integral is analogous to that in equation (7.56) and we have seen in section 7.6 that it can be non-vanishing for the water molecule when ψ_m^0 and ψ_n^0 belong to the symmetry combinations A_1/A_2 or B_1/B_2. Consequently, when ψ_m^0 and ψ_n^0 have the same symmetry, $\hat{H}_{\text{Stark}}$ can have non-vanishing matrix elements from the first term in equation (7.64) but when ψ_m^0 and ψ_n^0 belong to the symmetry combinations A_1/A_2 or B_1/B_2, $\hat{H}_{\text{Stark}}$ can have non-vanishing matrix elements from the second term of this equation. In the case of a Stark field, the Hamiltonian matrix factorizes into two blocks, an A_1/A_2 block and a B_1/B_2 block. We can understand this result by noting that (12) is a symmetry operation for $\hat{H}_{\text{Stark}}$ since it commutes with $\hat{H}$ and with μ_Z (we have seen in section 7.6 that $(12)\mu_Z = \mu_Z$). However, E^* is not a symmetry operation for $\hat{H}_{\text{Stark}}$ since $E^*\mu_Z = -\mu_Z$ (section 7.6) so that E^* does not commute with μ_Z. Therefore, matrix elements of $\hat{H}_{\text{Stark}}$ vanish between basis functions with the different values of $c^{(12)}$ [see equation (7.25)] whereas there is no similar selection rule for c^{E^*}. Thus, $\hat{H}_{\text{Stark}}$ can mix states of opposite parity. It can also mix states having total angular momentum quantum number differing by one.

7.10   The symmetry of H_3^+

The water molecule is a simple example for introducing symmetry operations, symmetry labels and a symmetry group, as well as for showing applications of the vanishing integral rule. It is simple for two reasons: (a) The symmetry operations (12), E^* and (12)* are each self-reciprocal, and (b) The states are non-degenerate, so that the effect of a symmetry operation is given by equation (7.25). To show what happens when these simplifying features are not present, we use the equilateral triangular two-electron molecule H_3^+ as an example.

The Hamiltonian of H_3^+ that has three identical nuclei, which we label 1, 2

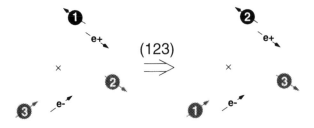

Figure 7.5. The effect of the symmetry operation (123) on an H_3^+ molecule. The molecular centre of mass is indicated by a $\times$.

and 3, will be invariant to the three transpositions[5] (12), (23) and (31) as well as to the cyclic permutations (123) and (132), where (123) means '1 is replaced by 2, 2 is replaced by 3, and 3 is replaced by 1'. The effect of (123) is depicted in figure 7.5.

Using the format of equations (7.3)–(7.5), we can write

$$(123)[\mathbf{R}_1, \sigma_1, \mathbf{R}_2, \sigma_2, \mathbf{R}_3, \sigma_3, \mathbf{R}_4, \sigma_4, \mathbf{R}_5, \sigma_5]$$
$$\underbrace{\qquad}_{①} \quad \underbrace{\qquad}_{②} \quad \underbrace{\qquad}_{③} \quad \underbrace{\qquad}_{④} \quad \underbrace{\qquad}_{⑤}$$
$$= [\mathbf{R}_1, \sigma_1, \mathbf{R}_2, \sigma_2, \mathbf{R}_3, \sigma_3, \mathbf{R}_4, \sigma_4, \mathbf{R}_5, \sigma_5] \qquad (7.65)$$
$$\underbrace{\qquad}_{②} \quad \underbrace{\qquad}_{③} \quad \underbrace{\qquad}_{①} \quad \underbrace{\qquad}_{④} \quad \underbrace{\qquad}_{⑤}$$
$$\equiv [\mathbf{R}_3, \sigma_3, \mathbf{R}_1, \sigma_1, \mathbf{R}_2, \sigma_2, \mathbf{R}_4, \sigma_4, \mathbf{R}_5, \sigma_5] \qquad (7.66)$$
$$\underbrace{\qquad}_{①} \quad \underbrace{\qquad}_{②} \quad \underbrace{\qquad}_{③} \quad \underbrace{\qquad}_{④} \quad \underbrace{\qquad}_{⑤}$$

or, omitting the circled numbers,

$$(123)(\mathbf{R}_1, \sigma_1, \mathbf{R}_2, \sigma_2, \mathbf{R}_3, \sigma_3, \mathbf{R}_4, \sigma_4, \mathbf{R}_5, \sigma_5)$$
$$= (\mathbf{R}_1', \sigma_1', \mathbf{R}_2', \sigma_2', \mathbf{R}_3', \sigma_3', \mathbf{R}_4', \sigma_4', \mathbf{R}_5', \sigma_5')$$
$$\equiv (\mathbf{R}_3, \sigma_3, \mathbf{R}_1, \sigma_1, \mathbf{R}_2, \sigma_2, \mathbf{R}_4, \sigma_4, \mathbf{R}_5, \sigma_5) \qquad (7.67)$$

where here $\mathbf{R}_i', \sigma_i'$ are the coordinates of particle i after having made the permutation (123).

The CNPI group of the H_3^+ molecule is:

$$\{E, (12), (23), (31), (123), (132), E^*, (12)^*, (23)^*, (31)^*, (123)^*, (132)^*\}. \qquad (7.68)$$

There are some obvious identities such as $(12) \equiv (21)$ and $(123) \equiv (231)$, and we must avoid including any such duplicates in the list of group operations. This CNPI group is called the $\mathbf{D}_{3h}(\mathrm{M})$ group. The effect of the operation E^* is shown

---

[5] By *transposition* we mean an interchange, such as (12), of *two* identical particles. A *cycle* or *cyclic permutation* is an operation such as (123) that involves more than two particles.

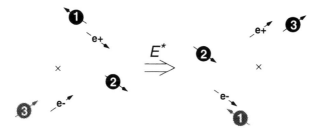

Figure 7.6. The effect of the symmetry operation E^* on an H_3^+ molecule. The molecular centre of mass is indicated by a $\times$.

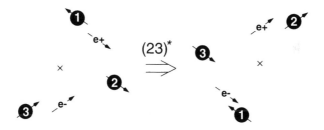

Figure 7.7. The effect of the symmetry operation $(23)^*$ on an H_3^+ molecule. The molecular centre of mass is indicated by a $\times$.

in figure 7.6 and that of $(23)^*$ is shown in figure 7.7. The multiplication table of the operations of this CNPI group contains results such as

$$(123)(123) = (132) \tag{7.69}$$
$$(123)(132) = E \tag{7.70}$$
$$(12)(123) = (23) \tag{7.71}$$

and

$$(123)(12) = (31). \tag{7.72}$$

Equation (7.70) shows that (123) and (132) are the reciprocals of each other, and equations (7.71) and (7.72) show that permutation multiplication is not necessarily commutative.

The effect of each of the operations of $D_{3h}(M)$ on a function is given by equation (7.7) so that, for example, if $f = X_1 + 3X_2 + 5X_3$, we have

$$
\begin{aligned}
(123)f(X_1, X_2, X_3) &= (123)(X_1 + 3X_2 + 5X_3) \\
&= (X_3 + 3X_1 + 5X_2) \\
&= f^{(123)}(X_1, X_2, X_3)
\end{aligned} \tag{7.73}
$$

so that $f^{(123)} = X_3 + 3X_1 + 5X_2$; the value of $f^{(123)}$ at the point $\{X_1, X_2, X_3\}$ is the same as the value of f at the point $\{X_1', X_2', X_3'\} = \{X_3, X_1, X_2\}$.

All the operations in equation (7.68) commute with the rovibronic Hamiltonian and, hence, convert any eigenfunction into another having the same eigenvalue according to equation (7.24) but now we must include the possibility of degeneracy. Suppose we have

$$\hat{H}_{rve}\psi_{ni} = E_n\psi_{ni} \tag{7.74}$$

where the level E_n is k-fold degenerate with an orthonormal set of eigenfunctions $\psi_{n1}, \psi_{n2}, \ldots, \psi_{nk}$. The effect of a symmetry operation R on the eigenfunction ψ_{ni} is to generate the function ψ_{ni}^R, which also has eigenvalue E_n. From equation (2.20), ψ_{ni}^R can only be a linear combination of the k functions ψ_{ni} and we can write

$$\psi_{ni}^R = \sum_{j=1}^{k} c_{ij}^R \psi_{nj}. \tag{7.75}$$

Instead of the constant c^R in equation (7.25), generated by a non-degenerate eigenfunction, the matrix $\mathbf{c}^R$, with elements c_{ij}^R, is generated for each symmetry operation R by a set of k-fold degenerate wavefunctions. These matrices multiply together [see equation (2.99)] in the same way as do the operations R and they form a *matrix representation* of the symmetry group of operations R. Such a set of k-fold degenerate matrices provides a symmetry label for k-fold degenerate states, in the same way as do the four (non-degenerate) symmetry labels in table 7.2 for the (non-degenerate) states of the water molecule. In applications, we rarely need the complete matrices and their *characters* $\chi^R[= \sum_i c_{ii}^R$, the trace of the matrix; see equation (2.93)] are usually all we need.

Tabulating the one-dimensional symmetry species together with the characters for the degenerate symmetry species produces the character table for the group. The character table for the $D_{3h}(M)$ group is given in table 7.3; there are six different symmetry species, or *irreducible* representations, in this group.

We now show how to multiply the degenerate symmetry species of the $D_{3h}(M)$ group together and this helps in understanding what a *reducible* representation is and in understanding why symmetry species are called irreducible representations.

The notation $\chi^\Gamma[R]$ is used for the character under the symmetry operation R in the representation Γ; for example, in the irreducible representations E' and E''

$$\chi^{E'}[E] = \chi^{E''}[E] = 2 \tag{7.76}$$

but

$$\chi^{E'}[E^*] = -\chi^{E''}[E^*] = 2. \tag{7.77}$$

The representation $E' \otimes E''$ is the four-dimensional matrix representation for which the character of the matrix representing the operation R is the product of numbers $\chi^{E'}[R]$ and $\chi^{E''}[R]$, i.e.

$$\chi^{E'\otimes E''}[R] = \chi^{E'}[R] \times \chi^{E''}[R]. \tag{7.78}$$

Table 7.3. The character table of $D_{3h}(M)$ for H_3^+. The protons are labelled 1, 2 and 3. For the non-degenerate symmetry species, each row of the table gives a possible combination of the constants c^R in equation (7.25), whereas for the degenerate ones the characters χ^R of the matrices in equation (7.75) are given. The last line gives the characters of the $E' \otimes E''$ product representation.

$R:$	E	(123) (132)	(12) (23) (31)	E^*	(123)* (132)*	(12)* (23)* (31)*
A_1' :	1	1	1	1	1	1
A_1'':	1	1	1	-1	-1	-1
A_2' :	1	1	-1	1	1	-1
A_2'':	1	1	-1	-1	-1	1
E' :	2	-1	0	2	-1	0
E'' :	2	-1	0	-2	1	0
$E' \otimes E''$:	4	1	0	-4	-1	0

Using this equation, the characters for the representation $E' \otimes E''$ are given at the bottom of table 7.3. This is a reducible representation of the group. It can be *reduced* to the sum of the irreducible representations A_1'', A_2'' and E'', which we write as

$$E' \otimes E'' = A_1'' \oplus A_2'' \oplus E''. \tag{7.79}$$

The characters satisfy

$$\chi^{E' \otimes E''}(R) = \chi^{A_1''}[R] + \chi^{A_2''}[R] + \chi^{E''}[R] \tag{7.80}$$

as can be verified from table 7.3. Because of equation (7.79), it is said that the reducible representation $E' \otimes E''$ 'contains' the irreducible representations A_1'', A_2'' and E''. In section 7.11, reducible and irreducible representations are discussed further.

To determine the result of multiplying representations and to reduce them to their irreducible components, we only need the characters of the representations involved. The product of two irreducible representations can give another irreducible representation. For example, in $D_{3h}(M)$, we have

$$A_2'' \otimes E' = E''. \tag{7.81}$$

The generalization of the vanishing integral rule (introduced in section 7.5 for the water molecule) which is applicable to all molecules and which includes the possibility of degenerate states is as follows.

The integral

$$I = \int \psi'^* \hat{O} \psi'' \, d\tau \qquad (7.82)$$

will vanish if the product of the symmetry species of ψ'^*, $\hat{O}$ and ψ'' does not *contain* the totally symmetric symmetry species $\Gamma^{(s)}$.

The vanishing integral rule can be written as

$$I = \int \psi'^* \hat{O} \psi'' \, d\tau = 0 \qquad \text{if } \Gamma(\psi'^*) \otimes \Gamma(\hat{O}) \otimes \Gamma(\psi'') \not\supset \Gamma^{(s)} \qquad (7.83)$$

or, equivalently,

$$I = \int \psi'^* \hat{O} \psi'' \, d\tau = 0 \qquad \text{if } \Gamma(\psi'^*) \otimes \Gamma(\psi'') \not\supset \Gamma(\hat{O}). \qquad (7.84)$$

If $\Gamma(\hat{O})$ is totally symmetric (of species $\Gamma^{(s)}$), as is the case if $\hat{O}$ is the field-free molecular Hamiltonian $\hat{H}$, then equation (7.84) reduces to the condition that I will vanish if

$$\Gamma(\psi'') \neq \Gamma(\psi') \qquad (7.85)$$

which is the same as equation (7.47) and this is true regardless of whether the state is degenerate, the wavefunction complex or whether the symmetry species concerned involve complex numbers.

Off-diagonal matrix elements of the Hamiltonian of a molecule in field-free space must vanish between states of different symmetry.

Thus, the Hamiltonian matrix in a basis set will block diagonalize into blocks for each symmetry species of the CNPI group of the molecule.

Using the vanishing integral rule, an electric dipole transition between the rovibronic states Φ'_{rve} and Φ''_{rve} of a molecule is forbidden if

$$\Gamma(\Phi'^*_{rve}) \otimes \Gamma(\mu_A) \otimes \Gamma(\Phi''_{rve}) \not\supset \Gamma^{(s)} \qquad (7.86)$$

where μ_A ($A = X, Y$ or Z) is a component of the electric dipole moment of the molecule. As stated at the end of section 7.6, we know how μ_A transforms in the CNPI group and, for any molecule:

Rovibronic transitions are forbidden if their symmetry species satisfy equation (7.86), where $\Gamma(\mu_A)$ is that one-dimensional irreducible representation of the CNPI group having character $+1$ for each permutation operation and character -1 for each permutation–inversion operation.

For H_3^+, $\Gamma(\mu_A) = A_1''$ and $\Gamma^{(s)} = A_1'$, so rovibronic transitions are forbidden between states having symmetries that satisfy

$$\Gamma(\Phi_{rve}'^*) \otimes A_1'' \otimes \Gamma(\Phi_{rve}'') \not\supset A_1'. \tag{7.87}$$

We can rewrite this condition as

$$\Gamma(\psi_{rve}'^*) \otimes \Gamma(\psi_{rve}'') \not\supset A_1''. \tag{7.88}$$

$E' \otimes E''$ does contain A_1'' and, hence, transitions between rovibronic states of symmetry E' and E'' are not forbidden by this symmetry rule. The transitions $A_1' \leftrightarrow A_2''$, $A_1'' \leftrightarrow A_2'$ and $E' \leftrightarrow E'$ are examples of forbidden rovibronic transitions for H_3^+.

7.11   Group theory

As stated by Fermi, group theory is basically a list of definitions. We use them when applying symmetry to understand molecules and we have introduced many of them above in applications involving the H_2O and H_3^+ molecules. The following is a list of definitions that we use and those that we have not yet defined in full are the subject of this section:

- Symmetry groups and subgroups.
- Irreducible and reducible representations.
- Character tables.
- Homomorphism and isomorphism.
- The reduction and generation of a representation.
- The symmetry of a product.
- Projection operators.
- The symmetrization of coordinates and wavefunctions.
- The correlation of the irreducible representations between a group and a subgroup.
- Even and odd permutations.

A subset of the operations of a group can be chosen in such a way that they satisfy the group axioms given on page 134; such a subset is a group and it is called a *subgroup*. The group

$$C_{3v}(M) = \{E, (123), (132), (12)^*, (23)^*, (31)^*\} \tag{7.89}$$

Table 7.4. The multiplication table for $C_{3v}(M)$. Each entry is the product of first applying the operation at the top of the column and then applying the operation at the left end of the row.

	E	(123)	(132)	(12)*	(23)*	(31)*
E	E	(123)	(132)	(12)*	(23)*	(31)*
(123)	(123)	(132)	E	(31)*	(12)*	(23)*
(132)	(132)	E	(123)	(23)*	(31)*	(12)*
(12)*	(12)*	(23)*	(31)*	E	(123)	(132)
(23)*	(23)*	(31)*	(12)*	(132)	E	(123)
(31)*	(31)*	(12)*	(23)*	(123)	(132)	E

is a subgroup of $D_{3h}(M)$. The multiplication table of the operations of $C_{3v}(M)$ are given in table 7.4 and the character table is given in table 7.5. A character table gives the characters of all the irreducible representations of the group; all representations can be expressed in terms of the irreducible ones. The irreducible representation E, whose characters $\chi^E[R]$ are given in table 7.5, consists of the six 2×2 matrices:

$$\{\mathbf{M}^E, \mathbf{M}^{(123)}, \mathbf{M}^{(132)}, \mathbf{M}^{(12)^*}, \mathbf{M}^{(23)^*}, \mathbf{M}^{(31)^*}\} \qquad (7.90)$$

where

$$\mathbf{M}^E = \begin{bmatrix} 1 & 0 \\ 0 & 1 \end{bmatrix} \qquad \mathbf{M}^{(12)^*} = \begin{bmatrix} 1 & 0 \\ 0 & -1 \end{bmatrix}$$

$$\mathbf{M}^{(123)} = \begin{bmatrix} -\frac{1}{2} & \frac{\sqrt{3}}{2} \\ -\frac{\sqrt{3}}{2} & -\frac{1}{2} \end{bmatrix} \qquad \mathbf{M}^{(23)^*} = \begin{bmatrix} -\frac{1}{2} & \frac{\sqrt{3}}{2} \\ \frac{\sqrt{3}}{2} & \frac{1}{2} \end{bmatrix}$$

$$\mathbf{M}^{(132)} = \begin{bmatrix} -\frac{1}{2} & -\frac{\sqrt{3}}{2} \\ \frac{\sqrt{3}}{2} & -\frac{1}{2} \end{bmatrix} \qquad \mathbf{M}^{(31)^*} = \begin{bmatrix} -\frac{1}{2} & -\frac{\sqrt{3}}{2} \\ -\frac{\sqrt{3}}{2} & \frac{1}{2} \end{bmatrix}. \qquad (7.91)$$

Replacing each operation R in table 7.4 by the matrix $\mathbf{M}^R$, it is seen that the multiplication table of the matrices represents that of the symmetry operations. This set of matrices is a group since it satisfies the group axioms with the definition of 'multiplication' being matrix multiplication [see equation (2.99)]. This is a *faithful* representation since each matrix represents only one symmetry operation; because of the 1:1 relation, it is said that the matrix group E and the group $C_{3v}(M)$ are *isomorphic*. If two groups, such as C_{3v} and $C_{3v}(M)$, are isomorphic, their irreducible representations are identical. In contrast to the E irreducible representation, the A_1 and A_2 irreducible representations of $C_{3v}(M)$ are *unfaithful* representations, and the groups $\{1\}$ and $\{1, -1\}$ are each said to be *homomorphic* to $C_{3v}(M)$.

Table 7.5. The character table for $C_{3v}(M)$.

R :	E	(123) (132)	(12)* (23)* (31)*
A_1 :	1	1	1
A_2 :	1	1	−1
E :	2	−1	0

The operations of a group form *classes* and the operations in a class all have the same character for a given representation; for example, the operations (123) and (132) are in the same class in $C_{3v}(M)$. The operations R and S in a group are in the same class if there exists a group operation Q such that

$$S = QRQ^{-1}. \tag{7.92}$$

The number of irreducible representations in a group is equal to the number of classes. Usually in the presentation of a character table only one operation from each class is given. In any CNPI group, E and E^* are in classes by themselves, permutations having the same structure (i.e. consisting of the same number of transpositions, cycles of three, cycles of four, etc) are in the same class and permutation inversions having the same structure are in the same class. However, in a subgroup of a CNPI group, the class structure does not necessarily follow from the permutation structure of the operations. If all the operations in a group commute with each other, the group is said to be Abelian and each operation is in a class by itself [$RQ = QR$ and equation (7.92) leads to $S = R$]; the $C_{2v}(M)$ group is Abelian. The number of operations in a group is called the *order* of the group and the sum of the squares of the dimensions of the irreducible representations is equal to the order of the group.

The reducible representation $A_1 \oplus E$ of $C_{3v}(M)$ is obtained by constructing the set of six 3×3 matrices $\mathbf{N}^R$, using the matrices $\mathbf{M}^R$ in equation (7.91) according to

$$\mathbf{N}^R = \begin{bmatrix} 1 & 0 \\ 0 & \mathbf{M}^R \end{bmatrix}. \tag{7.93}$$

These six matrices multiply like the operations of the group and, therefore, form a matrix representation of the group. If we subject each matrix $\mathbf{N}^R$ to the similarity transformation

$$\mathbf{A}\mathbf{N}^R\mathbf{A}^{-1} = \mathbf{O}^R \tag{7.94}$$

where $\mathbf{A}$ is a 2×2 matrix and $\mathbf{A}^{-1}$ is its inverse, the matrices $\mathbf{O}^R$ will not be block-diagonal like the matrices $\mathbf{N}^R$ and their 'origin' as $A_1 \oplus E$ is only apparent from their characters (the character of a matrix is unchanged by a similarity transformation; see problem 2.7 on page 41). The matrix representation O^R is said to be *equivalent* to the matrix representation N^R.

Any reducible representation Γ of $C_{3v}(M)$ can be written as

$$\Gamma = n_{A_1} A_1 \oplus n_{A_2} A_2 \oplus n_E E \tag{7.95}$$

where, for any operation R in the group, the characters satisfy

$$\chi^\Gamma [R] = n_{A_1} \chi^{A_1}[R] + n_{A_2} \chi^{A_2}[R] + n_E \chi^E [R]. \tag{7.96}$$

The reduction of the reducible representation Γ to its irreducible components Γ_i is achieved by determining the coefficients n_{Γ_i}; the n_{Γ_i} are given by

$$n_{\Gamma_i} = \frac{1}{h} \sum_R \chi^\Gamma [R] \chi^{\Gamma_i}[R]^* \tag{7.97}$$

where h is the order of the group.

Suppose that we have a set of n linearly independent functions

$$\phi_1, \phi_2, \ldots, \phi_n \tag{7.98}$$

that are transformed among each other by elements of the $C_{3v}(M)$ group, i.e. the effect of each $C_{3v}(M)$ symmetry operation R on each of the n functions ϕ_i is to produce a linear combination according to

$$R\phi_i = \sum_{j=1}^{n} D^\Gamma [R]_{ij} \phi_j \tag{7.99}$$

as in equation (7.75). In matrix notation, we can write this as

$$R\Phi = D^\Gamma [R]\Phi \tag{7.100}$$

where Φ is a column vector of the n functions ϕ_i, and the matrix $D^\Gamma [R]$ is an $n \times n$ matrix of the $D^\Gamma [R]_{ij}$. In this circumstance, the n-dimensional matrices $D^\Gamma [R]$ obtained for each operation R form an n-dimensional representation of the group $C_{3v}(M)$; it is said that the n functions ϕ_i *generate* the representation Γ. If the character of the representation matrix $D^\Gamma [R]$ for each operation R satisfies equation (7.96), we say that the n functions ϕ_i generate the reducible representation Γ in equation (7.95). It can be proved (see problem 7.7) that a set of n linearly independent functions ψ_k, related to the n functions ϕ_i in equation (7.98) by an orthogonal transformation

$$\psi_k = \sum_{i=1}^{n} A_{ki} \phi_i \tag{7.101}$$

where A is an orthogonal $n \times n$ matrix, will generate the same representation as that generated by the n functions ϕ_i. We obtain a combination of the functions ϕ_i that *transform irreducibly*, according to the mth row of the irreducible representation Γ_i, by applying the *projection operator* $P_{mm}^{\Gamma_i}$ to a particular one of the functions, ϕ_a say. A projection operator is a linear combination of the operations R defined by

$$P_{mm}^{\Gamma_i} = \frac{l_i}{h} \sum_R D^{\Gamma_i}[R]_{mm}^* R \tag{7.102}$$

where l_i is the dimension of Γ_i and $D^{\Gamma_i}[R]$ is the representation matrix in Γ_i for R. Replacing $D^{\Gamma_i}[R]_{mm}^*$ by $\chi^{\Gamma_i}[R]^*$ in equation (7.102) gives a projection operator that will generate a combination of the functions that transforms as Γ_i but its transformation properties under R will be unknown. However, we only need the characters, not the matrices, of the irreducible representations to construct the latter projection operator. An example, using a projection operator of this latter type for the purpose of constructing benzene molecular orbitals that transform irreducibly, is given in equations (10.19)–(10.32).

For the two pairs of degenerate functions (Φ_a, Φ_b) and (Ψ_a, Ψ_b), each of E symmetry in $C_{3v}(M)$, the product of them $(\Phi_a\Psi_a, \Phi_a\Psi_b, \Phi_b\Psi_a, \Phi_b\Psi_b)$ transforms as $E \otimes E = A_1 \oplus A_2 \oplus E$. By use of projection operators, we find that the combinations that transform irreducibly are

$$A_1 : (\Phi_a\Psi_a + \Phi_b\Psi_b) \tag{7.103}$$
$$A_2 : (\Phi_a\Psi_b - \Phi_b\Psi_a) \tag{7.104}$$

and

$$E : [(\Phi_a\Psi_a - \Phi_b\Psi_b), (\Phi_a\Psi_b + \Phi_b\Psi_a)]. \tag{7.105}$$

The three *symmetric product functions* $\Phi_a\Psi_a$, $\Phi_b\Psi_b$ and $(\Phi_a\Psi_b + \Phi_b\Psi_a)$ generate $A_1 \oplus E$ and the *antisymmetric product function* $(\Phi_a\Psi_b - \Phi_b\Psi_a)$ generates A_2.

The product of (Φ_a, Φ_b) with itself generates three independent functions transforming as the symmetric product $A_1 \oplus E$; the antisymmetric product combination vanishes since $\Phi_a\Phi_b = \Phi_b\Phi_a$. $A_1 \oplus E$ is the *symmetric product representation*, or *symmetric square*, of E with itself and we write

$$[E]^2 = [E \otimes E] = A_1 \oplus E. \tag{7.106}$$

A_2 is the *antisymmetric product representation*, or *antisymmetric square*, of E with itself, and we write

$$\{E\}^2 = \{E \otimes E\} = A_2. \tag{7.107}$$

The product $E \times E$ of any doubly degenerate representation E with itself is reducible to the sum of the symmetric product representation $[E \otimes E]$ and

Table 7.6. The correlation table for $D_{3h}(M)$ to $C_{3v}(M)$.

$D_{3h}(M)$	$C_{3v}(M)$
A_1'	A_1
A_1''	A_2
A_2'	A_2
A_2''	A_1
E'	E
E''	E

the antisymmetric product representation $\{E \otimes E\}$ where the characters in the symmetric product are given by

$$\chi^{[E \otimes E]}[R] = \tfrac{1}{2}((\chi^E[R])^2 + \chi^E[R^2]) \tag{7.108}$$

and the characters in the antisymmetric product are given by

$$\chi^{\{E \otimes E\}}[R] = \tfrac{1}{2}((\chi^E[R])^2 - \chi^E[R^2]). \tag{7.109}$$

The characters in the symmetric nth power of E (i.e. the symmetry of the set of $n + 1$ independent functions obtained by taking the nth power of a pair of E functions) can be obtained from the characters in the symmetric $(n - 1)$th power of E (where E is doubly degenerate) by using

$$\chi^{[E]^n}[R] = \tfrac{1}{2}(\chi^E[R]\chi^{[E]^{n-1}}[R] + \chi^E[R^n]). \tag{7.110}$$

In equations (7.108)–(7.110), $\chi^E[R^n]$ is the number obtained by determining the character in E under the operation $P = R^n$. The symmetric v_2th power of the E' irreducible representation of $D_{3h}(M)$ is used in equation (11.38) to obtain the symmetry of the vibrational wavefunctions of the H_3^+ molecule.

Any function that transforms as A_1' or A_2'' in $D_{3h}(M)$ generates characters $c^{(123)} = c^{(12)*} = +1$ and, hence, will transform as A_1 in $C_{3v}(M)$; we say that A_1' and A_2'' in $D_{3h}(M)$ each *correlate* with A_1 in $C_{3v}(M)$. As a result, we can set up the correlation table 7.6 of the irreducible representations of $D_{3h}(M)$ with those of its subgroup $C_{3v}(M)$. The correlation table 7.6 enables us to determine the symmetry of a function in $C_{3v}(M)$ if we know its symmetry in $D_{3h}(M)$. This is a rather simple example but to give a general discussion, we suppose we have a group C [in our example $D_{3h}(M)$] with operations $\{C_1, C_2, \ldots, C_c\}$ of order c and a subgroup P [in our example $C_{3v}(M)$] with operations $\{P_1, P_2, \ldots, P_p\}$ of order $p < c$, where $P_1 = C_1, P_2 = C_2, \ldots, P_p = C_p$. Any irreducible matrix representation Γ_α, say, of C will provide a matrix representation of P by considering only the matrices corresponding to the operations $C_1, C_2, \ldots, C_p$ of

Table 7.7. The reverse correlation table for $C_{3v}(M)$ to $D_{3h}(M)$.

$C_{3v}(M)$	$D_{3h}(M)$
A_1	$A_1' \oplus A_2''$
A_2	$A_2' \oplus A_1''$
E	$E' \oplus E''$

C. This will, in general, be a reducible representation of P which we can write as

$$\Gamma_\alpha = n_1^{(\alpha)}\Gamma_1 \oplus n_2^{(\alpha)}\Gamma_2 \oplus \cdots = \sum_i n_i^{(\alpha)}\Gamma_i \qquad (7.111)$$

where the Γ_i are the irreducible representations of P. This gives the correlation of Γ_α to the Γ_i. From equation (7.97), the $n_i^{(\alpha)}$ are given by

$$n_i^{(\alpha)} = \frac{1}{h}\sum_{r=1}^{p}\chi^{\Gamma_\alpha}[P_r]\chi^{\Gamma_i}[P_r]^*. \qquad (7.112)$$

We can also correlate the representations from the subgroup P to the larger group C. This will be useful later when we consider the effect of tunnelling on symmetry labels. The irreducible representation Γ_i of P correlates with the representation that we call $\Gamma(\Gamma_i)$ of the group C; the representation $\Gamma(\Gamma_i)$ of C is said to be *induced* by the representation Γ_i of the subgroup P. The representation $\Gamma(\Gamma_i)$ is given by

$$\Gamma(\Gamma_i) = \sum_i n_i^{(\alpha)}\Gamma_\alpha \qquad (7.113)$$

in terms of the irreducible representations Γ_α of C, where the $n_i^{(\alpha)}$ are as given in equation (7.112). Thus, given the irreducible representations of the subgroup P and the correlation table relating the irreducible representations of C to P, we can use that correlation table backwards to determine the symmetries of the levels in the larger group C. The reverse correlation table for $C_{3v}(M)$ to $D_{3h}(M)$ is given in table 7.7.

When we come to consider the determination of nuclear spin statistical weights in chapter 9, it will be necessary to understand the distinction between an *even* and an *odd* permutation. An odd(even) permutation can be written as the product of an odd(even) number of pair transpositions. A pair transposition such as (12) is an odd permutation. A cycle of three such as (123) is an even permutation, because no matter how we write it as the product of transpositions there will be an even number of them, e.g.

$$(123) = (23)(31) = (12)(23) = (31)(12) = (31)(12)(23)(23). \qquad (7.114)$$

In general, a cycle of length n will be even(odd) as n is odd(even).

7.12 Problems

7.1   Write down the CNPI groups of the molecules formaldehyde (CH_2O), ketene (CH_2CO), acetylene (C_2H_2), hydrogen peroxide (H_2O_2), difluoromethane (CH_2F_2), the vinylidene radical (CH_2C), *cis* and *trans* difluoroethylene (CHFCHF) and ethylene (C_2H_4). Why would you not like to write down all the elements of the CNPI group of the benzene molecule (C_6H_6)?

7.2   The CNPI group for methylfluoride is given in equation (7.68) and it is the group $D_{3h}(M)$ as for H_3^+. The elements of $D_{3h}(M)$ can be divided into two sets: The set of elements given in equation (7.89) which form the subgroup $C_{3v}(M)$, and the remaining elements, which do not form a group: $[(12),(23),(31),E^*,(123)^*,(132)^*]$. We call this latter set O(rest). The protons in the methylfluoride molecule are numbered 1, 2 and 3. One sees that there are two forms: One with a clockwise numbering (looking from the F nucleus) and one with an anticlockwise numbering. How do the elements of O(rest) differ from the elements of $C_{3v}(M)$ in relation to their effect on the clockwise and anticlockwise numbered forms?

7.3   Show that the groups C_{3v} and $C_{3v}(M)$ are isomorphic [see equations (6.9) and (7.89)].

7.4   Here we give the characters of several reducible representations of $D_{3h}(M)$. Reduce them to their irreducible components:

E	(123)	(12)	E^*	(123)*	(12)*	
4	4	0	0	0	0	
4	1	0	4	1	0	
8	2	0	0	0	0	(7.115)
8	−4	0	−8	4	0	
12	0	0	0	0	0	
16	−2	0	−8	4	0	
40	10	0	0	−30	0.	

7.5   For the group $D_{3h}(M)$, form the product of each irreducible representation with each of the rest and express each product in terms of the irreducible representations. List all symmetry pairs that represent forbidden electric dipole transitions for the H_3^+ molecule.

7.6*   The three protons in the PH$_3$ molecule [figure 6.10] are labelled by $i = 1, 2, 3$. The bond length between the P nucleus and proton i is denoted r_i. Determine the representation of $C_{3v}(M)$ [table 7.5] generated by r_1, r_2, r_3. Use projection operators in the form given in equation (7.102) to determine the linear combinations of r_1, r_2, r_3 that transform irreducibly. In forming the projection operators for the irreducible representation E, use the elements of the matrices in equation (7.91). Verify that the transformation properties of the coordinates with E symmetry are described by the matrices in equation (7.91).

7.7 Prove that the set of n functions ψ_k, related to the n functions ϕ_i by an orthogonal transformation, generates the same representation as that generated by the ϕ_i. Begin by applying the symmetry operation R to both sides of equation (7.101), use equation (7.99) for $R\phi_i$ on the right-hand side, and then use the inverse of equation (7.101) to express ϕ_i in terms of the ψ_r. Inserting the result of equation (2.107) will complete the proof.

Chapter 8

The symmetry groups of rigid molecules

In this chapter, we concentrate exclusively on *rigid molecules* in isolated electronic states. For this important, but limited, class of molecules the vibronic states can be symmetry classified using the appropriate point group. We show here how the point group can be derived from the complete nuclear permutation inversion (CNPI) group by introducing the molecular symmetry (MS) group.

> A rigid molecule is one for which there are no observable *tunnelling splittings* caused by wavefunction penetration between minima on the potential surface.

This is not to be confused with a rigid rotor. Tunnelling and tunnelling splittings are discussed in section 4.5. A rigid molecule has small amplitude vibrational displacements and it can suffer from the effects of anharmonicity, centrifugal distortion and Coriolis coupling. For an isolated electronic state, we do not have to consider the possibility of the breakdown of the Born–Oppenheimer approximation.

8.1 The CNPI group

The definition of the complete nuclear permutation inversion (CNPI) group is given on page 135 and the CNPI groups for the H_2O and H_3^+ molecules have already been discussed. The number of operations in a CNPI group is given by

$$h_{CNPI} = 2 \times n_1! \times n_2! \times n_3! \cdots \times n_r! \qquad (8.1)$$

where there are r different types of nucleus in the molecule, and n_i nuclei of type i. There are $n_i!$ ways of permuting n_i nuclei, and the factor 2 in equation (8.1) allows for the presence of the E^* operation and all the permutation–inversion operations. The acetic acid molecule CH_3COOH has three different types of nucleus, with $n_C = 2$, $n_H = 4$ and $n_O = 2$, so that $h_{CNPI} = 2 \times 2! \times 4! \times 2! = 192$. Alternatively, for ethylene C_2H_4, we have $h_{CNPI} = 2 \times 2! \times 4! = 96$.

We can set up the CNPI group for any molecule once we know its chemical formula. The structure of the molecule does not enter and the operations in the group do not reflect the structural symmetry of the molecule. For example, the order of the CNPI group of C_2H_4, which has a lot of structural symmetry at its equilibrium configuration in the ground electronic state, is less than the order of the CNPI group for CH_3COOH which has little such structural symmetry.

The CNPI group does not presuppose the Born–Oppenheimer approximation or the existence of an equilibrium structure. It is entirely appropriate as the group to use in the numerical approach (neglecting spin) that is mentioned in the first paragraph of chapter 3; we call this the 'big-computer little-understanding' approach. The CNPI group, just like the elementary rovibronic Hamiltonian in equation (2.75), can be set up immediately once we know the chemical formula of a molecule. In this approach, the CNPI group could, in principle, be used to block diagonalize the Hamiltonian matrix and to determine forbidden transitions. In the alternative approach adopted here, we usually find that we should use a subgroup of the CNPI group, and for a rigid molecule the choice of this subgroup depends on the equilibrium structure of the molecule. In the next section we state the criteria used for setting up this subgroup, but first we point out an interesting feature of the symmetry labelling that results from using the CNPI group.

The CNPI group for the CH_3F molecule is the $D_{3h}(M)$ group introduced in chapter 7 for the H_3^+ molecule; the character table is given in table 7.3 on page 147. Suppose we could use the direct numerical approach for solving the rovibronic Schrödinger equation with block diagonalization of the Hamiltonian matrix using the six irreducible representations of the CNPI group. For reasons that you will come to appreciate, all the energy levels up to energies of at least $20\,000$ cm^{-1} would come (apparently 'magically') in pairs. The eigenvalues of the A_1' and A_2'' blocks would be equal to each other, similarly the eigenvalues of the A_1'' and A_2' blocks would be equal to each other, and the eigenvalues of the E' and E'' blocks would be equal to each other. Such degeneracies that are not forced by symmetry are called 'accidental' but clearly these degeneracies are systematic. The observed levels of CH_3F would be labelled using the CNPI group as being of one of the three symmetries $A_1' \oplus A_2''$, $A_1'' \oplus A_2'$ or $E' \oplus E''$. These degeneracies, which are called *structural degeneracies*, are caused by the fact that there is more than one *version* of the equilibrium structure of the CH_3F molecule. We first define the term 'version' and then show how their multiple occurrence causes structural degeneracy.

For the CH_3F molecule in its ground electronic state, there is only one conformer but there are two symmetrically equivalent potential energy minima, one for each of the two versions shown in figure 8.1, where we have labelled the protons 1, 2 and 3. Different versions of a molecule have the same structure and can only be distinguished by labelling identical nuclei. By deforming the

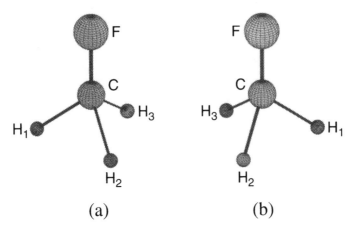

(a)   (b)

Figure 8.1. The two versions of a CH$_3$F molecule in its ground electronic state.

CH$_3$F molecule through the planar configuration we can interconvert the versions. Looking at the proton numbering from the F nucleus, we say that the version in figure 8.1(a) is the anticlockwise labelled version and that in figure 8.1(b) is the clockwise version. There are no other versions. By *ab initio* calculation, it is found that the barrier at the planar configuration is nearly the same as the dissociation energy of the C–F bond at around 35 000 cm^{-1}. Thus, the two versions of CH$_3$F, on the ground state potential energy surface, are separated by a very high barrier.

A complicated molecular system having a potential energy surface with many minima can have different conformers and each conformer can have several versions.

- Different conformers have different structures.
- Different versions of a particular conformer have the same structure and they can only be distinguished by labelling identical nuclei.

Versions are distinguished by labelling the identical nuclei. To determine all the versions for a molecule in its equilibrium configuration, one needs to label the nuclei and to determine how many distinct forms can be obtained by permuting the labels on identical nuclei with and without inverting the molecule; distinct forms cannot be interconverted by a mere rotation of the molecule in space.

> The number of versions of the equilibrium structure of a rigid molecule is given by dividing the order of the CNPI group by the order of the point group of the equilibrium configuration.

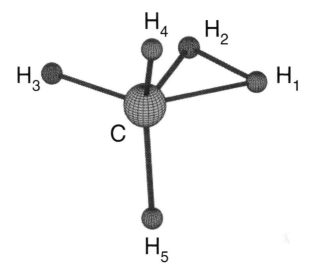

Figure 8.2. The CH_5^+ molecule at its calculated equilibrium geometry. The nuclei C, H_1, H_2 and H_3 are coplanar, and this plane bisects the H_4CH_5 angle.

Thus, a molecule will have many versions if it has many identical nuclei (to make the order of the CNPI group large) and if it has a very unsymmetrical equilibrium configuration (to make the order of its point group small). The CH_5^+ molecule is shown at its calculated equilibrium geometry in figure 8.2; its CNPI group has an order of $2 \times 5! = 240$. The point group symmetry of its equilibrium configuration is C_s with an order of 2; the only point group symmetry element is the reflection plane which passes through the nuclei C, H_1, H_2 and H_3. Thus, there are 120 versions of its equilibrium structure. Very large unsymmetrical molecules can have an astronomical number of versions. Small symmetrical molecules, such as H_2O and H_3^+ in their ground electronic states, have only one version.

For a rigid molecule that has n versions, there will be n identically shaped potential energy minima on the potential energy surface and each supports the same energy level pattern[1] and each observed level will have a structural degeneracy equal to the number of versions. In the ground electronic state of CH_3F the energy levels have a structural degeneracy of two, and every energy level is doubly degenerate (if both are allowed by the nuclear spin statistical formulae; see chapter 9).

In summary, although the CNPI group is a symmetry group and gives symmetry labels on energy levels, it is often a very large group and it often produces a labelling with systematic accidental degeneracies that we do not need to know about. It is possible to define a subgroup of the CNPI group

---

[1] There being no observable tunnelling splittings by definition for a rigid molecule.

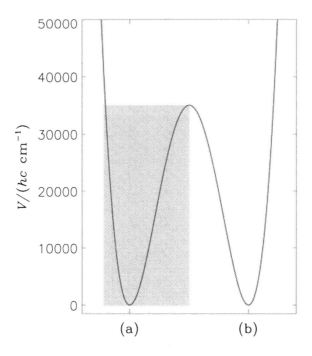

Figure 8.3. A schematic cut through the potential surface of the CH_3F molecule connecting the minima (a) and (b) shown in figure 8.1.

that provides all the useful symmetry labelling information that the CNPI group provides. Such symmetry information is used, for example, when we implement the vanishing integral rule for the purpose of determining selection rules and for block diagonalizing the Hamiltonian matrix. The subgroup can also be used for determining the nuclear spin statistical weights as discussed in chapter 9. We call this CNPI subgroup the *molecular symmetry* (MS) group.

8.2   The molecular symmetry (MS) group

In figure 8.3, we show a schematic cut through the potential energy surface V for the CH_3F molecule in its ground electronic state along the coordinate that connects the two minima (a) and (b) shown in figure 8.1. The barrier between the minima is *insuperable* (i.e. so high that there is no observed tunnelling) and so to calculate the lower vibrational energy levels, we only need focus on that part of V around one minimum; we have indicated such a region around the left-hand minimum with shading. The *ab initio* calculation of V would be made for nuclear geometries around this single minimum and the Taylor's series expansion around this minimum would lead to the determination of the harmonic force constants, normal coordinates and anharmonic force constants. The calculation of the rotation–vibrational energies and wavefunctions would be made using a region of nuclear coordinate space around a single minimum. The correctness of this procedure depends on the fact that we only consider energy levels sufficiently below the barrier maximum that tunnelling splittings can be neglected (see section 4.5). Thus, we neglect the structural degeneracy since it is of no consequence when there are no observable tunnelling splittings.

For a rigid molecule the calculation of energies is done in the coordinate space of one minimum of V, and to symmetry label the energy levels we fashion the symmetry group so that it only operates within that coordinate space. To accomplish this, we use a subgroup of the CNPI group obtained by deleting symmetry operations that cause a coordinate change which moves the molecule from one version to another. These deleted operations are called "unfeasible" operations of the CNPI group. They are useless in the same way that an *ab initio* calculation done to determine the energy at a geometry point on V around another version's minimum is useless.

> By deleting unfeasible operations from the CNPI group of a rigid molecule, we obtain a subgroup that acts within one minimum of V and does not interconvert versions; this is called the molecular symmetry (MS) group. For a rigid nonlinear molecule the MS group is isomorphic to the point group and we name it $G(M)$, where G is the name of the point group.

The CNPI group for CH_3F is given in equation (7.68). The unfeasible symmetry operations in the group are those that interconvert the clockwise and anticlockwise forms (see problem 7.2) and they are

$$[(12), (23), (31), E^*, (123)^*, (132)^*]. \qquad (8.2)$$

If we remove these operations from the CNPI group, we obtain the subgroup $C_{3v}(M)$ given in equation (7.89). This is the MS group for CH_3F. The symmetry operations in this group are *feasible* symmetry operations for CH_3F in its ground electronic state.

The ethylene molecule with its 12 versions provides a more complicated example which we work through in section 8.3.5. The H_2O and H_3^+ molecules are simpler since they each have only one version and, therefore, do not have structural degeneracy. This means that there are no unfeasible operations in their CNPI groups. For each of these molecules, the CNPI group and the MS group are the same group.

8.3   The MS group and the point group

As already stated, for a rigid nonlinear molecule the MS group is isomorphic to its point group. In this section, we explain the relationship between the two types of group. We will start by explaining things in a simple way using the water molecule as an example and we will neglect spin. We will then show the relationship of the MS group to the point group for the H_3^+ molecule in order to amplify and justify some of the statements we make when discussing the water molecule. For H_3^+, we include nuclear and electron spin in the discussion. After having treated the water and H_3^+ molecules, we make some important general statements about the effects of point group operations, and about the relationship between MS and point group operations for rigid molecules.

8.3.1   The H_2O molecule

Above we have presented the MS group as being deduced from the CNPI group after identifying and eliminating unfeasible operations; the operations of permuting identical nuclei and the inversion E^* being considered as fundamental symmetry operations that commute with the molecular Hamiltonian. This follows the work of Longuet-Higgins[2] who brought the permutation and inversion operations to the fore, who thought up the concept of feasibility, and who thereby showed how to define the MS group of any molecule, rigid or non-rigid. However, the first detailed description of the correct symmetry group to use for classifying the molecular states of rigid molecules was made by Hougen[3] who started from the point group. We will follow Hougen's procedure for the water molecule since it is an instructive way to understand the relationship between point group and MS group symmetry operations.

Labelling the protons 1 and 2, and the oxygen nucleus 3, the vibrational displacement coordinates of a water molecule are the nine coordinates $\Delta\alpha_r$ where $\alpha = x, y$ or z, and $r = 1, 2$ or 3. The three rotational coordinates θ, ϕ and χ are the Euler angles that define the orientation of the molecule-fixed xyz axes in space, where the xyz axes are attached to a water molecule in its equilibrium configuration as shown in figure 5.4 on page 102. Labelling the ten electrons 4 through 13, there are 30 electronic coordinates (x_i, y_i, z_i), where $i = 4$ to 13. We neglect spin in this example. The C_{2v} point group operations, which we define below, are E, C_{2x}, σ_{xz} and σ_{xy}.

Point group operations do not transform the Euler angles; they only transform the vibronic coordinates (i.e. the vibrational displacement coordinates and electronic coordinates). We first look at the point group operation C_{2x} which

[2] Longuet-Higgins H C 1963 *Mol. Phys.* **6** 445.
[3] Hougen J T 1962 *J. Chem. Phys.* **37** 1433; 1963 *J. Chem. Phys.* **39** 358.

rotates the vibronic coordinates about the x axis through π radians:

$$C_{2x}(\Delta x_1, \Delta x_2, \Delta x_3, x_i) = (\Delta x_1', \Delta x_2', \Delta x_3', x_i')$$
$$= (\Delta x_2, \Delta x_1, \Delta x_3, x_i) \tag{8.3}$$
$$C_{2x}(\Delta y_1, \Delta y_2, \Delta y_3, y_i) = (\Delta y_1', \Delta y_2', \Delta y_3', y_i')$$
$$= (-\Delta y_2, -\Delta y_1, -\Delta y_3, -y_i) \tag{8.4}$$

and

$$C_{2x}(\Delta z_1, \Delta z_2, \Delta z_3, z_i) = (\Delta z_1', \Delta z_2', \Delta z_3', z_i')$$
$$= (-\Delta z_2, -\Delta z_1, -\Delta z_3, -z_i) \tag{8.5}$$

where $i = 4$ through 13 labels the electrons. Similar equations can be written for the point group operation σ_{xz}, which reflects the vibronic variables through the xz plane. The Δx_r, x_i, Δz_r and z_i are unaffected by σ_{xz} but

$$\sigma_{xz}(\Delta y_1, \Delta y_2, \Delta y_3, y_i) = (\Delta y_1', \Delta y_2', \Delta y_3', y_i')$$
$$= (-\Delta y_1, -\Delta y_2, -\Delta y_3, -y_i). \tag{8.6}$$

For the point group operation σ_{xy}, which reflects the vibronic variables through the xy plane, we have

$$\sigma_{xy}(\Delta x_1, \Delta x_2, \Delta x_3, x_i) = (\Delta x_1', \Delta x_2', \Delta x_3', x_i')$$
$$= (\Delta x_2, \Delta x_1, \Delta x_3, x_i) \tag{8.7}$$
$$\sigma_{xy}(\Delta y_1, \Delta y_2, \Delta y_3, y_i) = (\Delta y_1', \Delta y_2', \Delta y_3', y_i')$$
$$= (\Delta y_2, \Delta y_1, \Delta y_3, y_i) \tag{8.8}$$

and

$$\sigma_{xy}(\Delta z_1, \Delta z_2, \Delta z_3, z_i) = (\Delta z_1', \Delta z_2', \Delta z_3', z_i')$$
$$= (-\Delta z_2, -\Delta z_1, -\Delta z_3, -z_i). \tag{8.9}$$

In figure 8.4, we show the effect of the three point group operations on the vibrational displacement coordinates of a water molecule that we depict in an arbitrarily displaced configuration, and on the coordinates of an arbitrary electron e_i. The display labelled 'E' represents the initial configuration. We also show the equilibrium configuration and the xyz axes. The equilibrium structure and the xyz axes are not transformed by the point group operations. In actuality, the molecule-fixed xyz axes for a vibrationally displaced water molecule are always positioned so that the three nuclei are in the xz plane and this means that the three Δy_i would always be zero. However, to show more clearly the effect of the point group operations on the coordinates, we have positioned the xyz axes in figure 8.4 so that the Δy_i are non-zero.

Following Hougen, we now apply a specially chosen bodily rotation operation about a molecule-fixed axis to the molecule after having applied each

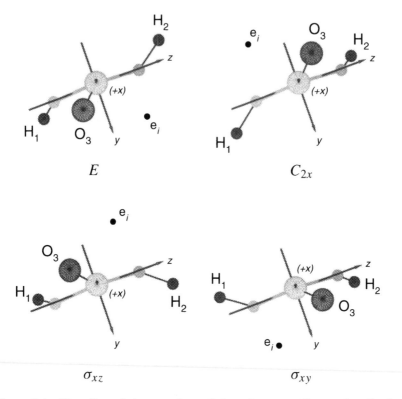

Figure 8.4. The effect of the operations of the point group C_{2v} on the vibrational displacements of the nuclei and on the coordinates (x_i, y_i, z_i) of electron e_i in the H_2O molecule. In all four displays, the electron is above the yz plane and the x axis points up out of the plane of the page. The display labelled 'E' shows the initial configuration and the equilibrium structure of H_2O is drawn as a grey-shaded structure.

of these point group operations. A bodily rotation operation does not affect the vibrational displacement coordinates $\Delta\alpha_r$ or electron coordinates α_i in the xyz axis system since the axes are rotated with the molecule, but it does affect the Euler angles θ, ϕ and χ that specify the orientation of the molecule-fixed axes in space.

After having applied the point group operation C_{2x} we apply the bodily rotation R_x^π, which is a bodily rotation about the x axis through π radians; the result of this combined operation is shown in the top right-hand display of figure 8.5. This display is obtained from the display labelled C_{2x} in figure 8.4 by rotating it bodily about the x axis through π radians. Comparing the top right-hand display in figure 8.5 with the display labelled E, we see that the combined operation $R_x^\pi C_{2x}$ simply permutes the protons numbered 1 and 2, and this is a

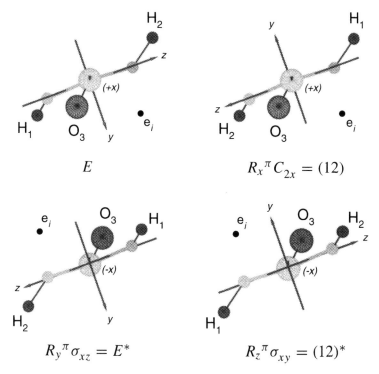

Figure 8.5. The effect of the product operations $R_x{}^\pi C_{2x}$, $R_y{}^\pi \sigma_{xz}$ and $R_z{}^\pi \sigma_{xy}$ on the vibrational displacements of the nuclei and the spatial coordinates of electron e_i in the H_2O molecule. As indicated, the three operations $R_x{}^\pi C_{2x}$, $R_y{}^\pi \sigma_{xz}$ and $R_z{}^\pi \sigma_{xy}$ have the same effect as the operations (12), E^*, and (12)*, respectively, of the molecular symmetry group $C_{2v}(M)$. The display labelled 'E' shows the initial configuration and the equilibrium structure of H_2O is drawn as a grey-shaded structure. In the two top displays, the electron is above the yz plane and the x axis points out of the plane of the page, whereas in the two bottom displays, the electron is below the yz plane and the x axis points into the plane of the page.

symmetry operation of the Hamiltonian. The top right-hand display in figure 8.5 is thus labelled '$R_x{}^\pi C_{2x} = (12)$'. Figure 8.5 also shows the effects of the products $R_y{}^\pi \sigma_{xz}$ and $R_z{}^\pi \sigma_{xy}$, which are identical to the effects of E^* and (12)*, respectively.

For the water molecule, we have obtained a symmetry group consisting of the operations $\{E, R_x{}^\pi C_{2x}, R_y{}^\pi \sigma_{xz}, R_z{}^\pi \sigma_{xy}\}$. When this group was initially obtained, it was called the *full* point group of the water molecule (to distinguish it from the point group which only acts on vibronic variables). After the invention of the MS group, it is now called the MS group $C_{2v}(M)$ and, with permutations and

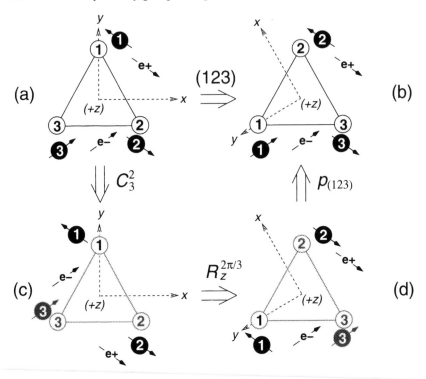

Figure 8.6. The effect of the successive application of the operations C_3^2, $R_z^{2\pi/3}$ and $P_{(123)}$ on a distorted H_3^+ ion and the equivalence of this to the permutation (123); $(123) = p_{(123)} R_z^{2\pi/3} C_3^2$.

the inversion brought to the fore, its operations are written $\{E, (12), E^*, (12)^*\}$. In presenting the character table of the MS group of a rigid molecule, it is useful to identify under each MS group operation the point group operation and overall bodily rotation operation (called the 'equivalent rotation') whose product gives the effect of the MS group operation. This makes it easier to appreciate the effects of the operations on the molecular coordinates.

8.3.2   The H_3^+ molecule

In figures 8.6–8.8, we show the effects of the MS group operations (123), E^* and (23)* on the coordinates of a vibrationally distorted H_3^+ molecule; the MS group operations take the molecule from part (a) to part (b) in each figure. The figures also show the transformations caused by the related point group operation, the related bodily rotation operation, and (where it is not the identity) the related nuclear spin permutation. In this way, we show how each operation of the MS group for the H_3^+ molecule is broken down into the product of a point group

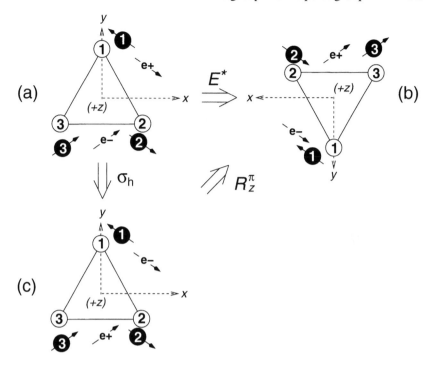

Figure 8.7. The effect of the successive application of the operations σ_h and R_z^{π} on a distorted H_3^+ ion and the equivalence of this to the inversion operation E^*; $E^* = R_z^{\pi}\sigma_h$. The molecule-fixed axis system xyz is right-handed in all figures.

operation that only transforms the vibronic coordinates, a bodily rotation that only transforms the rotational coordinates, and a nuclear spin permutation operation.

In part (a) of each figure, the instantaneous 'starting' distorted nuclear configuration is shown using filled circles for the positions of the nuclei. The electrons have instantaneous coordinates (x_i, y_i, z_i), where $i = 4$ or 5, and they are indicated using an 'e+' and an 'e−' to denote an electron above and below the page respectively. The small arrows on each particle indicate its spin. We can view these instantaneous particle positions as a snap-shot view of the molecule as the nuclear framework rotates and vibrates and as the electrons orbit. We show with open circles an appropriately oriented H_3^+ molecule at equilibrium with attached xyz axes. The (xyz) axes (with origin at the nuclear centre of mass) are located in the equilibrium configuration with the x and y axes in the molecular plane so that the y axis passes though proton number 1, the x axis passes through the bond connecting protons 1 and 2 and the z axis so that (xyz) is right-handed. The orientation of the (xyz) axes within the space-fixed axis system defines the Euler angles (θ, ϕ, χ). The xyz axes are located so that for all three nuclei $z_i = 0$ and $\Delta z_i = 0$; this means that the three nuclei are in the plane of the page.

In part (b) of each figure, we show the result of the MS group operation on the spatial coordinates and spins of the particles. It is vitally important to appreciate that the xyz axes, and the equilibrium H_3^+ structure, are positioned in each of these (b) figures *after* the symmetry operation has been performed on the instantaneous coordinates of the electrons and nuclei. In figure 8.6(b) the electron coordinates in space are unaffected by the nuclear permutation (123) but the molecule-fixed xyz axes, being tied to the nuclei, are rotated and, as a result, the electron coordinates (x_i, y_i, z_i) are transformed. In figure 8.7(b), the spatial coordinates of all particles are inverted through the molecular centre of mass from their locations in the starting distorted configuration. The spins are not affected by E^*. In figure 8.8(b), the effect of doing $(23)^*$ is depicted. Note that E^* and $(23)^*$ do not cause the xyz axes to become left-handed (i.e. inverted); an MS group operation cannot invert the molecule-fixed axes.

In each of these figures the coordinate changes caused by the MS group operation, and depicted in part (a)→(b), is broken down into successive coordinate changes:[4] (a)→(c), (c)→(d) and (d)→(b). Each of these parts involves only changing some of the coordinates:

- (a)→(c): The change in the *vibronic* coordinates, i.e. the change in the vibrational displacement coordinates $(\Delta x_i, \Delta y_i, \Delta z_i)$ of the nuclei, where $i = 1, 2$ and 3, and the change in the electronic coordinates (x_i, y_i, z_i), where $i = 4$ and 5.
- (c)→(d): The change in the rotational coordinates θ, ϕ and χ. This is a bodily rotation of the whole molecule about a molecule-fixed direction.
- (d)→(b): The permutation of the nuclear spins σ_i, for $i = 1, 2$ and 3.

The part (a)→(c) in each figure is a rotation of the vibronic variables about an axis, or a reflection of the vibronic coordinates through a plane. Each is a point group operation and it does not transform the molecule-fixed axes or the nuclear spins. The part (c)→(d) is a bodily rotation of the molecule about a molecule-fixed axis; the vibronic coordinates and nuclear spins are not transformed. Finally, the operation (d)→(b) is a permutation of the nuclear spins (but for E^* the nuclear spins are not permuted).

For the operations depicted in figures 8.6–8.8, we can write

$$(123) = P_{(123)} \, R_z^{2\pi/3} \, C_3^2 \tag{8.10}$$

$$E^* = P_0 \, R_z^{\pi} \, \sigma_h \tag{8.11}$$

$$(23)^* = P_{(23)} \, R_x^{\pi} \, \sigma_{yz} \tag{8.12}$$

where C_3^2, σ_h and σ_{yz} are point group operations and they only transform the vibronic variables. The operation C_3^2 rotates the vibronic variables about the z axis through $4\pi/3$ radians in a right-handed sense; the xyz axes are not rotated

---

[4] For the operation E^*, there is no nuclear spin permutation and we, therefore, omit part (d) in figure 8.7 since it would be identical to part (c).

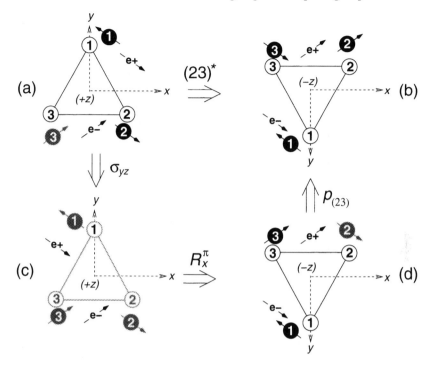

Figure 8.8. The effect of the successive application of the operations σ_{yz}, R_x^π and $P_{(23)}$ on a distorted H_3^+ ion and the equivalence of this to the permutation–inversion operation $(23)^*$; $(23)^* = P_{(23)} R_x^\pi \sigma_{yz}$. The molecule-fixed axis system xyz is right-handed in all figures.

by a point group operation such as C_3^2. The operations $R_z^{2\pi/3}$ and R_z^π are bodily rotations of the whole molecule about the z axis in a right-handed sense through $2\pi/3$ and π radians, respectively. The operation R_x^π is a bodily rotation of the whole molecule through π radians about the x axis. These rotation operations only affect the Euler angles. $p_{(123)}$ is the cyclic permutation of the nuclear spins and $p_{(23)}$ is the interchange of the spins of nuclei 2 and 3; p_0 is the identity. Note that after the operation $p_{(123)}$, nucleus 2 has the spin that 1 had, 3 has the spin that 2 had, and 1 has the spin that 3 had.

One can break down each of the 12 operations of the MS group $D_{3h}(\text{M})$ as the product of such vibronic, rotational and nuclear-spin operations. In this way, one finds out that the set of 12 vibronic operations of the type (a)→(c) for $D_{3h}(\text{M})$ form the point group D_{3h}.

8.3.3 General rules for rigid molecule symmetry groups

We can now state some general rules for rigid nonlinear molecules:

• The MS group and the point group are isomorphic and we use the same irreducible representation labels for each.
• Each operation R of the MS group can be written as the product:

$$R = R_{ns} R_{rot} R_{ve} \qquad (8.13)$$

where R_{ve} transforms the vibronic coordinates in the same way as R but it does not transform rotational coordinates or nuclear spins, R_{rot} transforms the rotational coordinates in the same way as R but it does not transform vibronic coordinates or nuclear spins, and R_{ns} transforms the nuclear spins in the same way as R but it does not transform vibronic or rotational coordinates. The three operations commute with each other and any could be the identity operation.
• The set of operations R_{ve} is the molecular point group of the molecule and this actually defines the operations of the molecular point group.
• The operations of a molecular point group R_{ve} only transform the vibronic variables (i.e. vibrational displacements and electronic coordinates).
• The molecular point group is not a symmetry group of the complete Hamiltonian but it is a symmetry group of the vibronic Hamiltonian. As a result, the molecular point group of a nonlinear rigid molecule can be used to classify the vibronic states but not the rovibronic or nuclear spin states.
• The MS group can be used to classify rovibronic, vibronic and nuclear spin states since it is the symmetry group of the complete Hamiltonian. The symmetry classification of the vibronic states of a rigid nonlinear molecule in the MS group duplicates that obtained using the molecular point group.

8.3.4 Linear rigid molecules

Rigid molecules that have a linear equilibrium configuration are special. The point group of a linear molecule is $D_{\infty h}$ if the molecule is centrosymmetric, like H_2 or HCCH, and it is $C_{\infty v}$ if it is not, like HF or HCN. These groups have an infinite number of operations in them and, like any point group, they only transform the vibronic variables. The MS group for a $C_{\infty v}$ linear molecule is $C_{\infty v}(M) = \{E, E^*\}$ and that for a $D_{\infty h}$ linear molecule is $D_{\infty h}(M) = \{E, p, E^*, p^*\}$ where p is the permutation operation that interchanges all pairs of identical nuclei symmetrically located about the molecular midpoint. The vibrational and electronic states of a linear molecule can be classified in the point group but the rovibronic states (often called the rotational levels) are classified in the MS group. As a result, the rotational levels of a $C_{\infty v}$ linear molecule are labelled $+$ (or Σ^+) and $-$ (or Σ^-) using $C_{\infty v}(M)$, and the rotational levels of a $D_{\infty h}$ linear molecule are labelled $+s$ (or Σ_g^+), $+a$ (or Σ_u^+), $-s$ (or Σ_u^-) and $-a$ (or Σ_g^-) using $D_{\infty h}(M)$; see tables B.14 and B.15 in Appendix B.

Using the isomorphic Hamiltonian of a linear molecule (see the footnote on page 94) the Euler angle χ is introduced as a variable and we can develop *extended molecular symmetry groups* $C_{\infty v}(EM)$ and $D_{\infty h}(EM)$ that are isomorphic with

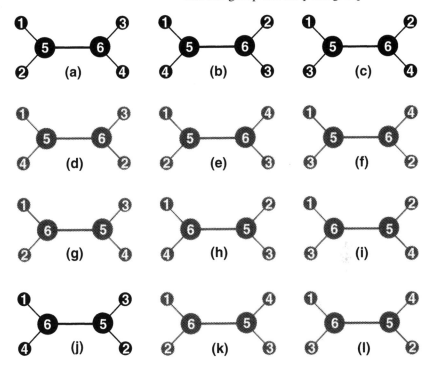

Figure 8.9. The 12 versions of the ethylene molecule C_2H_4.

the point groups but which transform the three Euler angles (and the nuclear spins); see tables B.16 and B.17 in Appendix B. These groups can be used to classify rovibronic, vibronic and nuclear spin states of a linear rigid molecule just like the MS groups are used for nonlinear rigid molecules.

8.3.5   The ethylene molecule C_2H_4

Here we determine the MS group $\boldsymbol{D}_{2h}(M)$ of the ethylene molecule as an example. We draw out its versions and discuss its structural degeneracy. We then consider which operations of the CNPI group are feasible in order to obtain the operations of the MS group.

As previously mentioned the order of the CNPI group for the ethylene molecule C_2H_4 is $2 \times 2! \times 4! = 96$. The order of its point group $\boldsymbol{D}_{2h}$ is 8 and so there are $96/8 = 12$ versions of its equilibrium structure. Numbering the four protons 1 through 4, and the carbon nuclei 5 and 6, we draw one version in part (a) of figure 8.9. Starting from the version drawn in figure 8.9(a) one determines the other versions by permuting the numerical labels of identical nuclei and then checking that the numbered configuration obtained cannot be rotated in space to coincide with version (a). The 12 versions so obtained are shown in figure 8.9.

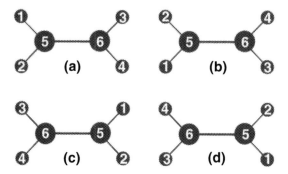

Figure 8.10. Four views of version (a) of the ethylene molecule shown in figure 8.9(a) that can be interconverted by an overall rotation in space.

In figure 8.10(a), we draw version (a) again but, by way of contrast, in figures 8.10(b)–(d), we draw three other numbered configurations, obtained by permuting the numerical labels of identical nuclei on (a), that *can* be made to coincide with version (a) by rotating it in space. In this way, we see that figure 8.10 merely shows four 'views' of version (a). As well as helping us to understand what a version is, figure 8.10 will help us determine the MS group of version (a) of the ethylene molecule.

Since there are 12 versions, there are 12 identically shaped minima in the potential energy surface of the ethylene molecule in its ground electronic state. The barriers between all the versions are high; the lowest barrier [the barrier to internal rotation between (a) and (e), (b) and (d), (c) and (f), (g) and (k), (h) and (j), or (i) and (l)] is about $25\,000$ cm^{-1}. Thus, for the lower rotation–vibration energy levels, tunnelling splittings will not be resolved and every level will have a structural degeneracy of 12.

To calculate the rotation–vibration energies, we would focus on one version and use the potential energy surface around that version. Let us choose version (a) in figure 8.9. To symmetry label the rotation–vibration levels of this version, we must determine its MS group; i.e. we must determine the feasible operations for version (a) from among the 96 operations in the CNPI group of ethylene. For a nonlinear rigid molecule, there must be the same number of operations in the MS group as in the point group and so we know that there are only eight operations (including the identity E) in the MS group of ethylene. We can now use the numbered configurations in figure 8.10, which show the four possible rotated arrangements of version (a), to help in determining the feasible operations. The nuclear permutations that connect these arrangements are feasible (since they do not correspond to passage over an insuperable potential energy barrier) and they are

$$E, (12)(34), (13)(24)(56) \quad \text{and} \quad (14)(23)(56). \quad (8.14)$$

We also have to determine the feasible permutation–inversion operations and

Table 8.1. The character table for D_{2h}(M): The MS group for the version of ethylene given in figure 8.9(a).

R:	E	$(12)(34)$	$(13)(24)(56)$	$(14)(23)(56)$	E^*	$(12)(34)^*$	$(13)(24)(56)^*$	$(14)(23)(56)^*$
A_g:	1	1	1	1	1	1	1	1
A_u:	1	1	1	1	-1	-1	-1	-1
B_{1g}:	1	1	-1	-1	-1	-1	1	1
B_{1u}:	1	1	-1	-1	1	1	-1	-1
B_{2g}:	1	-1	1	-1	-1	1	-1	1
B_{2u}:	1	-1	1	-1	1	-1	1	-1
B_{3g}:	1	-1	-1	1	1	-1	-1	1
B_{3u}:	1	-1	-1	1	-1	1	1	-1

again we look at figure 8.10 and see which permutation–inversion operations interconnect these arrangements of version (a). We see that E^* interconverts arrangements (a) and (d) [and also (b) and (c)] in figure 8.10, so E^* is a feasible operation; in this way we determine that the feasible permutation–inversion operations of the MS group of version (a) in figure 8.9 for ethylene are

$$E^*, (12)(34)^*, (13)(24)(56)^* \qquad \text{and} \qquad (14)(23)(56)^*. \qquad (8.15)$$

The MS group for version (a) in figure 8.9 of the ethylene molecule consists of the eight operations listed in equations (8.14) and (8.15); this group is called D_{2h}(M) and its character table is given in table 8.1.

For a planar rigid molecule E^* is always feasible and E^* has the same effect on the vibronic variables as the point group operation of reflection in the molecular plane.

8.4 Problems

8.1 Determine the MS group of the acetylene molecule HCCH.

8.2 Determine the MS group of 1,2 difluoroethylene CF_2CH_2.

8.3 Determine the MS group of formaldehyde CH_2O.

8.4 Determine the MS group of ketene CH_2CO.

8.5 Determine the MS group of *cis*-difluoroethylene CHFCHF.

8.6 Determine the MS group of *trans*-difluoroethylene CHFCFH.

8.7 Map the elements of the MS groups of *cis* and *trans* difluoroethylene onto the appropriate elements of their point groups.

8.8 The allene molecule is shown in its equilibrium configuration in figure 6.13. Determine its MS group and map each element onto the appropriate element of its point group D_{2d}; see problem 6.3.

8.9 Determine the MS groups of *ortho* and *para* difluorobenzene. In each case map the elements onto the appropriate elements of the point group.

8.10 Determine the MS group of the methane molecule CH_4.

8.11 Assuming the molecule to be a rigid molecule with no observable tunnelling, determine the MS group of the numbered form of the CH_5^+ molecule shown in figure 8.2.

PART 3

APPLICATIONS OF SYMMETRY

Chapter 9

Nuclear spin, statistical weights and hyperfine structure

9.1 The fifth postulate of quantum mechanics

In section 3.2, we stated the Pauli exclusion principle and it is part of the more general *fifth postulate of quantum mechanics*. To state this postulate, it is necessary to differentiate between two classes of particles: *fermions* and *bosons*. A fermion is a particle for which the spin angular momentum quantum number (called s for an electron and I for a nucleus in section 3.2) has a half-integer value, and a boson is a particle for which the spin angular momentum quantum number has an integer value. Electrons have $s = 1/2$ and are fermions. The nuclei ^{1}H, ^{3}He, ^{13}C, ^{15}N and ^{19}F with $I = 1/2$, the nuclei ^{7}Li, ^{9}Be, ^{11}B and ^{21}Ne with $I = 3/2$, and the ^{17}O nucleus with $I = 5/2$, are all fermions. In contrast, the nuclei ^{4}He, ^{12}C, ^{16}O, ^{18}O, ^{20}Ne and ^{22}Ne with $I = 0$, ^{2}H ($=$ D), ^{6}Li and ^{14}N with $I = 1$, and ^{10}B with $I = 3$, are all bosons.

It is an empirical fact that the complete internal wavefunction Φ (including spin) of a system of particles is changed in sign by an interchange of two identical fermions in the system but is unchanged by the interchange of two identical bosons. The statistical-mechanics treatment of many-body systems is affected by this (particularly the calculation of entropy) and it is said that fermions obey *Fermi–Dirac statistics*, whereas bosons obey *Bose–Einstein statistics*. This is the full statement of the fifth postulate of quantum mechanics.

No completely acceptable proof of the connection between particle spin and its statistics exists and we discuss experimental tests of it in section 15.6.

9.2   Statistical weights

A complete internal wavefunction Φ_{H_2O} for a water molecule depends on the coordinates $(R_1, \sigma_1, R_2, \sigma_2, R_3, \sigma_3, R_4, \sigma_4, R_5, \sigma_5, \ldots R_{13}, \sigma_{13})$ given in equation (7.1)[1], where the two protons are labelled 1 and 2, particle 3 is the oxygen nucleus and the electrons are labelled 4–13. Φ_{H_2O} is an eigenfunction of the complete Hamiltonian $\hat{H}_{int}$ for the internal dynamics of H_2O and the general expression for $\hat{H}_{int}$ is given in equation (2.77). Protons are fermions and, from the fifth postulate, any complete internal wavefunction for the H_2O molecule changes sign when the coordinates of the two protons are interchanged, i.e.

$$(12)\Phi_{H_2O} = -\Phi_{H_2O}. \tag{9.1}$$

Inspection of table 7.2 on page 135 shows that equation (9.1) restricts the symmetry of the complete internal wavefunctions for H_2O to be B_1 or B_2 in $C_{2v}(M)$; we use the symbol Γ_{tot} for the allowed symmetries of the complete internal wavefunction of a molecule. For D_2O the behaviour of a complete internal wavefunction Φ_{D_2O} under the interchange of the two deuterons is governed by Bose–Einstein statistics so that

$$(12)\Phi_{D_2O} = +\Phi_{D_2O} \tag{9.2}$$

and, for D_2O, Γ_{tot} can be only A_1 or A_2 in $C_{2v}(M)$.

In the Born–Oppenheimer approximation, and including electron spin, the rovibronic eigenfunctions for an H_2O or D_2O molecule can be written as

$$\begin{aligned}
\Phi_{rve,nj}&(R_1, R_2, R_3, R_4, \sigma_4, R_5, \sigma_5, \ldots, R_{13}, \sigma_{13})\\
&= \Phi_{elec,n}(R_1, R_2, R_3, R_4, \sigma_4, R_5, \sigma_5, \ldots, R_{13}, \sigma_{13})\\
&\quad \times \Phi_{rv,nj}(R_1, R_2, R_3).
\end{aligned} \tag{9.3}$$

from equation (3.8). $\Phi_{elec,n}$ is antisymmetric with respect to exchange of a pair of electrons. An approximation for the complete internal wavefunction of H_2O or D_2O is constructed by multiplying $\Phi_{rve,nj}$ by a *nuclear spin function* $\Phi_{ns,t}(\sigma_1, \sigma_2, \sigma_3)$, where t numbers the different spin functions. The $\Phi_{ns,t}$ are products of the spin functions of the individual nuclei. For a single nucleus labelled i, the spin wavefunctions satisfy

$$\hat{i}^2|I_i, \sigma_i\rangle = I_i(I_i + 1)\hbar^2|I_i, \sigma_i\rangle \tag{9.4}$$

and

$$\hat{i}_Z|I_i, \sigma_i\rangle = \sigma_i\hbar|I_i, \sigma_i\rangle \tag{9.5}$$

where, for a single nucleus, $\hat{i}^2$ is the operator for the square of the nuclear spin angular momentum and $\hat{i}_Z$ that for its space-fixed Z component. As explained in

---

[1] The R_i are the spatial coordinates and the σ_i are the spin labels; see remark after equation (7.2).

section 3.2, the quantum number I_i is a constant for a given nucleus and the spin projection σ_i has one of the $2I_i + 1$ values $-I_i, -I_i + 1, -I_i + 2, \ldots, I_i$.

In the H_2O molecule, $I_1 = I_2 = 1/2$, and (for ^{16}O) $I_3 = 0$. A proton has two possible spin functions $|1/2, \sigma_i\rangle$ with $\sigma_i = -1/2$ or $+1/2$; we write these two functions as

$$\alpha = |1/2, 1/2\rangle \quad \text{and} \quad \beta = |1/2, -1/2\rangle. \tag{9.6}$$

A ^{16}O nucleus has one possible spin function with $I_3 = \sigma_3 = 0$; we write this function

$$\delta = |0, 0\rangle. \tag{9.7}$$

Four different products of the one-particle spin functions exist: $\alpha\alpha\delta$, $\beta\beta\delta$, $\alpha\beta\delta$ and $\beta\alpha\delta$, where the first of the three factors is the spin function of proton 1, the second factor is the spin function of proton 2 and the third factor is the spin function of the oxygen nucleus. If we form the following linear combinations of these products

$$\Phi_{ns,1} = \alpha\alpha\delta \tag{9.8}$$

$$\Phi_{ns,2} = \beta\beta\delta \tag{9.9}$$

$$\Phi_{ns,3} = \frac{1}{\sqrt{2}}[\alpha\beta + \beta\alpha]\delta \tag{9.10}$$

and

$$\Phi_{ns,4} = \frac{1}{\sqrt{2}}[\alpha\beta - \beta\alpha]\delta \tag{9.11}$$

then the functions $\Phi_{ns,1}$, $\Phi_{ns,2}$ and $\Phi_{ns,3}$ each belong to the symmetry species A_1 of $C_{2v}(M)$ (table 7.2), whereas $\Phi_{ns,4}$ belongs to the symmetry species B_2. Taken together, the four nuclear spin functions $\Phi_{ns,1}$, $\Phi_{ns,2}$, $\Phi_{ns,3}$ and $\Phi_{ns,4}$ generate the representation

$$\Gamma_{ns} = 3A_1 \oplus B_2. \tag{9.12}$$

The three functions $\Phi_{ns,1}$, $\Phi_{ns,2}$ and $\Phi_{ns,3}$ of A_1 symmetry are components of an $I = 1$ total nuclear spin state, with $I_Z = +1, -1$ and 0, respectively; the B_2 function is an $I = 0$ total nuclear spin state.

The product functions

$$\Phi_{H_2O}(\boldsymbol{R}_1, \sigma_1, \boldsymbol{R}_2, \sigma_2, \boldsymbol{R}_3, \sigma_3, \boldsymbol{R}_4, \sigma_4, \boldsymbol{R}_5, \sigma_5, \ldots \boldsymbol{R}_{13}, \sigma_{13})$$
$$= \Phi_{rve,nj}(\boldsymbol{R}_1, \boldsymbol{R}_2, \boldsymbol{R}_3, \boldsymbol{R}_4, \sigma_4, \boldsymbol{R}_5, \sigma_5, \ldots, \boldsymbol{R}_{13}, \sigma_{13})\Phi_{ns,t} \tag{9.13}$$

are complete internal wavefunctions for H_2O; they include nuclear and electron spin. These functions automatically change sign when we interchange any two electrons in the molecule because $\Phi_{rve,nj}$ changes sign and $\Phi_{ns,t}$ remains unchanged (it does not depend on the electronic coordinates). The Fermi–Dirac statistical formulas, however, further require that Φ_{H_2O} change sign when we

Table 9.1. Spin statistical weights for H_2O and D_2O.

$H_2\ ^{16}O$				$D_2\ ^{16}O$			
Γ_{rve}	$\Gamma_{ns,t}$	Γ_{tot}	g_{ns}	Γ_{rve}	$\Gamma_{ns,t}$	Γ_{tot}	g_{ns}
A_1	B_2	B_2	1	A_1	$6A_1$	A_1	6
A_2	B_2	B_1	1	A_2	$6A_1$	A_2	6
B_1	$3A_1$	B_1	3	B_1	$3B_2$	A_2	3
B_2	$3A_1$	B_2	3	B_2	$3B_2$	A_1	3

interchange the two protons in H_2O [equation (9.1)] and this puts restrictions on the combinations of $\Phi_{rve,nj}$ and $\Phi_{ns,t}$ that we can use in equation (9.13).

If $\Phi_{rve,nj}$ has the symmetry Γ_{rve} and $\Phi_{ns,t}$ has the symmetry $\Gamma_{ns,t}$, then the function Φ_{H_2O} in equation (9.13) has the symmetry

$$\Gamma_{tot} = \Gamma_{rve} \otimes \Gamma_{ns,t} \qquad (9.14)$$

and, for a given Γ_{rve}, we must choose $\Gamma_{ns,t}$ such that $\Gamma_{tot} = B_1$ or B_2 for H_2O, in order to satisfy equation (9.1). On the left-hand side of table 9.1, we give the possible combinations of Γ_{rve}, $\Gamma_{ns,t}$ and Γ_{tot}. The table also introduces the *spin statistical weight* g_{ns}, which is simply the number of complete internal wavefunctions of the allowed Γ_{tot} symmetry that we can construct for a given Γ_{rve}. For $\Gamma_{ns,t} = A_1$, $g_{ns} = 3$ and for $\Gamma_{ns,t} = B_2$, $g_{ns} = 1$ from equation (9.12). For a molecule such as H_2O with two *spin modifications* (i.e. two different values of g_{ns}), the states having the higher g_{ns} are called *ortho states* and the states having the lower g_{ns} are called *para states*. For H_2O, rovibronic states of symmetry B_1 or B_2 have $g_{ns} = 3$ and are *ortho* states; states of symmetry A_1 or A_2 have $g_{ns} = 1$ and are *para* states (see figure 7.3 on page 136).

For a deuteron D, there are three possible spin functions $|1, \sigma\rangle$ with $\sigma = -1, 0$ and $+1$, respectively:

$$\lambda = |1, -1\rangle, \qquad \mu = |1, 0\rangle \qquad \text{and} \qquad \nu = |1, 1\rangle. \qquad (9.15)$$

By analogy with equations (9.8)–(9.11), we form nine symmetrized nuclear spin functions for D_2O:

$$\Phi_{ns,1} = \lambda\lambda\delta \qquad \Phi_{ns,2} = \mu\mu\delta \qquad \Phi_{ns,3} = \nu\nu\delta \qquad (9.16)$$

$$\Phi_{ns,4} = \frac{1}{\sqrt{2}}[\lambda\mu + \mu\lambda]\delta \qquad \Phi_{ns,5} = \frac{1}{\sqrt{2}}[\lambda\nu + \nu\lambda]\delta \qquad (9.17)$$

$$\Phi_{ns,6} = \frac{1}{\sqrt{2}}[\mu\nu + \nu\mu]\delta \qquad (9.18)$$

$$\Phi_{ns,7} = \frac{1}{\sqrt{2}}[\lambda\mu - \mu\lambda]\delta \qquad \Phi_{ns,8} = \frac{1}{\sqrt{2}}[\lambda\nu - \nu\lambda]\delta \qquad (9.19)$$

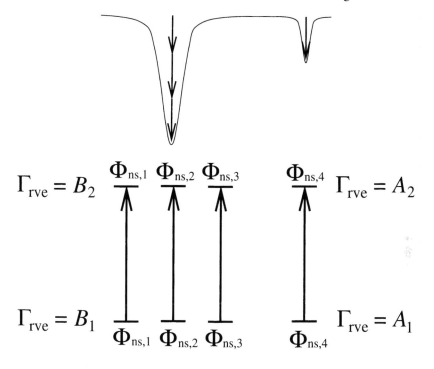

Figure 9.1. The effect of spin statistical weights on the intensities of H_2O absorption transitions.

and

$$\Phi_{ns,9} = \frac{1}{\sqrt{2}}[\mu\nu - \nu\mu]\delta. \tag{9.20}$$

The functions $\Phi_{ns,1}, \Phi_{ns,2}, \ldots, \Phi_{ns,6}$ each have A_1 symmetry in $C_{2v}(M)$, whereas $\Phi_{ns,7}, \Phi_{ns,8}$, and $\Phi_{ns,9}$ each have B_2 symmetry; the nine spin functions for D_2O generate the representation

$$\Gamma_{ns} = 6A_1 \oplus 3B_2 \tag{9.21}$$

of $C_{2v}(M)$. The allowed complete internal wavefunctions for D_2O can only have $\Gamma_{tot} = A_1$ or A_2 [from equation (9.2)], so that the spin statistical weights for D_2O are as given on the right-hand side of table 9.1. For D_2O, rovibronic states of symmetry B_1 or B_2 are *para* states and states of symmetry A_1 or A_2 are *ortho* states. This is the opposite of the situation for H_2O.

Spin statistical weights manifest themselves in a spectrum by giving rise to *intensity alternations* and we now explain what this means using the H_2O molecule as an example. Consider a transition from a lower rovibronic state $\Phi_{rve,nj''}$ with $\Gamma_{rve} = B_1$ to an upper rovibronic state $\Phi_{rve,nj'}$ with $\Gamma_{rve} = B_2$. Each rovibronic state in this transition can be combined with the three spin functions

$\Phi_{ns,1}$, $\Phi_{ns,2}$ and $\Phi_{ns,3}$ in equations (9.8)–(9.10) to produce three different states allowed by Fermi–Dirac statistics. An absorption transition induced by the electric dipole operator (which does not involve nuclear spin) will conserve the nuclear spin state. Thus, using simple rovibronic–nuclear spin product basis functions, as indicated in figure 9.1, with the neglect of the hyperfine Hamiltonian $\hat{H}_{hfs}$ [see equation (2.77)], the transition from $\Phi_{rve,nj''}$ to $\Phi_{rve,nj'}$ will consist of three overlapping transitions and the intensities will add up as indicated in the figure. For a transition from a lower rovibronic state $\Phi_{rve,nj''}$ with $\Gamma_{rve} = A_1$ to an upper rovibronic state $\Phi_{rve,nj'}$ with $\Gamma_{rve} = A_2$, only one nuclear spin function is available for each state. If the two transitions in figure 9.1 have the same absolute value of the transition moment integral $\int \Phi_{rve}'^{*} \mu_A \Phi_{rve}'' \, d\tau$ entering into equation (2.87) for the line strength, then [neglecting the Boltzmann and frequency factors in equation (1.9)] the transition involving the B_1 and B_2 rovibronic states will have an intensity three times larger than the transition involving the A_1 and A_2 rovibronic states because of the spin statistical weights. Such intensity alternations are often clearly visible in experimental spectra and examples are given in figures 13.4 and 13.10. The hyperfine Hamiltonian will cause a $B_2 - B_1$ spectral line for the water molecule to split into hyperfine components, but to see them requires very high spectral resolution. The effect of the hyperfine Hamiltonian is discussed in section 9.5.

9.3 Missing levels

9.3.1 CO_2

For $^{12}C^{16}O_2$, we label the two oxygen nuclei 1 and 2 and the carbon nucleus 3; the MS group is the $D_{\infty h}(M)$ group given in table B.15[2] in appendix B. ^{16}O and ^{12}C nuclei have $I_i = \sigma_i = 0$. We denote the one possible spin function for ^{16}O as $\delta_{(O)} = |0, 0\rangle_{(O)}$ and the one possible spin function for ^{12}C as $\delta_{(C)} = |0, 0\rangle_{(C)}$. The $^{12}C^{16}O_2$ molecule has the one possible spin function

$$\Phi_{ns,1} = \delta_{(O)}\delta_{(C)}\delta_{(O)} \tag{9.22}$$

which has $+s$ (Σ_g^+) symmetry. The two ^{16}O nuclei are bosons and the complete internal wavefunction of $^{12}C^{16}O_2$ must have $\Gamma_{tot} = \pm s$ (Σ_g^+ or Σ_u^-). Rovibronic states $\Phi_{rve,nj}$ with $\Gamma_{rve} = \pm s$ (Σ_g^+ or Σ_u^-) can be combined with the one allowed spin function $\Phi_{ns,1}$ in equation (9.22) to produce allowed complete internal wavefunctions with $\Gamma_{tot} = \pm s$ (Σ_g^+ or Σ_u^-), respectively, with spin statistical weight $g_{ns} = 1$ (see the left-hand side of table 9.2). For rovibronic states with $\Gamma_{rve} = \pm a$ (Σ_g^- or Σ_u^+), we cannot satisfy the requirements imposed by Bose–Einstein statistics. This is because in order to generate allowed complete internal wavefunctions for such states, we would have to combine them with a

---

[2] There are two possible notations for the irreducible representations: Either $+s$, $-s$, $+a$ and $-a$; or Σ_g^+, Σ_u^-, Σ_u^+ and Σ_g^-, respectively.

Table 9.2. Spin statistical weights for $^{12}C\,^{16}O_2$ and $^{12}C\,^{17}O_2$.

$^{12}C\,^{16}O_2$				$^{12}C\,^{17}O_2$			
Γ_{rve}	$\Gamma_{ns,t}$	Γ_{tot}	g_{ns}	Γ_{rve}	$\Gamma_{ns,t}$	Γ_{tot}	g_{ns}
Σ_g^+	Σ_g^+	Σ_g^+	1	Σ_g^+	$15\Sigma_u^+$	Σ_u^+	15
Σ_u^-	Σ_g^+	Σ_u^-	1	Σ_u^-	$15\Sigma_u^+$	Σ_g^-	15
Σ_g^-	—	—	0	Σ_g^-	$21\Sigma_g^+$	Σ_g^-	21
Σ_u^+	—	—	0	Σ_u^+	$21\Sigma_g^+$	Σ_u^+	21

spin function of $\pm a$ (Σ_g^- or Σ_u^+) symmetry and no such spin functions are available. Consequently, for levels with $\Gamma_{rve} = \pm a$ (Σ_g^- or Σ_u^+), $g_{ns} = 0$. These levels do not occur in the molecule and are said to be *missing*. If we solve the rovibronic Schrödinger equation for $^{12}C^{16}O_2$ (neglecting nuclear spin), we obtain solutions with $\pm s$ and $\pm a(\Sigma_g^+, \Sigma_u^-, \Sigma_g^-$ and Σ_u^+) symmetry but only the levels with $\Gamma_{rve} = \pm s$ (Σ_g^+ or Σ_u^-) exist.

Isotopomers of CO_2 involving oxygen isotopes with non-zero spin have no missing levels. For example, the isotope ^{17}O has $I = 5/2$ and we give the spin statistical weights of $^{12}C^{17}O_2$ on the right-hand side of table 9.2.

9.3.2 H_3^+

The H_3^+ molecule also has missing levels but they are imposed by Fermi–Dirac statistics rather than by Bose–Einstein statistics. The MS group of H_3^+ is $D_{3h}(M)$ whose character table is given in table 7.3 on page 147. The protons in H_3^+ are fermions with $I = 1/2$ so the complete wavefunction must change sign under the odd permutations (12), (23) and (31) and it must be invariant to the even permutations (123) and (132). Thus, the complete wavefunction of H_3^+ can only have the symmetries A_2' and A_2''.

The nuclear spin functions of H_3^+ involve the proton spin functions α and β from equation (9.6) and we can organize them as follows

$$
\begin{aligned}
(m_I = 3/2): &\quad \alpha\alpha\alpha = \Phi_{ns}^{(1)} \\
(m_I = 1/2): &\quad \alpha\alpha\beta = \Phi_{ns}^{(2)} \quad \alpha\beta\alpha = \Phi_{ns}^{(3)} \quad \beta\alpha\alpha = \Phi_{ns}^{(4)} \\
(m_I = -1/2): &\quad \alpha\beta\beta = \Phi_{ns}^{(5)} \quad \beta\alpha\beta = \Phi_{ns}^{(6)} \quad \beta\beta\alpha = \Phi_{ns}^{(7)} \\
(m_I = -3/2): &\quad \beta\beta\beta = \Phi_{ns}^{(8)}
\end{aligned}
\tag{9.23}
$$

where the nuclei are in the order 1, 2 and 3 in these functions, and $m_I = \sigma_1 + \sigma_2 + \sigma_3$ is the total projection quantum number for the proton spins. $\Phi_{ns}^{(1)}$ and $\Phi_{ns}^{(8)}$ each have symmetry A_1'. The three $m_I = \frac{1}{2}$ spin functions are transformed among themselves by the elements of the group and they generate the reducible

Table 9.3. Spin statistical weights for H_3^+.

Γ_{rve}	Γ_{ns}	Γ_{tot}	g_{ns}
A_1'	—	—	0
A_1''	—	—	0
A_2'	$4A_1'$	A_2'	4
A_2''	$4A_1'$	A_2''	4
E'	$2E'$	A_2'	2
E''	$2E'$	A_2''	2

representation

$$A_1' \oplus E'. \tag{9.24}$$

The three $m_I = -\frac{1}{2}$ spin functions generate the same representation of the MS group. Thus, the representation generated by the eight nuclear spin functions in equation (9.23) is

$$\Gamma_{ns} = 4A_1' \oplus 2E'. \tag{9.25}$$

The spin statistical weights for H_3^+ are given in table 9.3. Levels with $\Gamma_{rve} = A_1'$ or A_1'' are missing as there are no spin functions available to satisfy the requirements of Fermi–Dirac statistics. For levels with $\Gamma_{rve} = A_2'$ or A_2'', we can combine $\Phi_{rve,nj}$ with one of the four nuclear spin functions of A_1' symmetry and, for levels with $\Gamma_{rve} = E'$ or E'', we can use the two nuclear spin function pairs of E' symmetry since

$$E' \otimes E' = A_1' \oplus A_2' \oplus E' \quad \text{and} \quad E'' \otimes E' = A_1'' \oplus A_2'' \oplus E''. \tag{9.26}$$

A level with $\Gamma_{rve} = E'$ has associated with it *two* wavefunctions $\Phi_{rve,nj,a}$ and $\Phi_{rve,nj,b}$, say, that transform according to the E' irreducible representation. Equation (9.26) shows, however, that when we combine these two wavefunctions with two nuclear spin functions transforming according to E', then we obtain only *one* allowed complete wavefunction of A_2' symmetry. According to equation (9.25), we have two E' nuclear spin function pairs available and the spin degeneracy (i.e. the value of g_{ns}) for a level with $\Gamma_{rve} = E'$ is, therefore, 2. Similar arguments apply to levels with $\Gamma_{rve} = E''$.

9.4 Statistical weights for CH₃F

Methyl fluoride $^{12}CH_3{}^{19}F$ is a molecule for which the MS group is a subgroup of the CNPI group (see section 8.2) and it provides an example of such a case for determining nuclear spin statistical weights. In CH₃F, we label the three protons

Table 9.4. Spin statistical weights for CH_3F.

Γ_{rve}	Γ_{ns}	Γ_{tot}	g_{ns}
A_1	$8A_1$	A_1	8
A_2	$8A_1$	A_2	8
E	$4E$	A_1/A_2	8

1, 2 and 3, the ^{12}C nucleus 4 and the ^{19}F nucleus 5. The character table of $C_{3v}(M)$ is given in table 7.5 on page 151. Having three protons, it has the same eight proton spin states as H_3^+ given in equation (9.23). These functions transform as

$$4A_1 \oplus 2E. \tag{9.27}$$

Multiplying by the one ^{12}C nuclear spin state $\delta_{(C)}$ of A_1 symmetry and the two possible ^{19}F nuclear spin states with $\sigma_5 = -1/2$ or $+1/2$ that each have A_1 symmetry, the 16 possible nuclear spin functions transform according to

$$\Gamma_{ns} = 8A_1 \oplus 4E. \tag{9.28}$$

The complete internal wavefunction is invariant to the even permutation (123) of the protons and so it can have symmetry A_1 and A_2.

We can combine a rovibronic wavefunction of symmetry A_1 or A_2 with each of the eight A_1 spin functions to get eight allowed complete internal wavefunctions of symmetry A_1 or A_2, respectively. A level with $\Gamma_{rve} = E$ has associated with it two wavefunctions $\Phi_{rve,nj,a}$ and $\Phi_{rve,nj,b}$ that transform according to the E irreducible representation. In $C_{3v}(M)$,

$$E \otimes E = A_1 \oplus A_2 \oplus E \tag{9.29}$$

and so, when we combine the two wavefunctions $\Phi_{rve,nj,a}$ and $\Phi_{rve,nj,b}$ with the two nuclear spin functions transforming according to E, we obtain *two* allowed complete internal wavefunctions, one of A_1 symmetry and one of A_2 symmetry (and two, of E symmetry, that are *not* allowed). There are four E-pairs of spin functions available and we can use these spin functions to form a total of eight allowed complete internal wavefunctions. Therefore, each rovibronic level with $\Gamma_{rve} = E$ has a spin statistical weight of eight as given in table 9.4.

9.5   Nuclear spin hyperfine structure

The effects of nuclear spin are often visible in a spectrum as intensity alternations (or missing levels) due to the requirements imposed by Fermi–Dirac and Bose–Einstein statistics. Here we briefly discuss how nuclear spin manifests itself through the effect of the nuclear hyperfine Hamiltonian $\hat{H}_{hfs}$ [see equations (2.75)–(2.77)].

$\hat{H}_{hfs}$ causes the molecular rovibronic energy levels that have a nuclear spin statistical weight greater than unity to split into closely spaced sub-levels called *hyperfine structure* and, as a result, transitions between such rovibronic levels are split into *hyperfine components*. In figure 9.1, the rovibronic $B_2 \leftarrow B_1$ transition can split into hyperfine components but the $A_2 \leftarrow A_1$ transition cannot. In a singlet electronic state, the hyperfine structure on a particular rovibronic state can be calculated by diagonalizing the Hamiltonian $\hat{H}_{rve} + \hat{H}_{hfs}$ (see section 2.3). In setting up the Hamiltonian matrix, we do not use the 'uncoupled' basis set functions $\Phi_{rve,nj}\Phi_{ns,t}$ indicated above, where the $\Phi_{rve,nj}$ are eigenfunctions of $\hat{H}_{rve}$, and the $\Phi_{ns,t}$ are simple nuclear spin functions. Instead we use special 'coupled' linear combinations of such functions that are chosen to be eigenfunctions of the total angular momentum operator $\hat{F}^2$ and its space-fixed Z component $\hat{F}_Z$ (see sections 2.7 and 14.5).

Concentrating on the angular momentum properties of the basis functions, we can write an uncoupled basis function product as $|J, m_J\rangle|I, m_I\rangle$, where (J, m_J) and (I, m_I) are the rovibronic and spin angular momentum quantum numbers, respectively. The linear combinations of such functions that are eigenfunctions of $\hat{F}^2$ and $\hat{F}_Z$ are given by

$$|F, m_F\rangle = \sum_{m_J, m_I} C(JIF; m_J m_I m_F)|J, m_J\rangle|I, m_I\rangle$$

$$= \sum_{m_J, m_I} (-1)^{J-I+m_F} \sqrt{2F+1}$$

$$\times \begin{pmatrix} J & I & F \\ m_J & m_I & -m_F \end{pmatrix} |J, m_J\rangle|I, m_I\rangle \qquad (9.30)$$

where $C(JIF; m_J m_I m_F)$ and

$$\begin{pmatrix} J & I & F \\ m_J & m_I & -m_F \end{pmatrix}$$

are each an analytic function of the six quantum numbers occurring in them; they are called a Clebsch–Gordan coefficient and a $3j$-symbol, respectively. As stated in connection with equation (14.38), the possible values of F are $J + I, J + I - 1, \ldots, |J - I|$. The matrix of the Hamiltonian $\hat{H}_{rve} + \hat{H}_{hfs}$ in such a basis will diagonalize into blocks for each F value (and MS group symmetry species) and electric dipole transitions between the eigenfunctions will satisfy the selection rule $\Delta F = 0$ or ± 1 (but $F = 0 \leftrightarrow 0$ is forbidden); see section 14.5. Hyperfine structure components with $\Delta F = \Delta J$ are the most intense but transitions with $\Delta F \neq \Delta J$ are allowed. As an example of a $B_2 - B_1$ transition in the water molecule, the hyperfine structure on the $J_{K_a K_c} = 6_{16} \leftarrow 5_{23}$ transition has been measured[3] and it is shown in figure 9.2.

---

[3] Kukolich S G 1968 *J. Chem. Phys.* **50** 3751.

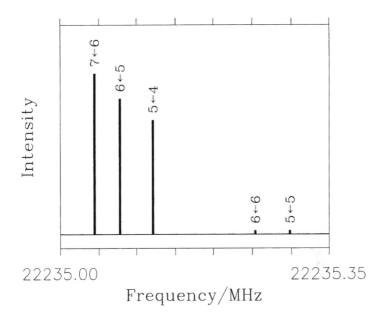

Figure 9.2. The hyperfine structure of the $J_{K_a K_c} = 6_{16} \leftarrow 5_{23}$ transition for the water molecule, with each transition labelled by the $F' \leftarrow F''$ assignment. The line positions are from Kukolich S G [1968 *J. Chem. Phys.* **50** 3751] and the relative intensities have been calculated for us by Kukolich from his data. The only other allowed hyperfine component ($F = 5 \leftarrow 6$) is too weak to show.

$\hat{H}_{\text{hfs}}$ can cause an interaction between an *ortho* state and a *para* state. Consider as an example a *para* state of H_2O with a rovibronic wavefunction $\Phi_{\text{rve,p}}$ of symmetry A_1 in $C_{2v}(M)$ and an *ortho* state with rovibronic wavefunction $\Phi_{\text{rve,o}}$ of symmetry B_2. We assume that these two states are accidentally close in energy, $E_{\text{rve,p}} \approx E_{\text{rve,o}}$, and that their J values do not differ by more than 1 so that rovibronic-spin states having the same value of F occur. We combine $\Phi_{\text{rve,p}}$, having angular momentum quantum number J_p say, with an $I = 0$ nuclear spin function $\Phi_{\text{ns,p}}$ of B_2 symmetry [i.e. $\Phi_{\text{ns,p}} = \Phi_{\text{ns,4}}$ given in equation (9.11)] so that the product $\Phi_p = \Phi_{\text{rve,p}}\Phi_{\text{ns,p}}$ has the total symmetry $A_1 \otimes B_2 = B_2$ and $F = J_p$. Similarly, we combine $\Phi_{\text{rve,o}}$, having $J = J_o$, with each of the three $I = 1$ nuclear spin functions $\Phi_{\text{ns,o}}$ of A_1 symmetry [i.e. the three functions $\Phi_{\text{ns,1}}, \Phi_{\text{ns,2}}$ and $\Phi_{\text{ns,3}}$ in equations (9.8)–(9.10)] so that the products each have total symmetry $B_2 \otimes A_1 = B_2$. Forming three coupled functions having $F = J + 1, J$ and $J - 1$, one, Φ_o say, has an F value equal to J_p. Since the two functions Φ_p and Φ_o have the same total symmetry B_2 in $C_{2v}(M)$ and the same F value, then $\hat{H}_{\text{hfs}}$ can have a non-vanishing matrix element between them. We write the 2×2 matrix of the Hamiltonian $\hat{H}_{\text{rve}} + \hat{H}_{\text{hfs}}$ as

$$\begin{bmatrix} H_{\text{oo}} & H_{\text{op}} \\ H_{\text{po}} & H_{\text{pp}} \end{bmatrix} \tag{9.31}$$

with

$$H_{qq} = E_{\text{rve},q} + \langle \Phi_q | \hat{H}_{\text{hfs}} | \Phi_q \rangle \tag{9.32}$$

with $q = \text{o}$ or p and (assuming, for simplicity, that the matrix elements are real)

$$H_{\text{po}} = H_{\text{op}} = \langle \Phi_p | \hat{H}_{\text{hfs}} | \Phi_o \rangle. \tag{9.33}$$

The matrix in equation (9.31) describes *ortho–para interaction* since the off-diagonal matrix elements connect *ortho* and *para* states. From equations (2.35)–(2.40), the eigenfunctions are

$$\Phi_{\text{o-p}}^{(1)} = c^+ \Phi_{\text{o}} - c^- \Phi_p \tag{9.34}$$

and

$$\Phi_{\text{o-p}}^{(2)} = c^+ \Phi_p + c^- \Phi_{\text{o}}. \tag{9.35}$$

We expect $|H_{\text{pp}} - H_{\text{oo}}| \gg |H_{\text{po}}|$ so that $|c^+|^2 \approx 1$ and $|c^-|^2 \ll 1$ [from equations (2.39) and (2.40)]. Thus, $\Phi_{\text{o-p}}^{(1)}$ is essentially an *ortho* state with a small admixture of *para* character, and $\Phi_{\text{o-p}}^{(2)}$ is essentially a *para* state with a small admixture of *ortho* character.

The creation of the two *ortho–para*-mixed states $\Phi_{\text{o-p}}^{(1)}$ and $\Phi_{\text{o-p}}^{(2)}$ by interaction between Φ_p and Φ_{o} is shown in figure 9.3. Suppose $\Phi_{\text{rve,o}}''$ is a pure *ortho* state of symmetry B_1 and that the transition from $\Phi_{\text{rve,o}}''$ to $\Phi_{\text{rve,o}}$ is allowed. Thus, the absolute value of the integral

$$\left| \int \Phi_{\text{rve,o}}''^* \mu_A \Phi_{\text{rve,o}} \, d\tau \right| \neq 0; \tag{9.36}$$

this integral enters into equation (2.87) and determines the intensity of the transition between $\Phi_{\text{rve,o}}''$ and $\Phi_{\text{rve,o}}$. However, the transition from $\Phi_{\text{rve,o}}''$ to $\Phi_{\text{rve,p}}$ is forbidden and

$$\left| \int \Phi_{\text{rve,o}}''^* \mu_A \Phi_{\text{rve,p}} \, d\tau \right| = 0 \tag{9.37}$$

because the transition from $\Phi_{\text{rve,o}}''$ to $\Phi_{\text{rve,p}}$ satisfies equation (7.59). However, we see that the two transitions from $\Phi_{\text{rve,o}}''$ to $\Phi_{\text{o-p}}^{(1)}$ and $\Phi_{\text{o-p}}^{(2)}$, respectively, both have non-vanishing line strengths since

$$I_{\text{TM}}^{(a)} = \left| \int \Phi_{\text{rve,o}}''^* \mu_A \Phi_{\text{o-p}}^{(1)} \, d\tau \right|^2 = |c^+|^2 \left| \int \Phi_{\text{rve,o}}''^* \mu_A \Phi_{\text{rve,o}} \, d\tau \right|^2 \neq 0 \tag{9.38}$$

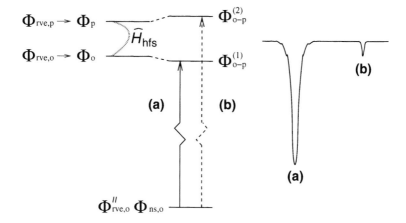

Figure 9.3. The principle of *ortho–para* interaction and intensity stealing. On the left-hand side, we show how the basis functions Φ_p and Φ_o mix to form the two *ortho–para* mixed states $\Phi_{o-p}^{(1)}$ and $\Phi_{o-p}^{(2)}$. The transitions to these states from a pure *ortho* level $\Phi_{rve,o}''$ are indicated, and the right-hand side is a schematic representation of the resulting transitions in the absorption spectrum.

and

$$I_{TM}^{(b)} = \left| \int \Phi_{rve,o}''^{*} \mu_A \Phi_{o-p}^{(2)} \, d\tau \right|^2 = |c^-|^2 \left| \int \Phi_{rve,o}''^{*} \mu_A \Phi_{rve,o} \, d\tau \right|^2 \neq 0. \quad (9.39)$$

The quantities $I_{TM}^{(a)}$ and $I_{TM}^{(b)}$ determine the intensities of the transitions marked (a) and (b), respectively, in figure 9.3. As indicated in the figure, we have $I_{TM}^{(a)} \gg I_{TM}^{(b)}$ because $|c^+|^2 \gg |c^-|^2$. We see that, for both transitions (a) and (b), the intensity originates in the integral given in equation (9.36). If there were no mixing of the two states $\Phi_{rve,p}\Phi_{ns,p}$ and $\Phi_{rve,o}\Phi_{ns,o}$, then transition (b) would have zero intensity. However, because of the *ortho–para* interaction, the state $\Phi_{o-p}^{(2)}$ contains a small amount of the state $\Phi_{rve,o}$ according to equation (9.35) and transition (b) has some intensity. We say that transition (b) *steals* intensity from transition (a) between $\Phi_{rve,o}''$ and $\Phi_{rve,o}$. Transition (b) takes place between the pure *ortho* state $\Phi_{rve,o}''$ and the predominantly *para* state $\Phi_{o-p}^{(2)}$; it is an *ortho–para* transition.

9.6 Problems

9.1   Confirm that the nuclear spin statistical weights of $^{12}C^{17}O_2$ are as given in table 9.2.

9.2   Determine the nuclear spin statistical weights of $^{12}C_2H_4$. The character table for the appropriate MS group, $D_{2h}(M)$, is given in table 8.1.

9.3   Determine the nuclear spin statistical weights of all three possible conformers of the ethylene isotopomer $^{12}C_2H_2D_2$.

9.4   For the eight H_3^+ proton spin functions in equation (9.23), determine the transformation matrices $D[R]$ for all operations R in $D_{3h}(M)$. Confirm that the characters satisfy equation (9.25).

9.5   Pretend that the ozone molecule has a D_{3h} equilibrium structure like H_3^+. In this circumstance, what would be the nuclear spin statistical weights of the rotational levels for the isotopomers $^{16}O_3$, $^{16}O_2{}^{18}O$, $^{17}O_2{}^{16}O$ and $^{16}O^{17}O^{18}O^2$.

Chapter 10

The symmetry of electronic wavefunctions

Molecular orbitals play a central role in the understanding of electronic structure and, in the present chapter, we discuss their symmetries, and the symmetries of the Slater determinants and electronic state wavefunctions that are constructed from them. We make use of the theory and equations developed in section 3.3.

As shown in section 8.3, for each of the (rigid) molecules discussed in this chapter, the molecular point group and the MS group are isomorphic and the operations of the point group have the same effect on the electronic coordinates as the corresponding MS group operations. Thus, either group could be used for classifying the electronic states and the same results would be obtained. However, it is simpler to use the point group since the orientation of the molecule-fixed axes is not affected by point group operations.

10.1   The water molecule

The water molecule is used here as a simple example having only non-degenerate symmetry species to show how the operations in the molecular point group affect the atomic orbitals and, through equation (3.37), the molecular orbitals. The molecular symmetry group for the water molecule is the group $C_{2v}(M)$ and the molecular point group is the group C_{2v}; see section 8.3.1. The character table for C_{2v}, and the correspondence between the operations in $C_{2v}(M)$ and C_{2v}, are shown in table 10.1.

Molecular orbitals for H_2O can be formed from the atomic orbitals 1s(O), 2s(O), $2p_x$(O), $2p_y$(O) and $2p_z$(O) centred on the oxygen nucleus, together with the atomic orbitals 1s(H$_1$) and 1s(H$_2$) centred on the protons. From figure 10.1, we see that

$$C_{2x}2p_y(O) = -2p_y(O). \tag{10.1}$$

Also the C_{2x} operation exchanges the atomic orbitals 1s(H$_1$) and 1s(H$_2$) centred on the protons. The following combinations of atomic orbitals transform

Table 10.1. The character table for $C_{2v}(M)$ and C_{2v}. The operations R_{MS} constitute the molecular symmetry group $C_{2v}(M)$ of the water molecule and the operations R_{PG} constitute the corresponding molecular point group C_{2v}. The xyz axes are defined in figure 10.1.

R_{MS}:	E	(12)	E^*	$(12)^*$
R_{PG}:	E	C_{2x}	σ_{xz}	σ_{xy}
A_1:	1	1	1	1
A_2:	1	1	−1	−1
B_1:	1	−1	−1	1
B_2:	1	−1	1	−1

irreducibly:

$$A_1 : \quad 1s(O), 2s(O), 2p_x(O), \frac{1}{\sqrt{2}}[1s(H_2) + 1s(H_1)]$$

$$B_1 : \quad 2p_y(O)$$

$$B_2 : \quad 2p_z(O), \frac{1}{\sqrt{2}}[1s(H_2) - 1s(H_1)]. \tag{10.2}$$

These symmetry adapted combinations of atomic orbitals are called *symmetry orbitals* (SOs).

In a Hartree–Fock calculation, SOs are used as basis functions ϕ_μ in setting up the trial MO functions ψ_i in equation (3.37) and these MOs have the symmetry of the SOs from which they are constructed. Orbitals of A_1 symmetry are labelled $(1a_1), (2a_1), (3a_1), \dots$, in order of ascending energy and, similarly, for other symmetries. The lowest lying trial MO is the 1s orbital on the oxygen nucleus and so $(1a_1) = 1s(O)$. The two SOs of B_2 symmetry form a bonding and an antibonding pair of MOs given by

$$(1b_2) = \frac{1}{\sqrt{2}} \left(2p_z(O) + \frac{1}{\sqrt{2}}[1s(H_2) - 1s(H_1)] \right) \tag{10.3}$$

and

$$(1b_2^*) = \frac{1}{\sqrt{2}} \left(2p_z(O) - \frac{1}{\sqrt{2}}[1s(H_2) - 1s(H_1)] \right). \tag{10.4}$$

The 2s(O) and $2p_x(O)$ orbitals each similarly form bonding and anti-bonding MOs with the $[1s(H_2) + 1s(H_1)]$ SO; the bonding MOs are given by

$$(2a_1) = \frac{1}{\sqrt{2}} \left(2s(O) + \frac{1}{\sqrt{2}}[1s(H_2) + 1s(H_1)] \right) \tag{10.5}$$

and

$$(3a_1) = \frac{1}{\sqrt{2}} \left(2p_x(O) + \frac{1}{\sqrt{2}}[1s(H_2) + 1s(H_1)] \right). \tag{10.6}$$

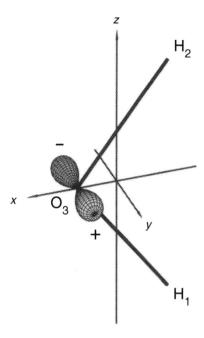

Figure 10.1. A water molecule in the equilibrium configuration of the ground electronic state with a $2p_y$ atomic orbital centred at the oxygen nucleus.

Finally, the $2p_y(O)$ orbital is an MO of symmetry B_1:

$$(1b_1) = 2p_y(O). \tag{10.7}$$

The MO $(1b_1)$ is centred entirely on the oxygen atom so it is neither bonding nor antibonding; it is a *non-bonding* orbital. There are ten electrons in a water molecule and so, filling the bonding and non-bonding MOs in energy order, we obtain the electronic configuration of the ground $\tilde{X}$ state as

$$(1a_1)^2(2a_1)^2(1b_2)^2(3a_1)^2(1b_1)^2 \tag{10.8}$$

where the orbitals are written in order of increasing energy.

The symmetry of the MO product

$$\phi_{1a_1}(r_1)\phi_{1a_1}(r_2)\phi_{2a_1}(r_3)\phi_{2a_1}(r_4)\phi_{1b_2}(r_5)$$
$$\times \phi_{1b_2}(r_6)\phi_{3a_1}(r_7)\phi_{3a_1}(r_8)\phi_{1b_1}(r_9)\phi_{1b_1}(r_{10}) \tag{10.9}$$

[where r_i denotes the coordinates of electron i] is given by

$$(A_1)^2 \otimes (A_1)^2 \otimes (B_2)^2 \otimes (A_1)^2 \otimes (B_1)^2 = A_1. \tag{10.10}$$

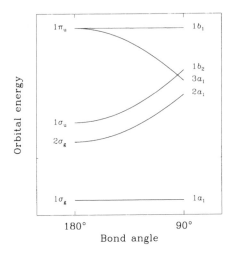

Figure 10.2. A Walsh diagram for an XH_2 molecule. It shows qualitatively how the orbital energies change as the molecule is bent.

The symmetry of a product of molecular orbitals is built up from the MO symmetries and the occupation numbers (0, 1 or 2). The electronic configuration for the first excited electronic state of a molecule is obtained by promoting an electron from the highest occupied molecular orbital (HOMO) of the electronic ground state to the lowest unoccupied molecular orbital (LUMO). For water, the LUMO is essentially an atomic 3s(O) orbital, $(4a_1)$. Thus, the MO product for the first excited $(\tilde{A})$ state of water is obtained by promoting an electron out of the $(1b_1)$ non-bonding orbital and into the $(4a_1)$ orbital to give the configuration

$$(1a_1)^2(2a_1)^2(1b_2)^2(3a_1)^2(1b_1)^1(4a_1)^1 \qquad (10.11)$$

which has symmetry B_1.

An electronic wavefunction $\Phi^0_{\text{elec},n}(Q^{(n)}, r_{\text{elec}}, \sigma)$, as given in equation (5.27), is normally obtained in an HF-SCF calculation followed by a CI calculation. It has the symmetry of the Slater determinants out of which it is constructed; these determinants all have the same symmetry. Each Slater determinant has the symmetry of the product of MOs entering into each of its $n!$ terms; these terms each have the same symmetry. An MO has the symmetry of the SOs out of which it is constructed. So, following the chain from AOs to SOs to MOs to Slater determinants and finally to CI wavefunctions, we can determine the symmetries of the CI wavefunctions. Thus, the electronic wavefunctions for the $\tilde{X}$ ground state and $\tilde{A}$ excited state of the water molecule have symmetries A_1 and B_1, respectively.

In figure 10.2, we show the *Walsh diagram* for the energies of the MOs of an XH_2 molecule as a function of the bond angle. This qualitative orbital energy

Table 10.2. The character table for the point group D_{6h}.

D_{6h}:	E	$2C_6$	$2C_3$	C_2	$3C_2'$	$3C_2''$	i	$2S_3$	$2S_6$	σ_h	$3\sigma_d$	$3\sigma_v$
A_{1g}:	1	1	1	1	1	1	1	1	1	1	1	1
A_{1u}:	1	1	1	1	1	1	−1	−1	−1	−1	−1	−1
A_{2g}:	1	1	1	1	−1	−1	1	1	1	1	−1	−1
A_{2u}:	1	1	1	1	−1	−1	−1	−1	−1	−1	1	1
B_{1g}:	1	−1	1	−1	1	−1	1	−1	1	−1	1	−1
B_{1u}:	1	−1	1	−1	1	−1	−1	1	−1	1	−1	1
B_{2g}:	1	−1	1	−1	−1	1	1	−1	1	−1	−1	1
B_{2u}:	1	−1	1	−1	−1	1	−1	1	−1	1	1	−1
E_{1g}:	2	1	−1	−2	0	0	2	1	−1	−2	0	0
E_{1u}:	2	1	−1	−2	0	0	−2	−1	1	2	0	0
E_{2g}:	2	−1	−1	2	0	0	2	−1	−1	2	0	0
E_{2u}:	2	−1	−1	2	0	0	−2	1	1	−2	0	0

diagram is based on simple ideas as to how the bonding energy of the MOs will vary with bond angle and can be used to predict the approximate geometry, linear or bent, of a given XH_2 molecule having a particular electron configuration. For example, the BeH_2 molecule has six electrons and the ground electronic state configuration is $(1a_1)^2(2a_1)^2(1b_2)^2$. From figure 10.2, we see that both the $(2a_1)$ and $(1b_2)$ orbitals have their strongest bonding (lowest energy) at linearity and so we would predict that BeH_2 would be linear in its ground electronic state. Very recently, this has been found to be the case[1]. In the ground $\tilde{X}$ state of the water molecule (with ten electrons), all the orbitals shown in figure 10.2 are full; the competition between the $(3a_1)$ orbital, which wants the molecule to be very strongly bent, and the $(1b_2)$ and $(2a_1)$ orbitals, which want the molecule to be linear, leads to a compromise bond angle of 105°. For larger molecules, orbital energy diagrams like figure 10.2 are also very useful in allowing one to understand how the electronic structure and electronic energy vary with molecular structure.

10.2 The benzene molecule

In section 3.4.3 on page 61, we introduced the Hückel approximation for describing the π electron system of benzene. We got as far as setting up the Hückel matrix in equation (3.63) and we will now use the vanishing integral rule to simplify its diagonalization. Figures 3.8 and 3.9 will be used here.

The molecular point group for the benzene molecule is D_{6h} whose character table is given in table 10.2. The symmetry elements are as follows.

[1] Bernath P F *et al* 2002 *Science* **297** 1323.

- One C_6 axis perpendicular to the molecular plane and passing through the common centre of the carbon nucleus and proton hexagons.
- Three C_2 axes, denoted C_2', which pass through opposite carbon atoms in the carbon-atom hexagon.
- Three C_2 axes, denoted C_2'', which bisect opposite edges of the carbon-atom hexagon.
- An inversion centre at the intersection of the C_6 axis with the molecular plane.
- A rotation–reflection axis S_3 coinciding with the C_6 axis.
- A rotation–reflection axis S_6 coinciding with the C_6 axis.
- A horizontal symmetry plane σ_h coinciding with the molecular plane.
- Three reflection symmetry planes denoted σ_d. Each of these planes contain the C_6 axis and one of the C_2'' axes. That is, the σ_d planes bisect opposite edges of the carbon-atom hexagon.
- Three reflection symmetry planes denoted σ_v. Each of these planes contain the C_6 axis and one of the C_2' axes. That is, the σ_v planes pass through opposite carbon atoms in the carbon-atom hexagon.

We want to determine symmetrized linear combinations of the ϕ_μ ($\mu = 1, 2, \ldots, 6$) C($2p_z$) atomic orbitals (pictured in figure 3.9) that transform irreducibly in $\boldsymbol{D}_{6h}$ and, to do this, we first determine the representation generated by the ϕ_μ orbitals using equations (7.99) and (7.100). As we show later, the characters of the reducible representation of $\boldsymbol{D}_{6h}$ that is generated by the six ϕ_μ orbitals are

$$
\begin{array}{ccccccccccccc}
E & 2C_6 & 2C_3 & C_2 & 3C_2' & 3C_2'' & i & 2S_3 & 2S_6 & \sigma_h & 3\sigma_d & 3\sigma_v \\
6 & 0 & 0 & 0 & -2 & 0 & 0 & 0 & 0 & -6 & 0 & 2
\end{array} \qquad (10.12)
$$

which reduces to

$$
\Gamma_\pi = A_{2u} \oplus B_{2g} \oplus E_{1g} \oplus E_{2u}. \qquad (10.13)
$$

To show how the characters in equation (10.12) are obtained, we consider, as an example of a C_2' operation, the operation $C_{2,1-4}'$, which is the $180°$ rotation about the C_2' axis that passes through the carbon nuclei 1 and 4 which coincides with the x axis. Figure 10.3 shows the effect of the $C_{2,1-4}'$ operation on the atomic orbitals ϕ_1 and ϕ_3; we see that $C_{2,1-4}'\phi_1 = -\phi_1$ and that $C_{2,1-4}'\phi_3 = -\phi_5$. Further, $C_{2,1-4}'\phi_2 = -\phi_6$, $C_{2,1-4}'\phi_4 = -\phi_4$, $C_{2,1-4}'\phi_5 = -\phi_3$ and $C_{2,1-4}'\phi_6 = -\phi_2$. Thus, we can write

$$
C_{2,1-4}'
\begin{bmatrix} \phi_1 \\ \phi_2 \\ \phi_3 \\ \phi_4 \\ \phi_5 \\ \phi_6 \end{bmatrix}
=
\begin{bmatrix} -\phi_1 \\ -\phi_6 \\ -\phi_5 \\ -\phi_4 \\ -\phi_3 \\ -\phi_2 \end{bmatrix}
=
\begin{bmatrix}
-1 & 0 & 0 & 0 & 0 & 0 \\
0 & 0 & 0 & 0 & 0 & -1 \\
0 & 0 & 0 & 0 & -1 & 0 \\
0 & 0 & 0 & -1 & 0 & 0 \\
0 & 0 & -1 & 0 & 0 & 0 \\
0 & -1 & 0 & 0 & 0 & 0
\end{bmatrix}
\begin{bmatrix} \phi_1 \\ \phi_2 \\ \phi_3 \\ \phi_4 \\ \phi_5 \\ \phi_6 \end{bmatrix}
$$
$$(10.14)$$

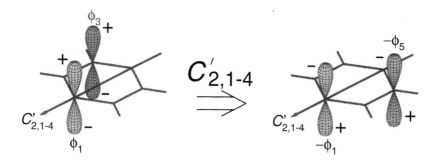

Figure 10.3. The effect of the $C'_{2,1-4}$ operation on the atomic orbitals ϕ_1 and ϕ_3; $C'_{2,1-4}\phi_1 = -\phi_1$ and $C'_{2,1-4}\phi_3 = -\phi_5$.

and the character of the representation matrix is -2 for C'_2. In section 10.6, the effect of the C_6 and of a σ_v symmetry operation are considered in detail.

It is rather easy to determine the characters by just picturing each rotation and reflection symmetry operation; only the ϕ_μ that are not sent to another position on the ring by the operation can contribute to the character (i.e. to the diagonal elements of the transformation matrix generated by the symmetry operation) and the sign of the contribution depends on whether the orbital is turned upside-down or not by the operation.

Each symmetry orbital for benzene can be written as

$$\psi_j = \sum_{\mu=1}^{6} c_{j\mu}\phi_\mu \tag{10.15}$$

where the $c_{j\mu}$ are expansion coefficients. To normalize the function ψ_j, we must consider the integral

$$\langle \psi_j | \psi_j \rangle = \int \psi_j^* \psi_j \, dx \, dy \, dz = \sum_{\mu=1}^{n} \sum_{\nu=1}^{n} c_{j\mu}^* c_{j\nu} \int \phi_\mu^* \phi_\nu \, dx \, dy \, dz = \sum_{\mu=1}^{n} |c_{j\mu}|^2 \tag{10.16}$$

where we have used equation (3.61). Thus, the condition for ψ_j to be normalized is

$$\sum_{\mu=1}^{n} |c_{j\mu}|^2 = 1. \tag{10.17}$$

Similarly, the overlap integral between two Hückel molecular orbitals ψ_j and $\psi_{j'}$

is

$$\langle \psi_j | \psi_{j'} \rangle = \int \psi_j^* \psi_{j'} \, dx \, dy \, dz$$

$$= \sum_{\mu=1}^{n} \sum_{\nu=1}^{n} c_{j\mu}^* c_{j'\nu} \int \phi_\mu^* \phi_\nu \, dx \, dy \, dz = \sum_{\mu=1}^{n} c_{j\mu}^* c_{j'\mu}. \quad (10.18)$$

When this is not zero, it is necessary to orthogonalize the MOs.

To form the SOs, we use the projection operators

$$P^{\Gamma_i} = \frac{l_i}{h} \sum_R \chi^{\Gamma_i}[R]^* R \quad (10.19)$$

where the sum runs over all operations R in D_{6h}, $\chi^{\Gamma_i}[R]$ is a character of the irreducible representation Γ_i, l_i is the dimension of this representation and h is the order of the group ($= 24$ for D_{6h}); see the discussion after equation (7.102). We derive

$$P^{A_{2u}} \phi_1 = \tfrac{1}{6}(\phi_1 + \phi_2 + \phi_3 + \phi_4 + \phi_5 + \phi_6) \quad (10.20)$$

and normalization of this function, using equations (10.16) and (10.17), yields the A_{2u} function,

$$\psi^{A_{2u}} = \frac{1}{\sqrt{6}}(\phi_1 + \phi_2 + \phi_3 + \phi_4 + \phi_5 + \phi_6). \quad (10.21)$$

Similarly,

$$P^{B_{2g}} \phi_1 = \tfrac{1}{6}(\phi_1 - \phi_2 + \phi_3 - \phi_4 + \phi_5 - \phi_6). \quad (10.22)$$

Upon normalization, we obtain the B_{2g} function

$$\psi^{B_{2g}} = \frac{1}{\sqrt{6}}(\phi_1 - \phi_2 + \phi_3 - \phi_4 + \phi_5 - \phi_6). \quad (10.23)$$

The irreducible representation E_{1g} of D_{6h} is doubly degenerate and, to derive two linear combinations of the ϕ_μ orbitals that transform according to it, we apply the projection operator $P^{E_{1g}}$ to ϕ_1 and ϕ_2, respectively:

$$P^{E_{1g}} \phi_1 = \tfrac{1}{6}(2\phi_1 + \phi_2 - \phi_3 - 2\phi_4 - \phi_5 + \phi_6) \quad (10.24)$$

and

$$P^{E_{1g}} \phi_2 = \tfrac{1}{6}(\phi_1 + 2\phi_2 + \phi_3 - \phi_4 - 2\phi_5 - \phi_6) \quad (10.25)$$

Normalization of these functions yields

$$\psi^{E_{1g},a} = \frac{1}{2\sqrt{3}}(2\phi_1 + \phi_2 - \phi_3 - 2\phi_4 - \phi_5 + \phi_6) \quad (10.26)$$

and

$$\psi^{E_{1g},b'} = \frac{1}{2\sqrt{3}}(\phi_1 + 2\phi_2 + \phi_3 - \phi_4 - 2\phi_5 - \phi_6) \tag{10.27}$$

and the overlap integral of the two normalized wavefunctions is obtained from equation (10.18) as

$$\langle \psi^{E_{1g},a} | \psi^{E_{1g},b'} \rangle = \tfrac{1}{12}(2 \times 1 + 1 \times 2 + (-1) \times 1 + (-2) \times (-1)$$
$$+ (-1) \times (-2) + 1 \times (-1)) = \tfrac{1}{2}. \tag{10.28}$$

The two functions are not orthogonal but the function

$$\psi^{E_{1g},b} = \psi^{E_{1g},b'} - \langle \psi^{E_{1g},a} | \psi^{E_{1g},b'} \rangle \psi^{E_{1g},a} = \psi^{E_{1g},b'} - \tfrac{1}{2}\psi^{E_{1g},a} \tag{10.29}$$

is orthogonal to $\psi^{E_{1g},a}$. Upon normalization,

$$\psi^{E_{1g},b} = \tfrac{1}{2}(\phi_2 + \phi_3 - \phi_5 - \phi_6). \tag{10.30}$$

The two functions $\psi^{E_{1g},a}$ and $\psi^{E_{1g},b}$ given in equations (10.26) and (10.30) are a pair of orthonormal functions of E_{1g} symmetry.

By applying the projection operator $P^{E_{2u}}$ to ϕ_1 and ϕ_2, respectively, and by normalizing and orthogonalizing the results, we obtain the pair of orthonormal E_{2u} functions

$$\psi^{E_{2u},a} = \frac{1}{2\sqrt{3}}(2\,\phi_1 - \phi_2 - \phi_3 + 2\,\phi_4 - \phi_5 - \phi_6) \tag{10.31}$$

and

$$\psi^{E_{2u},b} = \tfrac{1}{2}(\phi_2 - \phi_3 + \phi_5 - \phi_6). \tag{10.32}$$

The six SOs are pictured in figure 10.4, and they constitute an orthonormal basis of symmetrized trial MO functions in which to set up the matrix of the one-electron Hamiltonian $\hat{h}$. Each of them is a linear combination of the AOs ϕ_μ, and we obtain the matrix of the Hamiltonian in the symmetrized basis as

$$\langle \psi_j | \hat{h} | \psi_{j'} \rangle = \sum_{\mu=1}^{n} \sum_{\nu=1}^{n} c_{j\mu}^* c_{j'\nu} \langle \phi_\mu | \hat{h} | \phi_\nu \rangle \tag{10.33}$$

where the $\langle \phi_\mu | \hat{h} | \phi_\nu \rangle$ are the elements of the matrix $\mathbf{H}_\pi$ in equation (3.63). From equation (10.33), the symmetrized Hamiltonian matrix is

$$\mathbf{H}'_\pi = \begin{bmatrix} \alpha + 2\beta & 0 & 0 & 0 & 0 & 0 \\ 0 & \alpha - 2\beta & 0 & 0 & 0 & 0 \\ 0 & 0 & \alpha + \beta & 0 & 0 & 0 \\ 0 & 0 & 0 & \alpha + \beta & 0 & 0 \\ 0 & 0 & 0 & 0 & \alpha - \beta & 0 \\ 0 & 0 & 0 & 0 & 0 & \alpha - \beta \end{bmatrix} \tag{10.34}$$

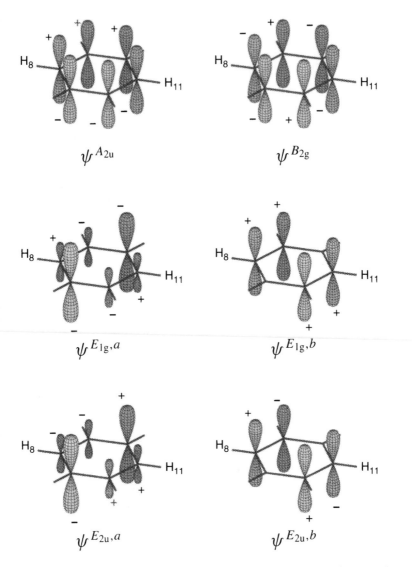

Figure 10.4. Pictorial representations of the six π MOs obtained in the Hückel approximation for the benzene molecule [see equations (10.21)–(10.32)]. For clarity, only the protons 8 and 11 are labelled since the other labels can be inferred from figure 3.8. Also, we only indicate the sign of the AO lobes either above or below the molecular plane, since all orbitals change sign when reflected in this plane.

where the basis functions are in the order $\psi^{A_{2u}}$, $\psi^{B_{2g}}$, $\psi^{E_{1g},a}$, $\psi^{E_{1g},b}$, $\psi^{E_{2u},a}$ and $\psi^{E_{2u},b}$; the matrix is diagonal. There are four different eigenvalues

$$
\begin{bmatrix} E^{A_{2u}} \\ E^{E_{1g}} \\ E^{E_{2u}} \\ E^{B_{2g}} \end{bmatrix} = \begin{bmatrix} \alpha + 2\beta \\ \alpha + \beta \\ \alpha - \beta \\ \alpha - 2\beta \end{bmatrix} = \begin{bmatrix} \alpha - 2|\beta| \\ \alpha - |\beta| \\ \alpha + |\beta| \\ \alpha + 2|\beta| \end{bmatrix} \tag{10.35}
$$

where, in the last equality, we have emphasized that fact that $\beta < 0$ so that $E^{A_{2u}} < E^{E_{1g}} < E^{E_{2u}} < E^{B_{2g}}$. The states with energy $E^{E_{1g}}$ and $E^{E_{2u}}$ are each doubly degenerate.

All of these orbitals have a nodal plane in the plane of the molecule and the orbital at lowest energy, $\psi^{A_{2u}}$ has no nodal planes perpendicular to this plane. The two orbitals $\psi^{E_{1g},a}$ and $\psi^{E_{1g},b}$ each have one nodal plane perpendicular to the plane of the molecule. For $\psi^{E_{1g},a}$, the nodal plane separates the carbon nuclei 2 and 3 and the carbon nuclei 5 and 6; for $\psi^{E_{1g},b}$, the nodal plane passes through the opposite carbon nuclei 1 and 4. The next two orbitals, $\psi^{E_{2u},a}$ and $\psi^{E_{2u},b}$, each have two nodal planes perpendicular to the plane of the molecule and the orbital at highest energy, $\psi^{B_{2g}}$, has three such nodal planes. In $\psi^{B_{2g}}$, each carbon nucleus is separated from each of its neighbouring nuclei by a nodal plane and this orbital is completely antibonding.

Each carbon atom in the benzene molecule contributes one π electron and, in the electronic ground state, these six electrons occupy the orbitals of lowest energy to give the electronic configuration

$$
\text{(doubly occupied } \sigma \text{ orbitals)} (\psi^{A_{2u}})^2 (\psi^{E_{1g},a})^2 (\psi^{E_{1g},b})^2. \tag{10.36}
$$

To determine the symmetry in D_{6h} of the electronic ground-state wavefunction of benzene corresponding to the electronic configuration in equation (10.36) is slightly tricky because the highest occupied molecular orbitals are components of the degenerate E_{1g} irreducible representation of D_{6h}. To do this, we first have to determine the transformation properties of the E_{1g} MOs. The effect of the operation C_6 on the atomic orbitals ϕ_μ ($\mu = 1, 2, \ldots, 6$) is given in equation (10.60), so

$$
C_6 \begin{bmatrix} \psi^{E_{1g},a} \\ \psi^{E_{1g},b} \end{bmatrix} = \begin{bmatrix} \frac{1}{2\sqrt{3}}(2\phi_2 + \phi_3 - \phi_4 - 2\phi_5 - \phi_6 + \phi_1) \\ \frac{1}{2}(\phi_3 + \phi_4 - \phi_6 - \phi_1) \end{bmatrix}
$$
$$
= \begin{bmatrix} \cos\left(\frac{\pi}{3}\right) & \sin\left(\frac{\pi}{3}\right) \\ -\sin\left(\frac{\pi}{3}\right) & \cos\left(\frac{\pi}{3}\right) \end{bmatrix} \begin{bmatrix} \psi^{E_{1g},a} \\ \psi^{E_{1g},b} \end{bmatrix} \tag{10.37}
$$

where $\cos(\pi/3) = 1/2$ and $\sin(\pi/3) = \sqrt{3}/2$. From equations (10.14) and (10.62), respectively, we determine in a similar manner that

$$
C'_{2,1-4} \begin{bmatrix} \psi^{E_{1g},a} \\ \psi^{E_{1g},b} \end{bmatrix} = \begin{bmatrix} -\psi^{E_{1g},a} \\ \psi^{E_{1g},b} \end{bmatrix} \qquad \text{and}
$$

$$\sigma_{v,1-4} \begin{bmatrix} \psi^{E_{1g},a} \\ \psi^{E_{1g},b} \end{bmatrix} = \begin{bmatrix} \psi^{E_{1g},a} \\ -\psi^{E_{1g},b} \end{bmatrix}. \tag{10.38}$$

The benzene molecule has 42 electrons and so the Slater determinant approximating the ground-state electronic wavefunction has 42! terms. Each of these terms can be written as

$$\text{(spin orbital product for inner electrons)}$$
$$\times |1/2, \alpha_i\rangle|1/2, \beta_j\rangle|1/2, \alpha_k\rangle|1/2, \beta_l\rangle$$
$$\times [\psi^{E_{1g},a}(r_i)\psi^{E_{1g},a}(r_j)\psi^{E_{1g},b}(r_k)\psi^{E_{1g},b}(r_l)]. \tag{10.39}$$

The 38 inner electrons comprise the 36 electrons in 18 doubly occupied σ orbitals and the two electrons in the $\psi^{A_{2u}}$ MO; thus the '(spin-orbital product for inner electrons)' in equation (10.39) has A_{1g} symmetry. The product $|1/2, \alpha_i\rangle|1/2, \beta_j\rangle|1/2, \alpha_k\rangle|1/2, \beta_l\rangle$ contains the spin part of the spin-orbitals from equation (3.23); electrons i and k both have α spin and j and l both have β spin. To determine the symmetry of the corresponding orbital product involving $\psi^{E_{1g},a}$ and $\psi^{E_{1g},b}$ in equation (10.39), we must take into account the fact that there are three other related terms in the Slater determinant that are obtained by (a) interchanging the coordinates of electrons i and k (which both have α spin), (b) interchanging the coordinates of electrons j and l (which both have β spin) or (c) making these two interchanges simultaneously. This gives four terms which we can write as

$$\text{(spin orbital product for inner electrons)}$$
$$\times |1/2, \alpha_i\rangle|1/2, \beta_j\rangle|1/2, \alpha_k\rangle|1/2, \beta_l\rangle$$
$$\times [+\psi^{E_{1g},a}(r_i)\psi^{E_{1g},a}(r_j)\psi^{E_{1g},b}(r_k)\psi^{E_{1g},b}(r_l)$$
$$+ \psi^{E_{1g},a}(r_k)\psi^{E_{1g},a}(r_l)\psi^{E_{1g},b}(r_i)\psi^{E_{1g},b}(r_j)$$
$$- \psi^{E_{1g},a}(r_k)\psi^{E_{1g},a}(r_j)\psi^{E_{1g},b}(r_i)\psi^{E_{1g},b}(r_l)$$
$$- \psi^{E_{1g},a}(r_i)\psi^{E_{1g},a}(r_l)\psi^{E_{1g},b}(r_j)\psi^{E_{1g},b}(r_k)]; \tag{10.40}$$

these terms all have the same spin part. The terms resulting from the interchanges (ik) or (jl) have signs opposite to that of the term in equation (10.39) because of the properties of a Slater determinant (i.e. the Pauli exclusion principle), whereas the term resulting from the simultaneous interchanges $(ik)(jl)$ has the same sign as the original term.

To determine the effect of the operation C_6 on the Slater determinant approximating the ground-state electronic wavefunction of benzene, it is necessary to consider the effect on sets of four terms in the determinant such as that given in equation (10.40) since C_6 mixes the four terms in the square brackets. The terms in equation (10.40) and, therefore, the complete Slater determinant[2], are unchanged by the operation C_6 and, therefore, also by the

---

[2] The 'spin-orbital product for inner electrons' has A_{1g} symmetry.

operations $C_6^2 = C_3, C_6^3 = C_2, C_6^4 = C_3^2$ and C_6^5 in D_{6h}. In fact, the Slater determinant is unchanged by all operations in D_{6h} and so the electronic wavefunction for the electronic ground state of benzene has A_{1g} symmetry.

In benzene, the HOMO is $\psi^{E_{1g},a}$ or $\psi^{E_{1g},b}$, since these two orbitals are degenerate, and the LUMO is $\psi^{E_{2u},a}$ or $\psi^{E_{2u},b}$, since these two orbitals are also degenerate. Thus, in benzene, there are four different ways of promoting an electron from the HOMO to the LUMO, leading to the four electronic configurations:

$$\ldots (\psi^{E_{1g},a})^1 (\psi^{E_{1g},b})^2 (\psi^{E_{2u},a})^1$$
$$\ldots (\psi^{E_{1g},a})^1 (\psi^{E_{1g},b})^2 (\psi^{E_{2u},b})^1$$
$$\ldots (\psi^{E_{1g},a})^2 (\psi^{E_{1g},b})^1 (\psi^{E_{2u},a})^1$$
$$\ldots (\psi^{E_{1g},a})^2 (\psi^{E_{1g},b})^1 (\psi^{E_{2u},b})^1.$$

In the Hückel approximation, these four configurations are all energetically degenerate. Their symmetries are given by

$$E_{1g} \otimes E_{2u} = B_{1u} \oplus B_{2u} \oplus E_{2u}. \tag{10.41}$$

However, an HF–SCF–CI calculation, which is more accurate than a Hückel calculation, shows that the B_{2u} electronic state is of lowest energy and so the first excited electronic state of benzene has B_{2u} symmetry. In the Hückel approximation, the electronic wavefunction for this state is

$$\Phi_{\text{elec},B_{2u}} = \tfrac{1}{2}(|\ldots \psi^{E_{1g},a} \psi^{E_{1g},b} \bar{\psi}^{E_{1g},b} \bar{\psi}^{E_{2u},b}\rangle$$
$$+ |\ldots \psi^{E_{1g},a} \bar{\psi}^{E_{1g},a} \psi^{E_{1g},b} \bar{\psi}^{E_{2u},a}\rangle$$
$$+ |\ldots \bar{\psi}^{E_{1g},a} \psi^{E_{1g},b} \bar{\psi}^{E_{1g},b} \psi^{E_{2u},b}\rangle$$
$$+ |\ldots \psi^{E_{1g},a} \bar{\psi}^{E_{1g},a} \bar{\psi}^{E_{1g},b} \psi^{E_{2u},a}\rangle) \tag{10.42}$$

where we have used the notation from equation (3.33) to indicate which orbitals are associated with α and β spin, respectively.

10.3   The butadiene molecule

In section 3.4.4, we determined the Hückel orbitals ψ_1, ψ_2, ψ_3 and ψ_4 which describe the π electron system of the buta-1,3-diene molecule in its *cis*-planar equilibrium configuration (figure 10.6). The orbitals are given in equations (3.68)–(3.71); in figure 10.5 we picture them. In the present section, we determine the symmetry properties of the butadiene orbitals in preparation for analysing the butadiene-cyclobutene conversion reaction in section 10.4.

The buta-1,3-diene molecule in its *cis*-planar equilibrium configuration has the point group C_{2v} whose character table is given in table 10.3. The operation σ is the reflection in the plane of the molecule, σ' is the reflection in the plane that is perpendicular to the plane of the molecule and which bisects the C_2–C_3

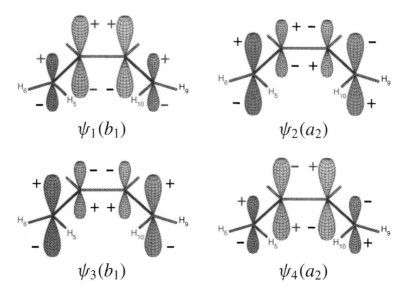

$\psi_1(b_1)$ $\psi_2(a_2)$

$\psi_3(b_1)$ $\psi_4(a_2)$

Figure 10.5. Pictorial representations of the four π orbitals ψ_1, ψ_2, ψ_3 and ψ_4, obtained in the Hückel approximation for the buta-1,3-diene molecule in its *cis*-planar equilibrium configuration [equations (3.68)–(3.71)]. For clarity only the protons 5, 6, 9 and 10 are labelled; the other labels can be inferred from figure 10.6.

Table 10.3. The character table for C_{2v}.

C_{2v}:	E	C_2	σ	σ'
A_1:	1	1	1	1
A_2:	1	1	−1	−1
B_1:	1	−1	−1	1
B_2:	1	−1	1	−1

bond and the line connecting C_1 and C_4 (see figure 10.6 for the labelling of the nuclei). The C_2 axis lies at the intersection of the two reflection symmetry planes. With these definitions, ψ_1 and ψ_3 have B_1 symmetry, whereas ψ_2 and ψ_4 have A_2 symmetry. Each carbon atom in buta-1,3-diene contributes one π electron so that the electronic ground state has the configuration

$$\text{(doubly occupied } \sigma \text{ orbitals)}(\psi_1)^2(\psi_2)^2. \tag{10.43}$$

The π-electron part of the electronic wavefunction has symmetry

$$(B_1)^2 \otimes (A_2)^2 = A_1 \tag{10.44}$$

and the product of doubly occupied σ orbitals also has A_1 symmetry. In the electronic ground state of buta-1,3-diene, ψ_2 is the HOMO and ψ_3 the LUMO, so that the first excited electronic state has the configuration

$$(\text{doubly occupied } \sigma \text{ orbitals})(\psi_1)^2(\psi_2)^1(\psi_3)^1 \qquad (10.45)$$

with symmetry

$$(B_1)^2 \otimes (A_2)^1 \otimes (B_1)^1 = B_2. \qquad (10.46)$$

As the orbital energy increases, the bonding character of the orbitals ψ_1, ψ_2, ψ_3 and ψ_4 (figure 10.5) changes from bonding to antibonding in a manner analogous to that described for benzene. The orbital ψ_2 has a nodal plane separating the carbon nuclei 2 and 3 and it gives rise to π bonds between nuclei 1 and 2 and between nuclei 3 and 4. Since ψ_2 is the HOMO, we draw the molecule with appropriate double bonds in figure 10.6. The LUMO ψ_3 describes a π bond between the carbon nuclei 2 and 3 but it has nodal planes separating nucleus 1 from nucleus 2 and nucleus 3 from nucleus 4.

10.4 Conservation of orbital symmetry

Point group symmetry can be used to predict the relative activation energies of competing chemical reactions by using the concept of *the conservation of orbital symmetry*. Woodward and Hoffmann[3] predicted that reactions in which orbital symmetry is conserved have activation energies much lower than reactions in which orbital symmetry is broken. To use this idea, it is necessary that the reactant(s) and the product(s), and all intermediate molecules, have at least one common symmetry element, such as a rotation symmetry axis or a reflection symmetry plane, that is conserved along the entire reaction path. The conserved symmetry elements define a point group and this point group is used to symmetry classify those orbitals of the reactant(s) and the product(s) that are involved in the reaction (the other orbitals being ignored). By symmetry analysis, correlations between the orbitals of the reactant(s) and the product(s) are established. From these correlations, it is possible to predict what happens to the bonding electrons of the reactant(s) in the course of the reaction and this, in turn, makes it possible to estimate qualitatively the part of the activation energy that results from conservation of orbital symmetry.

To illustrate the principles of the conservation of orbital symmetry, we use a simple electrocyclic reaction: The conversion of *cis*-planar buta-1,3-diene to cyclobutene. An electrocyclic reaction involves the formation of a σ-bond between the ends of a linear π-bonded molecule. Cyclo-addition reactions involve the formation of σ-bonds between the ends of two (or more) π-bonded molecule and they are another type of reaction to which the conservation of

---

[3] See Woodward R B and Hoffmann R 1970 *The Conservation of Orbital Symmetry* (Weinheim: Verlag Chemie GmbH).

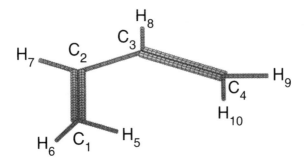

Figure 10.6. The buta-1,3-diene molecule $CH_2(CH)_2CH_2$ in its *cis*-planar configuration. Thin cylinders represent single bonds and double cylinders represent localized double bonds.

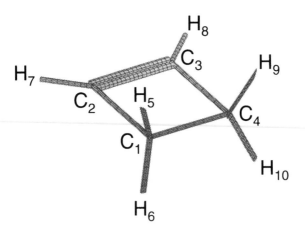

Figure 10.7. The cyclobutene molecule C_4H_6. Thin cylinders represent single bonds and double cylinders represent double bonds.

orbital symmetry applies; the Diels–Alder reaction is cyclo-addition reaction. In the example that we use here, the two competing reaction paths both lead to the same product (cyclobutene) and so the stereochemical predictions are hard to distinguish. However, the principles involved also apply to mildly substituted systems for which the competing reaction paths lead to different chemical products and for which the symmetry analysis leads to important and testable stereochemical predictions.

In the reaction of the *cis*-planar buta-1,3-diene molecule, shown in figure 10.6, to the cyclobutene molecule shown in figure 10.7, the two terminal CH_2 groups, along with the 2p orbitals on the sp^2 hybridized carbon atoms 1 and

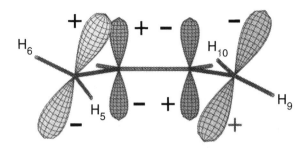

Figure 10.8. A conrotatory rotation about the C_1–C_2 and C_3–C_4 bonds of the buta-1,3-diene molecule in figure 10.6 with a pictorial representation of the HOMO ψ_2 in the electronic ground state. For clarity, only the protons 5, 6, 9 and 10 are labelled. During the conrotatory rotation converting buta-1,3-diene to cyclobutene, the molecule has C_2 symmetry throughout.

4, rotate by 90° about the adjacent C–C bonds, the π bonds connecting C_1 to C_2 and C_4 to C_3 are broken, the π bond connecting C_2 to C_3 is built and the σ bond between C_1 and C_4 is built from the $2p$ orbitals on C_1 and C_4 that are rotated to interact end-on. There is no change in the CH σ-bonds nor in the CC σ-bonds 1–2, 2–3 and 3–4. The orbitals involved in the reaction are the π orbitals ψ_1, ψ_2, ψ_3 and ψ_4 given in equations (3.68)–(3.71) and depicted in figure 10.5.

The conversion of buta-1,3-diene to cyclobutene involves the 90° rotation of the two 'end' CH_2 groups and reaction paths where these rotations happen in a concerted 'together' manner are the most important. There are two such paths: Either the two CH_2 rotate in the same direction (a *conrotatory* reaction path), as illustrated in figure 10.8, or they rotate in opposite directions (a *disrotatory* reaction path), as illustrated in figure 10.9. In the figures, the HOMO ψ_2 in the electronic ground state of buta-1,3-diene is depicted. If the reaction path is conrotatory, the two protons H_5 and H_{10} in buta-1,3-diene (figure 10.6) end up on opposite sides of the carbon-nucleus plane in cyclobutene, whereas a disrotatory reaction path puts these two protons on the same side of the carbon-nucleus plane[4]. If the direction of rotation for both CH_2 groups in figure 10.8 is reversed (so that the protons 6 and 10 move downwards while 5 and 9 move upwards), an alternative conrotatory reaction path is obtained that is equivalent to that shown in the figure. The alternative path leads to the same results as obtained using the path shown in figure 10.8. An alternative, and equivalent, disrotatory reaction path is obtained by reversing the direction of rotation for both CH_2 groups in figure 10.9.

During the conrotatory rotation of the two CH_2 groups (figure 10.8), the nuclear arrangement has, as its only symmetry element, a C_2 axis passing through

---

[4] The labelling of the nuclei in figure 10.7 is chosen arbitrarily. Nothing can be inferred from this figure about the conrotatory or disrotatory nature of the reaction path.

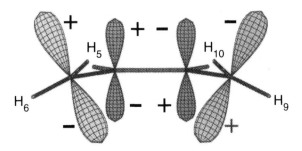

Figure 10.9.   A disrotatory rotation about the C_1–C_2 and C_3–C_4 bonds of the buta-1,3-diene molecule in figure 10.6 with a pictorial representation of the HOMO ψ_2 in the electronic ground state. For clarity, only the protons 5, 6, 9 and 10 are labelled. During the disrotatory rotation converting buta-1,3-diene to cyclobutene, the molecule has C_s symmetry throughout.

the mid point of the C_2–C_3 bond and through the mid point of the line connecting C_1 and C_4. Thus, the appropriate point group for investigating the conservation of orbital symmetry by a conrotatory reaction path is

$$C_2 = \{E, C_2\}. \tag{10.47}$$

During the disrotatory rotation (figure 10.9), the only symmetry element of the nuclear arrangement is a reflection symmetry plane perpendicular to the plane defined by the four carbon nuclei and passing through the mid point of the C_2–C_3 bond and through the mid point of the line connecting C_1 and C_4. Thus, the appropriate point group for investigating the conservation of orbital symmetry by a disrotatory reaction path is

$$C_s = \{E, \sigma'\}. \tag{10.48}$$

The reflection operation is denoted σ' for consistency with the point group C_{2v} of *cis*-planar buta-1,3-diene (see section 3.4.4 and table 10.3). The character tables of the two groups C_2 and C_s are given in table 10.4.

The σ orbitals in buta-1,3-diene are unchanged in cyclobutene but the π orbital system is changed. The two π orbitals ϕ_1 and ϕ_4 in figure 3.11 are rotated by 90° about the C_1–C_2 and C_3–C_4 bonds, respectively. They now lie in the carbon-nucleus plane and have turned into σ orbitals. The rotated orbitals interact to form a bonding orbital σ_{cb} and an antibonding orbital σ_{cb}^*; we show σ_{cb} in figure 10.10(a). The orbital σ_{cb}^* has a nodal plane coinciding with the reflection symmetry plane σ' of cyclobutene. From table 10.3, the bonding orbital σ_{cb} has A_1 symmetry in C_{2v} whereas σ_{cb}^* has B_2 symmetry. The two π atomic orbitals ϕ_2 and ϕ_3 centred on the carbon atoms 2 and 3 in *cis*-planar buta-1,3-diene (figure 3.11) are the only π basis functions available in cyclobutene. They

Table 10.4. The character tables of the groups C_2 and C_s.

	C_2			C_s	
	E	C_2		E	σ'
A	1	1	A'	1	1
B	1	-1	A''	1	-1

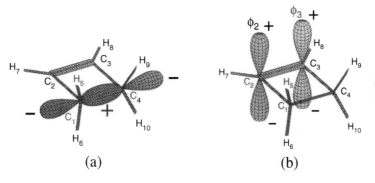

Figure 10.10. (a) The bonding orbital σ_{cb} for cyclobutene and (b) the two π orbitals ϕ_2 and ϕ_3 centred on the carbon atoms 2 and 3, respectively, in cyclobutene.

are shown in figure 10.10(b). They interact as given in equation (3.60) to produce bonding and antibonding π molecular orbitals

$$\pi_{cb} = \frac{1}{\sqrt{2}}[\phi_2 + \phi_3] \quad \text{and} \quad \pi_{cb}^* = \frac{1}{\sqrt{2}}[\phi_2 - \phi_3]. \tag{10.49}$$

The orbital π_{cb} has B_1 symmetry in C_{2v} while π_{cb}^* has A_2 symmetry.

Of the four π molecular orbitals of *cis*-planar buta-1,3-diene considered here, ψ_1 has the lowest energy and ψ_2, ψ_3 and ψ_4 have increasingly higher energies [see equation (3.67)]. In order to determine the correlations between these orbitals and the four orbitals σ_{cb}, σ_{cb}^*, π_{cb} and π_{cb}^* of cyclobutene, it is necessary to estimate the relative energies of the cyclobutene orbitals.

The two π orbitals π_{cb} and π_{cb}^* are obtained by interaction of the two atomic p_z orbitals ϕ_2 and ϕ_3 and in the Hückel approximation, the energies of the molecular orbitals are the eigenvalues of the matrix

$$\begin{bmatrix} \alpha & \beta \\ \beta & \alpha \end{bmatrix}; \tag{10.50}$$

these eigenvalues are

$$E^\pi = \alpha - |\beta| \quad \text{and} \quad E^{\pi^*} = \alpha + |\beta| \tag{10.51}$$

where the lower energy E^π corresponds to the bonding orbital π_{cb} and the higher energy $E^{\pi*}$ corresponds to the antibonding orbital π^*_{cb}. In the approximate theory developed here for obtaining E^π and $E^{\pi*}$, the same values of α and β are used as for *cis*-planar buta-1,3-diene in equation (3.67).

The two orbitals σ_{cb} and σ^*_{cb} are obtained by interaction of $90°$-rotated versions of the the two atomic p_z orbitals ϕ_1 and ϕ_4 in figure 3.11 and the energies of σ_{cb} and σ^*_{cb} are obtained as the eigenvalues of a matrix

$$\begin{bmatrix} W_\sigma & -Q_\sigma \\ -Q_\sigma & W_\sigma \end{bmatrix} \tag{10.52}$$

by analogy with equation (3.51). Since the atomic orbitals used as basis functions in obtaining this matrix are rotated versions of the orbitals ϕ_1 and ϕ_4, that give rise to the diagonal matrix elements of α in the Hückel theory for *cis*-planar buta-1,3-diene, we approximate $W_\sigma = \alpha$. With this approximation, the eigenvalues of equation (10.52) are obtained as

$$E^\sigma = \alpha - |Q_\sigma| \quad \text{and} \quad E^{\sigma*} = \alpha + |Q_\sigma| \tag{10.53}$$

where the lower energy E^σ corresponds to the bonding orbital σ_{cb} and the higher energy $E^{\sigma*}$ corresponds to the antibonding orbital σ^*_{cb}.

When one electron occupies the the rotated ϕ_1 orbital and the other electron occupies the rotated ϕ_4 orbital, the energy is 2α in this approximation. When both electrons occupy the bonding orbital σ_{cb}, the energy is $2\alpha - 2|Q_\sigma|$, and the energy lowering by forming the σ bond is $2|Q_\sigma|$. Similarly, the energy lowering by forming the π bond is $2|\beta|$. The energy lowering obtained by formation of a bonding σ orbital is, in general, larger than the energy lowering obtained by formation of a bonding π orbital (see section 3.4.2), so $|Q_\sigma| > |\beta|$. This relation, in conjunction with equations (10.51) and (10.53), yields the energy ordering $E^\sigma < E^\pi < E^{\pi*} < E^{\sigma*}$ for the cyclobutene orbitals.

In the electronic ground state of cyclobutene, the four available electrons occupy the orbitals of lowest energy to give the electronic configuration

$$\text{(doubly occupied } \sigma \text{ orbitals)}(\sigma_{cb})^2(\pi_{cb})^2. \tag{10.54}$$

The bonding orbital π_{cb} describes a π bond between the carbon nuclei 2 and 3 so we have indicated a double bond between these two nuclei in figures 10.7 and 10.10. The ground-state electronic wavefunction of cyclobutene has the symmetry

$$(A_1)^2 \otimes (B_1)^2 = A_1 \tag{10.55}$$

since the product of doubly occupied σ orbitals has A_1 symmetry. The electronic configuration for the first excited electronic state of cyclobutene has the configuration

$$\text{(doubly occupied } \sigma \text{ orbitals)}(\sigma_{cb})^2(\pi_{cb})^1(\pi^*_{cb})^1 \tag{10.56}$$

Table 10.5. Symmetries of molecular orbitals for buta-1,3-diene and cyclobutene in the point groups C_{2v}, C_2, and C_s.

	Buta-1,3-diene				Cyclobutene		
	C_{2v}	C_2	C_s		C_{2v}	C_2	C_s
ψ_1	B_1	B	A'	σ_{cb}	A_1	A	A'
ψ_2	A_2	A	A''	π_{cb}	B_1	B	A'
ψ_3	B_1	B	A'	π_{cb}^*	A_2	A	A''
ψ_4	A_2	A	A''	σ_{cb}^*	B_2	B	A''

with the symmetry

$$(A_1)^2 \otimes (B_1)^1 \otimes (A_2)^1 = B_2. \tag{10.57}$$

Correlations between the molecular orbitals of *cis*-planar buta-1,3-diene and those of cyclobutene can now be established. In the approximation employed here, the molecular orbitals involved in the interconversion between *cis*-planar buta-1,3-diene and cyclobutene are (in order of increasing energy for each molecule) ψ_1, ψ_2, ψ_3 and ψ_4 for *cis*-planar buta-1,3-diene and σ_{cb}, π_{cb}, π_{cb}^* and σ_{cb}^* for cyclobutene. Table 10.5 summarizes the symmetries of these molecular orbitals in C_{2v} and gives also their symmetries in the point groups C_2 and and C_s. These latter groups are subgroups of C_{2v} and the symmetries of the molecular orbitals in them can be sraightforwardly obtained from the symmetry in C_{2v} by correlation as discussed in connection with table 7.6.

In figure 10.11(a), the *orbital correlation diagram* for the conrotatory interconversion of buta-1,3-diene and cyclobutene is shown. On the left-hand side of the diagram, the energies of the four buta-1,3-diene orbitals ψ_1, ψ_2, ψ_3 and ψ_4 are schematically indicated with their symmetry labels (table 10.5) in the point group C_2 [equation (10.47)] which is appropriate for the symmetry analysis of the conrotatory reaction path. On the right-hand side of figure 10.11(a), the energies and C_2 symmetry labels of the cyclobutene orbitals σ_{cb}, π_{cb}, π_{cb}^* and σ_{cb}^* are shown. Broken lines (correlations) are drawn connecting orbitals of the same symmetry. Lines connecting two pairs of orbitals of the same symmetry can never cross because of the *non-crossing rule* (see section 10.5) and this rule uniquely defines the correlation diagram. The correlation lines depict the variation of the orbital energies along the reaction path. The orbital energies are the eigenvalues of matrices such as the Hückel matrix for buta-1,3-diene in equation (3.64) or, more generally, the more accurate matrices diagonalized in HF-SCF calculations.

Figure 10.11(b) is the analogous correlation diagram for the disrotatory interconversion of buta-1,3-diene and cyclobutene. The orbitals and their energies are the same as for the conrotatory interconversion but the orbitals are now symmetry classified in the point group C_s [equation (10.48)] which is appropriate

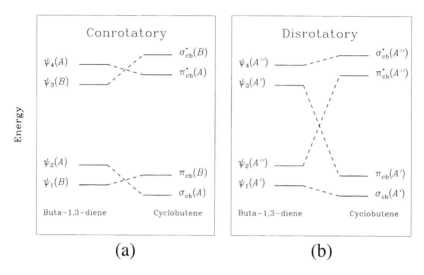

Figure 10.11. Orbital correlation diagrams for the (a) conrotatory and (b) disrotatory interconversion of buta-1,3-diene and cyclobutene. For the conrotatory reaction path, the molecular orbitals are symmetry classified in the point group C_2 and for the disrotatory reaction path, they are classified in C_s.

Table 10.6. Correlation of electronic states for the interconversion of buta-1,3-diene and cyclobutene.

Conrotatory		Disrotatory	
Buta-1,3-diene	Cyclobutene	Buta-1,3-diene	Cyclobutene
$\psi_1^2\psi_2^2$	$\sigma_{cb}^2\pi_{cb}^2$	$\psi_1^2\psi_2^2$	$\sigma_{cb}^2\pi_{cb}^{*2}$
$\psi_1^2\psi_2^1\psi_3^1$	$\sigma_{cb}^1\pi_{cb}^2\sigma_{cb}^{*1}$	$\psi_1^2\psi_2^1\psi_3^1$	$\sigma_{cb}^2\pi_{cb}^1\pi_{cb}^{*1}$
$\psi_1^1\psi_2^2\psi_4^1$	$\sigma_{cb}^2\pi_{cb}^1\pi_{cb}^{*1}$	$\psi_1^2\psi_3^2$	$\sigma_{cb}^2\pi_{cb}^2$

for the symmetry analysis of the disrotatory reaction path; as a result, the correlations are different from those in figure 10.11(a).

According to figure 10.11(a), a conrotatory reaction path takes the HOMO ψ_2 of buta-1,3-diene into the bonding orbital σ_{cb} of cyclobutene so that the two bonding electrons of buta-1,3-diene, that occupy ψ_2, are transformed into two bonding electrons of cyclobutene. However, a disrotatory reaction path [figure 10.11(b)] places the same two electrons in the orbital π_{cb}^* at much higher energy. Thus, if the interconversion reaction takes place with the molecules at thermal equilibrium at a moderate temperature (a *thermal reaction*) so that

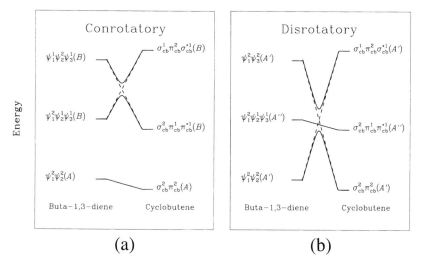

Figure 10.12. State correlation diagrams for the (a) conrotatory and (b) disrotatory interconversion of buta-1,3-diene and cyclobutene. For the conrotatory reaction path, the electronic states are symmetry classified in the point group C_2 and for the disrotatory reaction path, they are classified in C_s. The full curves result from application of the non-crossing rule (see section 10.5).

reactants and products are predominantly in their respective electronic ground states, then the conrotatory reaction path is favoured. If we excite the molecules to their first excited electronic states by irradiation with light (a *photochemical reaction*), the HOMO of buta-1,3-diene becomes ψ_3. A conrotatory reaction path places the electron occupying ψ_3 of buta-1,3-diene in the highly excited antibonding orbital σ^*_{cb} of cyclobutene, whereas a disrotatory reaction path places this electron in the bonding cyclobutene orbital π_{cb} at much lower energy. Therefore, for the photochemical reaction, the disrotatory reaction path is preferred.

A more complete understanding of the reaction mechanism is obtained by using the orbital correlation diagrams in figure 10.11 to determine the correlation diagram for the low-energy electronic state configurations of buta-1,3-diene and cyclobutene. These are called *state correlation diagrams*.[5] For example, the electronic ground state of buta-1,3-diene has the configuration[6] $(\psi_1)^2(\psi_2)^2$ [equation (10.43)]. From figure 10.11(a), a conrotatory reaction

---

[5] State correlation diagrams (and orbital correlation diagrams) for analysing the conservation of orbital symmetry were first introduced by Longuet-Higgins H C and Abrahamson E W 1965 *J. Am. Chem. Soc.* **87** 2045.

[6] For brevity, we omit in the configurations the doubly occupied σ orbitals which are assumed to be identical for buta-1,3-diene and cyclobutene.

path correlates this configuration with the ground-state configuration $(\sigma_{cb})^2(\pi_{cb})^2$
[equation (10.54)] of cyclobutene. Figure 10.11(b) shows that a disrotatory
reaction path correlates the electronic ground-state configuration of buta-1,3-
diene with a highly excited electronic state, $(\sigma_{cb})^2(\pi_{cb}^*)^2$, of cyclobutene.
Correlations of this type involving the electronic ground states [equations (10.43)
and (10.54)] and the first excited electronic states [equations (10.45) and (10.56)]
of buta-1,3-diene and cyclobutene are summarized in table 10.6. The correlations
in the table are presented graphically in the state correlation diagrams of
figure 10.12. These diagrams indicate the symmetry labels of the electronic
states in C_2 for the conrotatory reaction path and, in C_s, for the disrotatory
reaction path; these symmetries are determined as discussed in section 10.1; see
equation (10.10).

The state correlations shown for the conrotatory reaction path as dotted lines
in figure 10.12(a) are obtained from the orbital correlation diagram. Application
of the non-crossing rule (see section 10.5) produces the actual state correlations,
and these are given as full curves. These curves give an approximate indication
of the change in energy along the reaction path between buta-1,3-diene and
cyclobutene and it is seen that a buta-1,3-diene molecule in the first excited
electronic state must overcome an energy barrier to convert to cyclobutene in
the first excited electronic state. However, the state correlation shows that a buta-
1,3-diene molecule in the electronic ground state can convert to cyclobutene in its
electronic ground state without having to overcome an energy barrier.

In figure 10.11(b), the state correlation diagram for the disrotatory reaction
path is shown. For this reaction, a buta-1,3-diene molecule in the electronic
ground state must overcome an energy barrier, whereas a molecule in the first
excited electronic state need not. Comparison of figures 10.11(a) and 10.11(b)
demonstrates that for the thermal reaction, where the molecules start out in their
respective electronic ground states, the conrotatory reaction path is preferred since
the molecules can convert without having to overcome an energy barrier. For
the photochemical reaction, where the molecules start out in their first excited
electronic states, the disrotatory reaction path is favoured. This result is in
agreement with the conclusions we drew before using the orbital correlation
diagrams.

Woodward and Hoffmann have used these ideas for a systematic study of
many reactions. For example, the results obtained for the butadiene-cyclobutene
interconversion can be extended to the reaction of a general chain polyene
$H_2C(CH)_k CH_2$ with $(k+2)\pi$ electrons to a cyclic molecule with $k\pi$ electrons. It
is found that for $k = 4n+2$ $(n = 0, 1, 2, \ldots)$, the reaction is thermally conrotatory
and photochemically disrotatory. For $k = 4n$ $(n = 1, 2, 3, \ldots)$ the reaction is
thermally disrotatory and photochemically conrotatory. This is an example of a
Woodward–Hoffmann rule.

Conservation of orbital symmetry only applies to concerted reaction paths,
i.e. to paths that involve two or more parts of a molecule moving together in
concert. Molecular reactions need not follow such paths. For example, steric

constraints in the reactant and/or product molecules may make the concerted pathway unfavourable.

10.5   The non-crossing rule

In the development of the orbital and state correlation diagrams, mention was made of the *non-crossing rule* for orbital and state electronic energies. For orbital energies this rule states that if we vary the nuclear coordinates along the reaction path and diagonalize the appropriate sequence of matrices to obtain the orbital energies as functions of these nuclear coordinates, two eigenvalues corresponding to orbitals of the same symmetry cannot cross. The diagonal matrix elements corresponding to these two orbitals *can* cross but, when the diagonal elements are close, the corresponding eigenvalues are approximately given by equations (2.32) and (2.33); these two eigenvalues cannot become equal for a non-zero value of the off-diagonal matrix element H_{12} in equation (2.30). This matrix element is allowed by the vanishing integral rule to be non-zero and this rule presupposes that it will never be *exactly* zero since it is not forced by symmetry to vanish. As a result, the eigenvalues associated with two orbitals of the same symmetry suffer an *avoided crossing* when the corresponding diagonal elements cross as functions of the nuclear coordinates (it so happens, however, that this does not occur in the orbital correlation diagrams of figure 10.11).

Because of the non-crossing rule for orbital energies, the lowest buta-1,3-diene orbital of a given symmetry must correlate with the lowest cyclobutene orbital of the same symmetry. The next (in order of ascending energy) buta-1,3-diene orbital of the given symmetry must correlate with the next cyclobutene orbital of the same symmetry, and so on.

In the state correlation diagrams of figure 10.12, there are avoided crossings. For the conrotatory reaction path in figure 10.12(a), the orbital correlation diagram in figure 10.11(a) suggests the correlation indicated by broken lines in figure 10.12(a): Each of the first excited states correlates with a highly excited state of the other molecule. Indeed, if we were to carry out a series of Hartree–Fock calculations at nuclear geometries along the conrotatory reaction path, the computed electronic energies would vary approximately as indicated by the broken lines in figure 10.12(a). However, if we were to improve the accuracy of the calculated electronic energies by carrying out configuration-interaction (CI) calculations based on the Hartree–Fock calculations just described, there would be a drastic change. In the CI calculations, the Hartree–Fock energies indicated by the broken lines in figure 10.12(a) are diagonal elements of the CI matrix. Since the two electronic configurations corresponding to these diagonal elements have the same symmetry (B) in C_2, they are coupled by a non-vanishing off-diagonal matrix element. Therefore, the eigenvalues of the CI matrix will not cross along the reaction path as the diagonal elements do; the eigenvalues will suffer an avoided crossing analogous to that discussed for the orbital correlation diagrams. The avoided crossing is indicated by the full curves connecting the excited electronic states in figure 10.12(a).

Because of the avoided crossings caused by configuration interaction, potential energy curves connecting two pairs of electronic states of the same symmetry can never cross. Therefore, the first excited electronic state of buta-1,3-diene, which is the lowest buta-1,3-diene state of B symmetry in C_2, *must* correlate with the first excited electronic state of cyclobutene, which is the lowest cyclobutene state of B symmetry in C_2[7]. This leads to the correlations indicated by the full curves in figure 10.12(a). In figure 10.12(b), the state correlation diagram for the disrotatory path also has an avoided crossing.

10.6   The C_6 and σ_v operations for benzene

The operation C_6 is a rotation of $60°(=\pi/3$ radians) about the C_6 axis (the z axis in figure 3.8 on page 62) and so it rotates xyz electron coordinates about the z axis by $60°$, measured in a right-handed sense. If an electron initially has coordinates (x_i, y_i, z_i), then after the operation C_6 has been applied, its coordinates are transformed to (x_i', y_i', z_i') given by

$$
\begin{bmatrix} x_i' \\ y_i' \\ z_i' \end{bmatrix} = \begin{bmatrix} \cos\left(\frac{\pi}{3}\right) & -\sin\left(\frac{\pi}{3}\right) & 0 \\ \sin\left(\frac{\pi}{3}\right) & \cos\left(\frac{\pi}{3}\right) & 0 \\ 0 & 0 & 1 \end{bmatrix} \begin{bmatrix} x_i \\ y_i \\ z_i \end{bmatrix}
$$

$$
= \begin{bmatrix} \frac{1}{2} & -\frac{\sqrt{3}}{2} & 0 \\ \frac{\sqrt{3}}{2} & \frac{1}{2} & 0 \\ 0 & 0 & 1 \end{bmatrix} \begin{bmatrix} x_i \\ y_i \\ z_i \end{bmatrix}. \tag{10.58}
$$

The effect of a point group symmetry operation on a function is defined by equation (7.7); using this definition we obtain

$$
C_6\phi_1(x_i, y_i, z_i) = \phi_1(x_i', y_i', z_i') = \phi_2(x_i, y_i, z_i) \tag{10.59}
$$

where some thought is required to understand the second equality. If, for instance, the initial point (x_i, y_i, z_i) lies near carbon nucleus 2, where ϕ_2 has a significant amplitude, the transformed point (x_i', y_i', z_i') lies near carbon atom 1, where ϕ_1 has the same amplitude (see figures 3.8 and 3.9). Thus, as shown in figure 10.13, $C_6\phi_1 = \phi_2$. It follows analogously that $C_6\phi_2 = \phi_3, \ldots, C_6\phi_5 = \phi_6$ and

---

[7] We do not know with certainty if, for the highest energy curve drawn in figure 10.12(a), the configuration $(\psi_1)^1(\psi_2)^2(\psi_3)^1$ of buta-1,3-diene correlates with the configuration $(\sigma_{cb})^1(\pi_{cb})^2(\sigma_{cb}^*)^1$ of cyclobutene. There may be other configurations of B symmetry in C_2 that have lower energies than one or both of these high-lying configurations so that they change this particular correlation. However, the correlation of this pair of highly excited electronic states is of no interest to us.

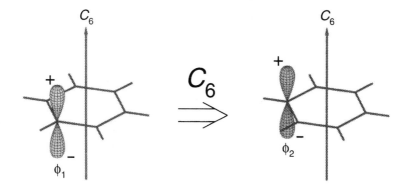

Figure 10.13. The effect of the C_6 operation on the atomic orbital ϕ_1: $C_6\phi_1 = \phi_2$. See figures 3.8 and 3.9 for the labelling of the nuclei and the definition of the atomic orbitals ϕ_μ ($\mu = 1, 2, \ldots, 6$).

$C_6\phi_6 = \phi_1$. Thus,

$$
C_6
\begin{bmatrix}
\phi_1 \\ \phi_2 \\ \phi_3 \\ \phi_4 \\ \phi_5 \\ \phi_6
\end{bmatrix}
=
\begin{bmatrix}
\phi_2 \\ \phi_3 \\ \phi_4 \\ \phi_5 \\ \phi_6 \\ \phi_1
\end{bmatrix}
=
\begin{bmatrix}
0 & 1 & 0 & 0 & 0 & 0 \\
0 & 0 & 1 & 0 & 0 & 0 \\
0 & 0 & 0 & 1 & 0 & 0 \\
0 & 0 & 0 & 0 & 1 & 0 \\
0 & 0 & 0 & 0 & 0 & 1 \\
1 & 0 & 0 & 0 & 0 & 0
\end{bmatrix}
\begin{bmatrix}
\phi_1 \\ \phi_2 \\ \phi_3 \\ \phi_4 \\ \phi_5 \\ \phi_6
\end{bmatrix} . \tag{10.60}
$$

The operation $\sigma_{v,1-4}$, the reflection in the σ_v plane containing the carbon nuclei 1 and 4, is a reflection in the xz plane (figure 3.8) and the electronic coordinates transform as

$$
\sigma_{v,1-4}
\begin{bmatrix} x_i \\ y_i \\ z_i \end{bmatrix}
=
\begin{bmatrix} x_i' \\ y_i' \\ z_i' \end{bmatrix}
=
\begin{bmatrix} x_i \\ -y_i \\ z_i \end{bmatrix}
=
\begin{bmatrix}
1 & 0 & 0 \\
0 & -1 & 0 \\
0 & 0 & 1
\end{bmatrix}
\begin{bmatrix} x_i \\ y_i \\ z_i \end{bmatrix} . \tag{10.61}
$$

We obtain $\sigma_{v,1-4}\phi_1(x_i, y_i, z_i) = \phi_1(x_i', y_i', z_i') = \phi_1(x_i, y_i, z_i)$ together with $\sigma_{v,1-4}\phi_2 = \phi_6$, $\sigma_{v,1-4}\phi_3 = \phi_5$, $\sigma_{v,1-4}\phi_4 = \phi_4$, $\sigma_{v,1-4}\phi_5 = \phi_3$ and $\sigma_{v,1-4}\phi_6 = \phi_2$ (see figure 10.14 for the effect of $\sigma_{v,1-4}$ on ϕ_1 and ϕ_3). Consequently,

$$
\sigma_{v,1-4}
\begin{bmatrix}
\phi_1 \\ \phi_2 \\ \phi_3 \\ \phi_4 \\ \phi_5 \\ \phi_6
\end{bmatrix}
=
\begin{bmatrix}
\phi_1 \\ \phi_6 \\ \phi_5 \\ \phi_4 \\ \phi_3 \\ \phi_2
\end{bmatrix}
=
\begin{bmatrix}
1 & 0 & 0 & 0 & 0 & 0 \\
0 & 0 & 0 & 0 & 0 & 1 \\
0 & 0 & 0 & 0 & 1 & 0 \\
0 & 0 & 0 & 1 & 0 & 0 \\
0 & 0 & 1 & 0 & 0 & 0 \\
0 & 1 & 0 & 0 & 0 & 0
\end{bmatrix}
\begin{bmatrix}
\phi_1 \\ \phi_2 \\ \phi_3 \\ \phi_4 \\ \phi_5 \\ \phi_6
\end{bmatrix} . \tag{10.62}
$$

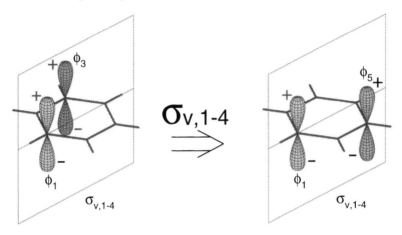

Figure 10.14. The effect of the $\sigma_{v,1-4}$ operation on the atomic orbitals ϕ_1 and ϕ_3; $\sigma_{v,1-4}\phi_1 = \phi_1$ and $\sigma_{v,1-4}\phi_3 = \phi_5$. See figures 3.8 and 3.9 for the labelling of the nuclei and the definition of the atomic orbitals ϕ_μ ($\mu = 1, 2, \ldots, 6$).

10.7 Problems

10.1 In the Walsh diagram for an XH_2 molecule, figure 10.2, the $(1b_1)$ and $(3a_1)$ orbitals are shown as having the same energy at linearity, whereas, on bending, the energy of the $(3a_1)$ strongly decreases. Explain why this is so.

10.2 Using the concept of sp hybrid orbitals [see equation (3.59)] explain why the $(1b_2)$ and $(2a_1)$ orbitals in figure 10.2 are most strongly bonding at linearity.

10.3 Extensive *ab initio* calculations[8] lead to the prediction that, in its ground electronic state, the CH_2^+ ion is bent at equilibrium but with only a small barrier in the potential energy function at the linear configuration and that in its first excited electronic state the equilibrium structure is linear but that at this linear structure the energy of this excited state is equal to that of the ground electronic state. Use figure 10.2 to explain these numerical results.

10.4 An *ab initio* calculation[9] for the CH_2 molecule predicts that in an excited electronic state of A_2 symmetry the equilibrium bond angle is less than $50°$. Explain this result.

10.5 The planar D_{3h} molecule borazine $B_3N_3H_6$ is sometimes called inorganic

---

[8] See, for example, figure 1 of Jensen P *et al* 2002 *Spectrochim. Acta* A **58** 763.
[9] Yamaguchi Y and Schaefer H F III 1997 *J. Chem. Phys.* **106** 1819.

benzene. What is its equilibrium structure? Use Hückel MO theory to explain why it is given this alternative name.

10.6   Is the photochemical cyclization of 1,3,5 hexatriene $CH_2(CH)_4CH_2$ to 1,3 cyclohexadiene a conrotatory or disrotatory process? Explain your answer using orbital and state correlation diagrams.

Chapter 11

The symmetry of rotation–vibration wavefunctions

11.1 The transformation properties of the Euler angles

The coordinates used to describe molecular rotation are the Euler angles (θ, ϕ, χ) defined in figures 5.1–5.3. The three angles determine the orientation of the molecule-fixed xyz axes relative to the $\xi\eta\zeta$ axes. The present section discusses the effect of MS group operations on Euler angles. To obtain the transformation properties of the Euler angles for a particular molecule, it is sufficient to consider the molecule at equilibrium but we must choose a convention for attaching the xyz axes. The origin of the xyz axes is always the nuclear centre of mass and the xyz axes are always right-handed.

Initially, with the xyz axes attached to the molecule according to the chosen convention, the Euler angles are (θ, ϕ, χ). After applying an MS group operation R, we again use the chosen convention to attach the molecule-fixed axes. The newly oriented molecule-fixed axes $x'y'z'$ have orientation in space given by the new Euler angle values (θ', ϕ', χ'). We write

$$R(\theta, \phi, \chi) = (\theta', \phi', \chi'). \tag{11.1}$$

The rotation that takes the xyz axes into the $x'y'z'$ axes is called the *equivalent rotation* of the MS group operation R. For the H_3^+ molecule, we see from figure 8.6 on page 168 that the effect of the permutation (123) on the xyz axes is to rotate them by $120°$ (or $2\pi/3$ radians) in a right-handed sense about the z axis. Consequently, the equivalent rotation of (123) is the rotation $R_z^{2\pi/3}$ whose effect is detailed in the (c)→(d) part of figure 8.6. Similarly, figure 8.7 shows that the equivalent rotation of E^* is R_z^{π}, a rotation of $180°$ about the z axis and figure 8.8 shows that the equivalent rotation of $(23)^*$ is R_x^{π}, a rotation of $180°$ about the x axis.

For symmetric and asymmetric top molecules, the MS operations give rise to equivalent rotations of two types: R_z^{β}, a rotation about the z axis through the

222

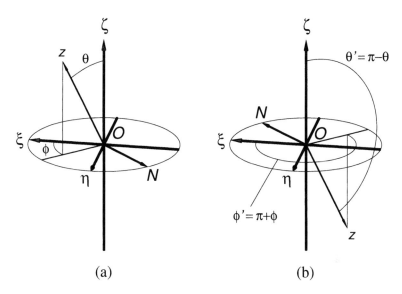

Figure 11.1. (a) The definition of the Euler angles θ and ϕ and (b) the effect on θ and ϕ of the equivalent rotation $R_\alpha{}^\pi$ which reverses the direction of the z axis.

angle β (in radians), where β is measured in a right-handed sense about the z axis, and $R_\alpha{}^\pi$, a rotation of π radians about an axis in the xy plane that forms the angle α (in radians) with the x axis, where α is measured in a right-handed sense about the z axis. Spherical top molecules have equivalent rotations other than $R_z{}^\beta$ and $R_\alpha{}^\pi$ but we do not consider these more complicated rotations here. Notice that $R_0{}^\pi = R_x{}^\pi$ and $R_{\pi/2}{}^\pi = R_y{}^\pi$.

The Euler angles θ and ϕ are the polar coordinates that define the orientation of the z axis in the $\xi\eta\zeta$ axis system [see figure 11.1(a)]. Since $R_z{}^\beta$ is a rotation about the z axis, it leaves the orientation of this axis and, therefore, the angles θ and ϕ, unchanged. It is seen from figure 11.2 that a rotation about the z axis through the angle β causes the change $\chi \rightarrow \chi' = \chi + \beta$ and so

$$R_z{}^\beta(\theta, \phi, \chi) = (\theta', \phi', \chi') = (\theta, \phi, \chi + \beta). \tag{11.2}$$

The equivalent rotation $R_\alpha{}^\pi$ reverses the direction of the z axis and, from figure 11.1(b), we see that

$$R_\alpha{}^\pi(\theta, \phi) = (\theta', \phi') = (\pi - \theta, \pi + \phi). \tag{11.3}$$

Figure 11.3 illustrates the effect of $R_\alpha{}^\pi$ on χ. Figure 11.3(a) shows the initial situation in which the y axis forms the angle χ with the node line ON and the rotation axis pq forms the angle α with the x axis. Rotating the xyz axes by 180° about the axis pq gives the orientation of the axes as depicted in figure 11.3(b).

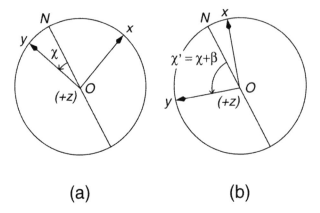

(a)   **(b)**

Figure 11.2. The molecule-fixed axes xyz drawn so that the xy plane coincides with the plane of the page. (a) An initial situation where the y axis forms the angle χ with the node line ON. (b) The situation resulting from applying the equivalent rotation $R_z{}^\beta$. In (a) and (b), the z axis points up (out of the page).

Since the z axis is reversed, the direction of the node line ON and the positive sense for measuring the angle χ are reversed (see figure 11.1). In figure 11.3(b),

$$\chi' = \angle(NOq) + \angle(qOp) + \angle(pOy). \tag{11.4}$$

Elementary trigonometry[1] gives

$$\chi' = \left(\frac{\pi}{2} - \alpha - \chi\right) + \pi + \left(\frac{\pi}{2} - \alpha\right) = 2\pi - 2\alpha - \chi \tag{11.5}$$

and, hence,

$$R_\alpha{}^\pi(\theta, \phi, \chi) = (\pi - \theta, \phi + \pi, 2\pi - 2\alpha - \chi). \tag{11.6}$$

The Euler angle transformations are summarized in table 11.1, where we also give the transformation properties of the angular momentum components $\hat{J}_\alpha$ which we have obtained by using the Euler angle transformations and equations (5.45)–(5.47) for the $\hat{J}_\alpha$.

11.2 The symmetry of rotational wavefunctions

To determine the effect of the equivalent rotation $R_z{}^\beta$ on the symmetric top rotational wavefunctions

$$|J, k, m\rangle = \Phi_{\text{rot}}(\theta, \phi, \chi) \tag{11.7}$$

---

[1] The angle $\angle(NOq)$ in figure 11.3(b) equals the angle $\angle(NOp)$ in figure 11.3(a), which is $(\pi/2 - \alpha - \chi)$, and the angle $\angle(pOy)$ in figure 11.3(b) equals the angle $\angle(pOy)$ in figure 11.3(a), which is $(\pi/2 - \alpha)$.

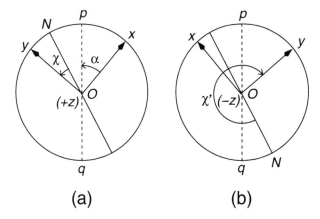

Figure 11.3. The molecule-fixed axes xyz drawn so that the xy plane coincides with the plane of the page. (a) An initial situation where the y axis forms the angle χ with the node line ON and the rotation axis (shown as a broken line pq) forms the angle α with the x axis. (b) The situation after applying the equivalent rotation $R_\alpha{}^\pi$. In (a), the z axis points up (out of the page) and, in (b), it points down.

Table 11.1. Transformation properties under[a] $R_z{}^\beta$ and $R_\alpha{}^\pi$.

	$R_z{}^\beta$	$R_\alpha{}^\pi$
θ	θ	$\pi - \theta$
ϕ	ϕ	$\phi + \pi$
χ	$\chi + \beta$	$2\pi - 2\alpha - \chi$
$\hat{J}_x$	$\hat{J}_x \cos\beta + \hat{J}_y \sin\beta$	$\hat{J}_x \cos 2\alpha + \hat{J}_y \sin 2\alpha$
$\hat{J}_y$	$-\hat{J}_x \sin\beta + \hat{J}_y \cos\beta$	$\hat{J}_x \sin 2\alpha - \hat{J}_y \cos 2\alpha$
$\hat{J}_z$	$\hat{J}_z$	$-\hat{J}_z$
$|J, k, m\rangle$	$e^{ik\beta}|J, k, m\rangle$	$(-1)^J e^{-2ik\alpha}|J, -k, m\rangle$

[a] $R_z{}^\beta$ is a rotation of the molecule-fixed xyz axes through β radians about the z axis (β is measured in a right-handed sense about the z axis) and $R_\alpha{}^\pi$ is a rotation of the molecule-fixed xyz axes through π radians about an axis in the xy plane making an angle α with the x axis (α is measured in a right-handed sense about the z axis).

given by equations (5.51)–(5.56), we insert the transformed coordinates from table 11.1 into equation (5.52) and this gives

$$R_z{}^\beta|J, k, m\rangle = e^{ik\beta}|J, k, m\rangle. \tag{11.8}$$

For $R_\alpha{}^\pi | J, k, m \rangle$, it can be shown (see problem 11.1) that the effect of inserting the transformed coordinates from table 11.1 is

$$R_\alpha{}^\pi | J, k, m \rangle = (-1)^J e^{-2ik\alpha} | J, -k, m \rangle. \tag{11.9}$$

To determine the MS group symmetries of the zero-order rotational wavefunctions for a given molecule, we first determine the equivalent rotations for the operations in the MS group and we then employ equations (11.8) and (11.9) to determine the effect of the equivalent rotations on the wavefunctions. We work through this procedure for H_3^+.

11.2.1 H_3^+

The MS group of H_3^+ is $D_{3h}(M)$ whose character table is given in table 7.3 on page 147. The convention for the way the xyz axes are attached in the equilibrium configuration is that the x and y axes are in the molecular plane with the y axis passing though proton number 1, the x axis passing through the bond connecting protons 1 and 2, and the z axis directed so that xyz is right-handed. To determine the representations spanned by the rotational functions of H_3^+, it is sufficient to determine their transformation properties under one selected operation from each of the classes in $D_{3h}(M)$ (see table 7.3). We select the operations

$$
\begin{array}{ccccccc}
D_{3h}(M): & E & (123) & (23) & E^* & (123)^* & (23)^* \\
\text{Equiv. rot.}: & R^0 & R_z^{2\pi/3} & R_{\pi/2}{}^\pi & R_z{}^\pi & R_z^{5\pi/3} & R_0{}^\pi
\end{array}
\tag{11.10}
$$

where we have also given the equivalent rotations. The equivalent rotations of (123), E^* and (23)* are from figures 8.6–8.8, and those for (23) and (123)* can be inferred from figure 11.4, where we show the effect on the orientation of the xyz axes of each of the operations in equation (11.10).

We use equations (11.8) and (11.9) to determine the effect of the equivalent rotations on the rotational functions $| J, k, m \rangle$ for H_3^+. We distinguish two cases: $K = |k| = 0$ and $K = |k| > 0$. For $K = 0$, the wavefunction $| J, 0, m \rangle$ spans a one-dimensional irreducible representation of $D_{3h}(M)$ whereas, for $K > 0$, the wavefunctions $| J, K, m \rangle$ and $| J, -K, m \rangle$ span a two-dimensional representation which may be reducible or irreducible. For $K > 0$, we write

$$\Phi_K = \left[\begin{array}{c} | J, K, m \rangle \\ | J, -K, m \rangle \end{array} \right]. \tag{11.11}$$

The effect of the operation (123) is

$$
(123)\Phi_K = R_z^{2\pi/3} \Phi_K = R_z^{2\pi/3} \left[\begin{array}{c} | J, K, m \rangle \\ | J, -K, m \rangle \end{array} \right]
$$

$$
= \left[\begin{array}{cc} e^{2iK\pi/3} & 0 \\ 0 & e^{-2iK\pi/3} \end{array} \right] \left[\begin{array}{c} | J, \ K, m \rangle \\ | J, -K, m \rangle \end{array} \right]
\tag{11.12}
$$

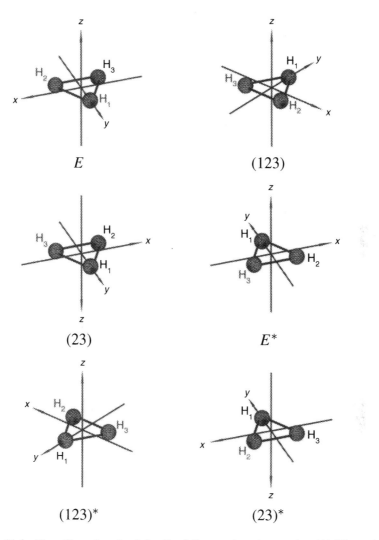

Figure 11.4. The effect of each of the $D_{3h}(M)$ operations in equation (11.10) on the orientation of the molecule-fixed xyz axes for the H_3^+ molecule.

where in the last equality we have used equation (11.8). The character of this representation matrix is

$$\chi_K[(123)] = e^{2iK\pi/3} + e^{-2iK\pi/3} = 2\cos\left(\frac{2K\pi}{3}\right). \qquad (11.13)$$

Table 11.2. The symmetry species of the rotational wavefunctions of H_3^+ in the $D_{3h}(M)$ group, and of CH_3F in the $C_{3v}(M)$ group. The $(2J + 1)$ m-degeneracy is ignored, n is integral, and $K = |k|$.

K		H_3^+	K		CH_3F
0	J even	A_1'	0	J even	A_1
	J odd	A_2'		J odd	A_2
$6n \pm 1$		E''	$3n \pm 1$		E
$6n \pm 2$		E'	$3n$		$A_1 \oplus A_2$
$6n \pm 3$		$A_1'' \oplus A_2''$			
$6n \pm 6$		$A_1' \oplus A_2'$			

The effect of the operation (23) is

$$(23)\Phi_K = R_{\pi/2}{}^{\pi}\, \Phi_K = R_{\pi/2}{}^{\pi} \begin{bmatrix} |J, K, m\rangle \\ |J, -K, m\rangle \end{bmatrix}$$

$$= \begin{bmatrix} 0 & (-1)^{J+K} \\ (-1)^{J+K} & 0 \end{bmatrix} \begin{bmatrix} |J, K, m\rangle \\ |J, -K, m\rangle \end{bmatrix} \qquad (11.14)$$

where in the last equality we have used equation (11.9); this representation matrix has character zero. By deriving the representation matrix for each of the operations in equation (11.10), we obtain the characters listed as '$K > 0$' in equation (11.15). For $K = 0$, we obtain the characters of the representation spanned by the single function $|J, 0, m\rangle$ by using equations (11.8) and (11.9); these are listed as '$K = 0$' in equation (11.15).

$D_{3h}(M)$:	E	(123)	(23)	E^*	$(123)^*$	$(23)^*$
$K > 0$:	2	$2\cos(\frac{2}{3}K\pi)$	0	$2(-1)^K$	$2\cos(\frac{5}{3}K\pi)$	0
$K = 0$:	1	1	$(-1)^J$	1	1	$(-1)^J$

$$\qquad (11.15)$$

Using equation (7.97), in conjunction with the characters of the irreducible representations (in table 7.3) and the identity

$$2\cos\left(\frac{5K\pi}{3}\right) = 2\cos\left(\frac{2K\pi}{3} + K\pi\right) = 2(-1)^K \cos\left(\frac{2K\pi}{3}\right) \qquad (11.16)$$

the representations in equation (11.15) are reduced to the rotational symmetries given in table 11.2. This table also gives the rotational symmetries for CH_3F in the $C_{3v}(M)$ group.

11.2.2 H_2O

H_2O is an asymmetric top with the MS group $C_{2v}(M)$, whose irreducible representations are given in table 7.2. Using molecule-fixed axes abc (as described in section 5.3.4), we first determine the equivalent rotations of the operations of the MS group.

We label the two protons in the H_2O molecule as 1 and 2, and the oxygen nucleus as 3. In the equilibrium configuration, the b principal axis coincides with the C_2 axis, the a axis is in the molecular plane and the c axis is perpendicular to the molecular plane. The convention we use for attaching the abc axes is such that the b coordinate of the oxygen nucleus is positive, the a coordinate of proton 2 is positive, and the c axis is perpendicular to the molecular plane with orientation such that abc is right-handed. The effect of the operations (12), E^* and $(12)^*$ on the orientation of the abc axes is shown in figure 11.5. From these results, the equivalent rotations are:

$$C_{2v}(M) : \quad E \quad (12) \quad E^* \quad (12)^*$$
$$\text{Equiv. rot.} : \quad R^0 \quad R_b{}^\pi \quad R_c{}^\pi \quad R_a{}^\pi$$

(11.17)

where $R_q{}^\pi$, $q = a, b$ or c, is a rotation of $180°$ about the q axis.

The zero-order rotational wavefunctions $|J_{K_a K_c}\rangle$ of any asymmetric top molecule are linear combinations of the $|J, k, m\rangle$ functions as described in section 5.3.4, and the states are labelled ee, eo, oe or oo depending on the evenness or oddness of K_a and K_c, respectively. The equivalent rotations of the elements of the MS group of any asymmetric top molecule can only be one of E, $R_a{}^\pi$, $R_b{}^\pi$ or $R_c{}^\pi$. Water is an example of a molecule for which all four of these types of equivalent rotation occur but, in the general case, they need not all occur. Detailed analysis (see problem 11.2) leads to the determination of the asymmetric top symmetry rule which states that:

The ee functions transform as the totally symmetric representation, the eo functions as the representation having $+1$ for $R_a{}^\pi$ (and -1 for $R_b{}^\pi$ and $R_c{}^\pi$), the oe functions as the representation having $+1$ for $R_c{}^\pi$ (and -1 for $R_a{}^\pi$ and $R_b{}^\pi$) and the oo functions as the representation having $+1$ for $R_b{}^\pi$ (and -1 for $R_a{}^\pi$ and $R_c{}^\pi$).

Using this rule, the symmetries of the rotational states of the water molecule are determined from the equivalent rotations given in equation (11.17) to be as in table 11.3.

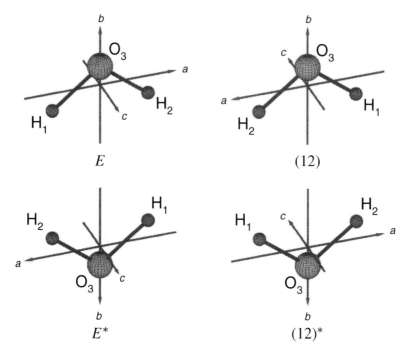

Figure 11.5. The effect of the operations of $C_{2v}(M)$ on the orientation of the abc axes for the H_2O molecule.

Table 11.3. The symmetry of the rotational states of H_2O: e(o) indicates even(odd) K_a or K_c value.

$K_a K_c$	Γ_{rot}	$K_a K_c$	Γ_{rot}
ee	A_1	oe	B_2
eo	B_1	oo	A_2

11.3 The symmetry of normal coordinates

To determine the symmetries of the zero-order vibrational wavefunctions given in equation (4.18):

$$\Phi_{vib} = \Phi_{v_1}(Q_1)\Phi_{v_2}(Q_2)\cdots\Phi_{v_{3N-6}}(Q_{3N-6}) \qquad (11.18)$$

where the functions $\Phi_{v_r}(Q_r)$ are the harmonic-oscillator eigenfunctions discussed in section 4.3 and N is the number of nuclei in the molecule, it is necessary to determine the symmetries of the vibrational normal coordinates Q_1

to Q_{3N-6} upon which the Φ_{vib} depend[2]. To do this, we include the translational normal coordinates $T_\alpha(T_x = Q_{3N-5}, T_y = Q_{3N-4}$ and $T_z = Q_{3N-3})$ and the rotational normal coordinates $R_\alpha(R_x = Q_{3N-2}, R_y = Q_{3N-1}$ and $R_z = Q_{3N})$ given in equations (4.11) and (4.12). Since the $3N$ normal coordinates Q_1 to Q_{3N} are related to the $3N$ mass-weighted Cartesian displacement coordinates $m_i^{1/2}\Delta\alpha_i$, where $\alpha = x, y$ or z and $i = 1$ to N, by an orthogonal transformation [see equation (4.9)], the symmetry of the $3N Q_r$ is the same as the symmetry of the $3N\ m_i^{1/2}\Delta\alpha_i$ [see equation (7.101)]. The symmetry of the $3N\ m_i^{1/2}\Delta\alpha_i$ is the same as the symmetry of the $3N$ Cartesian displacement coordinates $\Delta\alpha_i$. Thus,

> To determine the symmetry Γ_Q of the $(3N - 6)$ vibrational normal coordinates, we determine the symmetry Γ_{Car} of the $3N$ Cartesian displacement coordinates and subtract the symmetry Γ_{TR} of the translational and rotational normal coordinates.

We write

$$\Gamma_{\text{Car}} = \Gamma_Q \oplus \Gamma_{\text{TR}}. \tag{11.19}$$

The analytic expression for the translational and rotational normal coordinates T_α and R_α in terms of the Cartesian displacement coordinates $\Delta\alpha_i$ are given in equations (4.11) and (4.12). Thus, once the symmetry transformation properties of the $\Delta\alpha_i$ have been determined, these equations can be used to determine the transformation properties of the coordinates T_α and R_α. These coordinates transform as irreducible representations of the symmetry group and their symmetry species are usually listed in any set of character tables so they can be simply looked up there. The symmetry of the rotational normal coordinate R_α is the same as that of the rotational angular momentum component $\hat{J}_\alpha$. Having obtained Γ_Q as the sum of its irreducible components, we then have the symmetries of the vibrational normal coordinates, since the following can be proved.

> Each normal coordinate Q_r of a molecule transforms according to an irreducible representation of the point group or, equivalently, of the isomorphic MS group.

We treat the water and benzene molecules as examples below.

---

[2] Linear molecules are discussed in section 5.3.2; they have $(3N - 5)$ vibrational normal coordinates and two rotational normal coordinates.

In section 8.3, we showed that for nonlinear rigid molecules the molecular point group is isomorphic to the MS group, and that the operations in it have the same effect on the vibrational and electronic coordinates as the corresponding MS group operations. Thus, for nonlinear rigid molecules, we can use either the molecular point group or the MS group for the vibrational symmetry analysis. For the symmetry analysis of electronic wavefunctions described in chapter 10, we had the same choice and there we chose the molecular point group since it gave rise to a simpler theoretical description. In the investigation of vibrational symmetry, the situation is less clear-cut. For small molecules, it is straightforward to use the MS group, in particular if the symmetry properties or the rotational wavefunctions are also being investigated, whereas, for large molecules in situations where the rotational symmetry is not needed, the molecular point group is preferable. We consider here two examples: The water molecule, for which we use the MS group but show the connection to the results obtained using the molecular point group; and benzene, for which we use the molecular point group only.

11.3.1 H_2O

Equations (8.3)–(8.9) give the transformation properties of the $\Delta\alpha_i$ (where we use a I^r convention so that $xyz \equiv bca$) under the effect of the operations C_{2x}, σ_{xz} and σ_{xy} in the molecular point group C_{2v} and, in figure 8.5, we demonstrate the equivalence of these transformations to those resulting from the operations (12), E^* and $(12)^*$ in the MS group $C_{2v}(M)$. These results are summarized on the left-hand side of table 11.4, where we have also indicated the isomorphism between $C_{2v}(M)$ and C_{2v}.

To determine the representation of $C_{2v}(M)$ generated by the nine coordinates $\Delta\alpha_i$, we write them as a column vector and determine, for each operation $R(= E, (12), E^*$ or $(12)^*)$ in $C_{2v}(M)$, the 9×9 representation matrix that describes the effect of R on the $\Delta\alpha_i$ in an equation like equation (7.100). The characters of the representation generated are given in the line labelled 'χ_{Car}' in table 11.4, and this reduces to

$$\Gamma_{Car} = 3A_1 \oplus A_2 \oplus 2B_1 \oplus 3B_2. \tag{11.20}$$

From the right-hand side of table 11.4 (see also table B.4),

$$\Gamma_{TR} = A_1 \oplus A_2 \oplus 2B_1 \oplus 2B_2. \tag{11.21}$$

Using equation (11.19), we obtain Γ_Q, the symmetry of the vibrational normal coordinates for the water molecule, as

$$\Gamma_Q = 2A_1 \oplus B_2. \tag{11.22}$$

Thus, two normal coordinates, Q_1 and Q_2, are each of A_1 symmetry and the third, Q_3, is of B_2 symmetry.

Table 11.4. Transformation properties for water molecule coordinates in the $C_{2v}(M)$ and C_{2v} groups. The operations R_{MS} constitute the MS group $C_{2v}(M)$ of the water molecule and the operations R_{PG} constitute the molecular point group C_{2v}.

R_{MS}:	E	(12)	E^*	$(12)^*$	E	(12)	E^*	$(12)^*$	
R_{PG}:	E	C_{2x}	σ_{xz}	σ_{xy}	E	C_{2x}	σ_{xz}	σ_{xy}	
	Δx_1	Δx_2	Δx_1	Δx_2	T_x	T_x	T_x	T_x	: A_1
	Δx_2	Δx_1	Δx_2	Δx_1	T_y	$-T_y$	$-T_y$	T_y	: B_1
	Δx_3	Δx_3	Δx_3	Δx_3	T_z	$-T_z$	T_z	$-T_z$	: B_2
	Δy_1	$-\Delta y_2$	$-\Delta y_1$	Δy_2	R_x	R_x	$-R_x$	$-R_x$	: A_2
	Δy_2	$-\Delta y_1$	$-\Delta y_2$	Δy_1	R_y	$-R_y$	R_y	$-R_y$	: B_2
	Δy_3	$-\Delta y_3$	$-\Delta y_3$	Δy_3	R_z	$-R_z$	$-R_z$	R_z	: B_1
	Δz_1	$-\Delta z_2$	Δz_1	$-\Delta z_2$					
	Δz_2	$-\Delta z_1$	Δz_2	$-\Delta z_1$					
	Δz_3	$-\Delta z_3$	Δz_3	$-\Delta z_3$					
χ_{Car}:	9	-1	3	1					

Figure 11.6 shows the motion of the H_2O molecule as the three normal coordinates Q_1, Q_2 and Q_3 are varied. We show the displacements when each of the three normal coordinates is given a value of 1 $u^{1/2}$Å, while the other two are set equal to zero. The displacement coordinates are calculated using equation (4.9)[3]. Q_1 describes the *symmetric stretch*, Q_2 describes the *bending motion* and Q_3 describes the *antisymmetric stretch*.

An alternative way of determining the representation Γ_Q for a molecule involves *internal coordinates*. Internal coordinates are typically defined as bond lengths and bond angles whose values unambiguously define the instantaneous nuclear geometry. For the water molecule, these are the two internuclear distances r_1 and r_2 together with the bond angle $\gamma = \angle(HOH)$. For a nonlinear rigid molecule with N nuclei, $3N - 6$ independent internal coordinates are required to specify the nuclear geometry and these coordinates span the representation Γ_Q. For the H_2O molecule, the three coordinates (r_1, r_2, γ) transform according to the representation $\Gamma_Q = 2A_1 \oplus B_2$ of $C_{2v}(M)$ or C_{2v}. For small molecules, it is often easier to obtain Γ_Q directly using internal coordinates instead of generating Γ_{Car} and subtracting Γ_{TR}.

11.3.2 Benzene

The benzene molecule C_6H_6 has 30 vibrational normal coordinates and we determine their symmetry in the molecular point group D_{6h} whose irreducible

---

[3] The l matrix elements are from table 2 of Hoy A R *et al* 1972 *Mol. Phys.* **24** 1265.

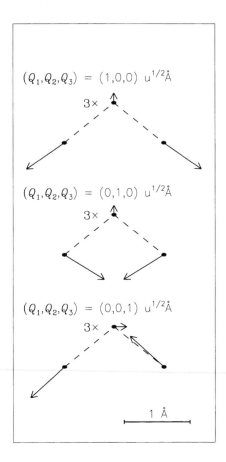

$(Q_1, Q_2, Q_3) = (1,0,0)\ u^{1/2}\text{Å}$

$3\times$

$(Q_1, Q_2, Q_3) = (0,1,0)\ u^{1/2}\text{Å}$

$3\times$

$(Q_1, Q_2, Q_3) = (0,0,1)\ u^{1/2}\text{Å}$

$3\times$

1 Å

Figure 11.6. The motion of the H_2O molecule as the three normal coordinates Q_1, Q_2 and Q_3 are varied. The bonds at equilibrium are represented by broken lines and the nuclear displacements by full arrows. The displacement vector for the oxygen nucleus is enlarged $3\times$ to be visible on the scale of the figure.

representations are given in table 10.2. The MS group $D_{6h}(M)$ is given in table B.9 and we see there that the six coordinates T_x, T_y, T_z, R_x, R_y and R_z generate the representation

$$\Gamma_{TR} = A_{2g} \oplus A_{2u} \oplus E_{1g} \oplus E_{1u}. \qquad (11.23)$$

The numbering of the nuclei in C_6H_6 and the molecule-fixed xyz axes are defined in figure 3.8. To determine the D_{6h} symmetry Γ_{Car} of the 36 displacement coordinates $\Delta x_1, \Delta y_1, \Delta z_1, \Delta x_2, \Delta y_2, \Delta z_2, \dots, \Delta x_{12}, \Delta y_{12}, \Delta z_{12}$ for C_6H_6, we must, at least in principle, construct the representation matrices $\mathbf{D}^{\Gamma_{Car}}[R]$ [see equation (7.100)] generated by the displacement coordinates. In practice, it suffices to consider the diagonal elements of the $\mathbf{D}^{\Gamma_{Car}}[R]$ matrices since they determine the characters. For most operations R in $\mathbf{D}_{6h}$, all diagonal elements of the matrices $\mathbf{D}^{\Gamma_{Car}}[R]$ vanish to produce zero characters. There are only four exceptions:

- The identity operation E. For this operation, all 36 diagonal elements have the value 1 and so the corresponding character is 36.
- The 180° rotations C_2' about axes that contain two carbon nuclei and two protons. For example, the operation $C_{2,1-4}'$ (see figure 10.3) changes the displacement coordinates of the nuclei $i = 1, 4, 7$ and 10 as $(\Delta x_i, \Delta y_i, \Delta z_i) \rightarrow (\Delta x_i, -\Delta y_i, -\Delta z_i)$. Hence, $\mathbf{D}^{\Gamma_{Car}}[C_{2,1-4}']$ has 12 non-vanishing diagonal elements with a sum of -4.
- The reflection σ_h in the molecular plane. It changes the displacement coordinates of all nuclei as $(\Delta x_i, \Delta y_i, \Delta z_i) \rightarrow (\Delta x_i, \Delta y_i, -\Delta z_i)$ and generates a character of 12.
- The reflections σ_v in planes that contain two carbon nuclei and two protons. The operation $\sigma_{v,1-4}$ (see figure 10.14) changes the displacement coordinates of the nuclei $i = 1, 4, 7$ and 10 as $(\Delta x_i, \Delta y_i, \Delta z_i) \rightarrow (\Delta x_i, -\Delta y_i, \Delta z_i)$ and so the matrix $\mathbf{D}^{\Gamma_{Car}}[\sigma_{v,1-4}]$ has 12 non-vanishing diagonal elements with a sum of 4.

Thus, Γ_{Car} has the following characters

$$
\begin{array}{ccccccccccc}
E & 2C_6 & 2C_3 & C_2 & 3C_2' & 3C_2'' & i & 2S_3 & 2S_6 & \sigma_h & 3\sigma_d & 3\sigma_v \\
36 & 0 & 0 & 0 & -4 & 0 & 0 & 0 & 0 & 12 & 0 & 4.
\end{array}
\tag{11.24}
$$

This representation reduces to

$$
\begin{aligned}
\Gamma_{Car} = {}& 2A_{1g} \oplus 2A_{2g} \oplus 2A_{2u} \oplus 2B_{1u} \oplus 2B_{2g} \oplus 2B_{2u} \\
& \oplus 2E_{1g} \oplus 4E_{1u} \oplus 4E_{2g} \oplus 2E_{2u}.
\end{aligned}
\tag{11.25}
$$

Subtracting the representation Γ_{TR}, given in equation (11.23), the representation generated by the 30 vibrational normal coordinates of benzene is

$$
\begin{aligned}
\Gamma_Q = {}& 2A_{1g} \oplus A_{2g} \oplus A_{2u} \oplus 2\,B_{1u} \oplus 2B_{2g} \oplus 2B_{2u} \\
& \oplus E_{1g} \oplus 3E_{1u} \oplus 4E_{2g} \oplus 2E_{2u}.
\end{aligned}
\tag{11.26}
$$

For large molecules like benzene, it can be difficult to determine the representation Γ_Q using internal coordinates. Although it is simple to choose and symmetry classify internuclear distances, such as the six C–C distances and the six C–H distances for benzene, it is often difficult to choose a non-redundant set of independent angular internal coordinates that, in combination with the internuclear distances, unambiguously define the molecular geometry.

11.4   The symmetry of vibrational wavefunctions

11.4.1 H_2O

For the H_2O molecule, equation (4.18) reduces to

$$\Phi_{vib} = \Phi_{v_1}(Q_1)\Phi_{v_2}(Q_2)\Phi_{v_3}(Q_3). \qquad (11.27)$$

From equation (11.22), the normal coordinates Q_1 and Q_2 have A_1 symmetry and, thus, the wavefunctions $\Phi_{v_1}(Q_1)$ and $\Phi_{v_2}(Q_2)$ each have A_1 symmetry independently of the values of v_1 and v_3. The normal coordinate Q_3 has B_2 symmetry. Φ_{v_3} is an even(odd) function of Q_3 for v_3 even(odd) (see section 4.3) so $\Phi_{v_3}(Q_3)$ generates a representation of C_{2v} with the characters

$$
\begin{array}{cccc}
E & C_{2x} & \sigma_{xy} & \sigma_{xz} \\
1 & (-1)^{v_3} & 1 & (-1)^{v_3}
\end{array}
\qquad (11.28)
$$

which means that $\Phi_{v_3}(Q_3)$ has A_1 symmetry for v_3 even and B_2 symmetry for v_3 odd. Thus, we can write the symmetry of the zero-order vibrational wavefunction Φ_{vib} for the water molecule [equation (11.27)] as

$$\Gamma_{vib} = (A_1)^{v_1} \otimes (A_1)^{v_2} \otimes (B_2)^{v_3}. \qquad (11.29)$$

11.4.2 H_3^+

The reader can determine that the vibrational normal coordinates of H_3^+ have symmetry

$$\Gamma_Q = A_1' \oplus E' \qquad (11.30)$$

in the MS group $D_{3h}(M)$ (see table 7.3 for the irreducible representations) or, equivalently, in the molecular point group D_{3h}.

The normal coordinate of A_1' symmetry is denoted Q_1 and the two normal coordinates of E' symmetry are denoted (Q_{2a}, Q_{2b}). Since (Q_{2a}, Q_{2b}) span a doubly degenerate irreducible representation their λ_r values [see equation (4.8)] must be equal (i.e. $\lambda_{2a} = \lambda_{2b}$) and (Q_{2a}, Q_{2b}) describe a two-dimensional harmonic-oscillator [see equations (4.34)–(4.40)]. For H_3^+, the zero-order vibrational wavefunctions are written as

$$\Phi_{vib} = \Phi_{v_1}(Q_1)\Psi_{v_2,l_2}(Q_2, \alpha_2) \qquad (11.31)$$

where $(Q_{2a}, Q_{2b}) = (Q_2 \cos\alpha_2, Q_2 \sin\alpha_2)$ and the function Ψ_{v_2,l_2} is defined in equation (4.38).

The normal coordinate Q_1 and, hence, the function $\Phi_{v_1}(Q_1)$, for any value of v_1, are of A_1' symmetry. For a given value of v_2, the $(v_2 + 1)$ functions Ψ_{v_2,l_2} with $l_2 = -v_2, -v_2 + 2, -v_2 + 4, \ldots, +v_2$ generate the same $(v_2 + 1)$-dimensional representation Γ_{v_2} of $D_{3h}(M)$ as the $(v_2 + 1)$ products $Q_{2a}^{v_2}$,

$Q_{2a}^{v_2-1} Q_{2b}, Q_{2a}^{v_2-2} Q_{2b}^2, Q_{2a}^{v_2-3} Q_{2b}^3, \ldots, Q_{2a} Q_{2b}^{v_2-1}, Q_{2b}^{v_2}$. For $v_2 = 0$, there is only one product, $Q_{2a}^0 Q_{2b}^0 = 1$, which is totally symmetric, so

$$\Gamma_0 = A_1'. \tag{11.32}$$

For $v_2 = 1$, there are two products $(Q_{2a}^1 Q_{2b}^0, Q_{2a}^0 Q_{2b}^1) = (Q_{2a}, Q_{2b})$, and so

$$\Gamma_1 = E'. \tag{11.33}$$

For $v_2 = 2$, there are three products Q_{2a}^2, $Q_{2a} Q_{2b}$ and Q_{2b}^2. It follows from the discussion of equation (7.106) that the symmetry of these three products is the symmetric square of E':

$$\Gamma_2 = [E']^2 = A_1' \oplus E'. \tag{11.34}$$

For an arbitrary value of v_2,

$$\Gamma_{v_2} = [E']^{v_2} \tag{11.35}$$

the symmetric v_2'th power of E'; the characters of this representation can be obtained from equation (7.110). For example, we have

$$\Gamma_3 = [E']^3 = A_1' \oplus A_2' \oplus E'. \tag{11.36}$$

In conclusion, the zero-order vibrational wavefunction Φ_{vib} for the H_3^+ ion [equation (11.31)] has the symmetry

$$\Gamma_{\text{vib}} = A_1' \otimes [E']^{v_2} = [E']^{v_2} = \begin{cases} A_1' & \text{for } v_2 = 0 \\ E' & \text{for } v_2 = 1 \\ A_1' \oplus E' & \text{for } v_2 = 2 \\ A_1' \oplus A_2' \oplus E' & \text{for } v_2 = 3 \\ \vdots & \vdots \end{cases} \tag{11.37}$$

We can rewrite equation (11.37) as

$$\Gamma_{\text{vib}} = (A_1')^{v_1} \otimes [E']^{v_2}. \tag{11.38}$$

Equations (11.29) and (11.38) are particular examples of the following general rule:

For a molecule with vibrational normal coordinates of symmetries $\Gamma^{(1)}, \Gamma^{(2)}, \ldots, \Gamma^{(f)}$ in a vibrational state with quantum numbers $v_1, v_2, \ldots, v_f$ the symmetry of the vibrational wavefunction is

$$\Gamma(v_1, v_2, \ldots, v_f) = [\Gamma^{(1)}]^{v_1} \otimes [\Gamma^{(2)}]^{v_2} \otimes \cdots \otimes [\Gamma^{(f)}]^{v_f} \tag{11.39}$$

where $[\ ]^v$ is the symmetric vth product for a degenerate species and the ordinary vth power for a non-degenerate species.

It follows from equation (11.39) that in the *vibrational ground state* of a molecule, i.e. the state with $v_1 = v_2 = \cdots = v_f = 0$, the vibrational wavefunction is totally symmetric, and that in a *fundamental level* of a molecule, i.e. where one $v_j = 1$ and $v_i = 0$ for $i \neq j$, the symmetry of the vibrational wavefunction is $\Gamma^{(j)}$, the symmetry of Q_j.

11.5   Rotation–vibration coupling

Neglecting nuclear spin and nuclear spin hyperfine structure [discussed in section 9.5], the zero-order molecular wavefunctions are products of LCAO-SCF-CI electronic spin-orbital wavefunctions harmonic-oscillator vibrational wavefunctions, and rigid-rotor rotational wavefunctions [see equation (5.27)]. These states can be coupled by non-vanishing off-diagonal matrix elements of the electronic fine structure Hamiltonian $\hat{H}_{es}$ [see equation (2.77)], the nuclear kinetic energy operator $\hat{T}_N$ [see equation (3.13)], the anharmonicity potential energy term V_N^{anh} [see equation (4.15)] and the rotational coupling term $\hat{H}'_{rot}$ [see equations (5.36)–(5.39)]. Symmetry is used, as already described in chapter 10 and in this chapter, to label the zero-order states. We can also use symmetry to determine which coupling terms in V_N^{anh} and $\hat{H}'_{rot}$ must vanish and to make a general determination of which zero-order states cannot be coupled by any of these terms in the full Hamiltonian.

Anharmonicity was discussed in section 4.4 and it involves the effect of the cubic terms $\Phi_{rst} Q_r Q_s Q_t$ and quartic terms $\Phi_{rstu} Q_r Q_s Q_t Q_u$. Like all terms in the Hamiltonian, V_N^{anh} must be totally symmetric and generate the totally symmetric irreducible representation $\Gamma^{(s)}$ of the MS group of the molecule. Because of this, only cubic terms for which $Q_r Q_s Q_t$ are of symmetry $\Gamma^{(s)}$ can occur. Similarly only quartic terms for which $Q_r Q_s Q_t Q_u$ are of symmetry $\Gamma^{(s)}$ can occur. This leads to the following result:

Φ_{rst} will vanish if

$$\Gamma(Q_r) \otimes \Gamma(Q_s) \otimes \Gamma(Q_t) \not\supset \Gamma^{(s)} \qquad (11.40)$$

and Φ_{rstu} will vanish if

$$\Gamma(Q_r) \otimes \Gamma(Q_s) \otimes \Gamma(Q_t) \otimes \Gamma(Q_u) \not\supset \Gamma^{(s)}. \qquad (11.41)$$

In section 4.4, we stated that for the water molecule, the cubic anharmonicity term Φ_{223} vanishes. Using equation (11.40), we now see why this must be so.

The terms in equations (5.36) and (5.37) are part of the rotational Hamiltonian $\hat{H}_{rot}$ and they involve both rotational and vibrational degrees of freedom; they also give rise to rotation–vibration interaction. This interaction occurs because molecular rotation introduces two forces: The centrifugal force and the Coriolis force. The centrifugal force causes centrifugal distortion and this stretches bonds and opens out bond angles to increase the instantaneous moments of inertia as a molecule rotates. As a result, the rotational term values do not follow an expression quadratic in the angular momenta, as they do in the rigid-rotor approximation, and terms of higher power in angular momenta are introduced. The term in equation (5.36) gives rise to the centrifugal distortion correction and it involves the parameters $a_r^{\alpha\beta}$ introduced in the expansion of the $\mu_{\alpha\beta}$ elements given in equation (5.29). A Coriolis force is experienced within the xyz rotating reference frame by an nucleus when it moves within that frame. The term in equation (5.37) expresses the Coriolis force correction and it involves the Coriolis coupling constant $\zeta_{r,s}^\alpha$ introduced in equation (5.31) for the vibrational angular momentum operator $\hat{p}_\alpha$.

Symmetry can be used to determine which coupling parameters $a_r^{\alpha\beta}$ in equation (5.30), and which Coriolis coupling constants $\zeta_{r,s}^\alpha$ in equation (5.32), must vanish. By substituting equations (5.29) and (5.31) into equation (5.28), one determines that the $a_r^{\alpha\beta}$ and $\zeta_{r,s}^\alpha$ occur in the terms

$$\hat{H}_1 = -\frac{1}{2hc} \sum_{\alpha,\beta,r} \mu_{\alpha\alpha}^e a_r^{\alpha\beta} \mu_{\beta\beta}^e Q_r \hat{J}_\alpha \hat{J}_\beta \tag{11.42}$$

and

$$\hat{H}_2 = -\frac{1}{hc} \sum_{\alpha,r,s} \mu_{\alpha\alpha}^e \zeta_{r,s}^\alpha Q_r \hat{P}_s \hat{J}_\alpha \tag{11.43}$$

respectively. The terms $\hat{H}_1$ and $\hat{H}_2$ must be totally symmetric in the MS group of the molecule since they are part of the molecular Hamiltonian. Thus, $a_r^{\alpha\beta}$ must vanish if the product of the symmetries of Q_r, $\hat{J}_\alpha$ and $\hat{J}_\beta$ does not contain $\Gamma^{(s)}$. Similarly, $\zeta_{r,s}^\alpha$ must vanish if the product of the symmetries of Q_r, $\hat{P}_s$ and $\hat{J}_\alpha$ does not contain $\Gamma^{(s)}$. The normal coordinate momentum $\hat{P}_s$ has the same symmetry as the conjugate normal coordinate Q_s, and we obtain the symmetry conditions:

$a_r^{\alpha\beta}$ will vanish if

$$\Gamma(Q_r) \otimes \Gamma(\hat{J}_\alpha) \otimes \Gamma(\hat{J}_\beta) \not\supset \Gamma^{(s)} \tag{11.44}$$

and $\zeta_{r,s}^\alpha$ will vanish if

$$\Gamma(Q_r) \otimes \Gamma(Q_s) \otimes \Gamma(\hat{J}_\alpha) \not\supset \Gamma^{(s)}. \tag{11.45}$$

For a symmetric top or linear molecule, the zero-order vibrational wavefunctions are

$$|V, L\rangle = \Phi_{v_1}(Q_1)\Phi_{v_2}(Q_2)\ldots\Phi_{v_t}(Q_t)\Psi_{v_{t+1},l_{t+1}}(Q_{t+1}, \alpha_{t+1})$$
$$\times \Psi_{v_{t+2},l_{t+2}}(Q_{t+2}, \alpha_{t+2})\ldots\Psi_{v_f,l_f}(Q_f, \alpha_f) \tag{11.46}$$

where there are t non-degenerate normal coordinates Q_i of symmetry $\Gamma^{(i)}$, $(f - t)$ doubly degenerate pairs (Q_{ja}, Q_{jb}) of symmetry $\Gamma^{(j)}$, $V = (v_1, v_2, \ldots, v_t, v_{t+1}, \ldots, v_f)$ and $L = (l_{t+1}, l_{t+2}, \ldots, l_f)$. The zero-order rotation–vibration wavefunctions are

$$|V, L\rangle|J, k, m\rangle \tag{11.47}$$

where, for a symmetric top molecule, $K = |k| \leq J$ and, for a linear molecule, $K = |k| = |\sum_{j=t+1}^{f} l_j| \leq J$; the zero-order rotational wavefunctions $|J, k, m\rangle$ are defined in equation (5.53). The functions in equation (11.47) are the eigenfunctions of the zero order harmonic-oscillator rigid-rotor Hamiltonian $\hat{H}_{vib}^0 + \hat{H}_{rot}^0$ from equations (4.14) and (5.35).

In the variational approach for calculating rotation–vibration energies, we diagonalize a matrix representation of the complete rotation–vibration Hamiltonian, including V_N^{anh} and the terms in equations (5.36) and (5.37), in a basis set of functions $|V, L\rangle|J, k, m\rangle$. Proceeding as outlined in section 2.3, the rotation–vibration wavefunctions are obtained as

$$\Phi_{rv}^{(J,\Gamma_{rv},j)} = \sum_{V,L} \sum_{k=-J}^{J} c_{V,L,k}^{(J,\Gamma_{rv},j)}|V, L\rangle|J, k, m\rangle \tag{11.48}$$

where the $c_{V,L,k}^{(J,\Gamma_{rv},j)}$ are expansion coefficients. Because of the vanishing integral rule, only products $|V, L\rangle|J, k, m\rangle$ of the same symmetry, Γ_{rv} say, in the MS group can be connected by the non-vanishing matrix elements of the rotation–vibration Hamiltonian. Therefore, as discussed in chapter 7, the wavefunction $\Phi_{rv}^{(J,\Gamma_{rv},j)}$ has the useful symmetry label Γ_{rv}. Off-diagonal matrix elements are zero between basis set functions having different values of the rotational quantum number J as a result of rotational symmetry (see section 14.5), so J is also a useful label. We introduce the running number j to distinguish between eigenstates having the same values of Γ_{rv} and J. If $|V, L\rangle$ has the symmetry Γ_{vib} and $|J, k, m\rangle$ has the symmetry Γ_{rot}, the product $|V, L\rangle|J, k, m\rangle$ occurs in the summation of equation (11.48) if and only if

$$\Gamma_{rv} = \Gamma_{vib} \otimes \Gamma_{rot}. \tag{11.49}$$

There will, however, be different combinations of Γ_{vib} and Γ_{rot} that all satisfy equation (11.49).

For H_3^+, the vibrational wavefunctions can have the symmetries $\Gamma_{vib} = A_1', A_2'$ or E' in the MS group $D_{3h}(M)$. In table (11.5), we list the Γ_{rv}

Table 11.5. The possible combinations of $D_{3h}(M)$ symmetries Γ_{rv}, Γ_{vib} and Γ_{rot} [equation (11.49)] for H_3^+.

Γ_{rv}	Γ_{vib}	Γ_{rot}	Γ_{rv}	Γ_{vib}	Γ_{rot}
A_1'	A_1'	A_1'	A_2''	A_1'	A_2''
	A_2'	A_2'		A_2'	A_1''
	E'	E'		E'	E''
A_1''	A_1'	A_1''	E'	A_1'	E'
	A_2'	A_2''		A_2'	E'
	E'	E''		E'	E'
A_2'	A_1'	A_2'	E''	A_1'	E''
	A_2'	A_1'		A_2'	E''
	E'	E'		E'	E''

symmetries resulting from all possible combinations of these Γ_{vib} values with the possible Γ_{rot} symmetries. We see, for example, that a rotation–vibration wavefunction $\Phi_{rv}^{(J,A_1',j)}$ of A_1' symmetry in $D_{3h}(M)$ can be a superposition of products $|V,L\rangle|J,k,m\rangle$ with $(\Gamma_{vib}, \Gamma_{rot}) = (A_1', A_1')$, (A_2', A_2') and (E', E'), respectively.

Asymmetric top molecules like H_2O have no l quantum numbers since all normal coordinates span non-degenerate irreducible representations and so we write the zero-order vibrational wavefunction as $|V\rangle$. The zero-order rotation–vibration wavefunctions are

$$|V\rangle|J_{K_a K_c}\rangle \tag{11.50}$$

where $|J_{K_a K_c}\rangle$ is an asymmetric top rotational wavefunction. To be coupled by the rotation–vibration Hamiltonian, the products $|V\rangle|J_{K_a K_c}\rangle$ must all have the same symmetry Γ_{rv} in the MS group. We use the H_2O molecule as an example. From equation (11.29), the possible values for Γ_{vib} are A_1 and B_2 and, in table 11.6, we show the possible combinations of Γ_{vib} and Γ_{rot} symmetries that produce all possible Γ_{rv} for H_2O in $C_{2v}(M)$. A rotation–vibration wavefunction $\Phi_{rv}^{(J,A_1,j)}$ of A_1 symmetry in $C_{2v}(M)$, for example, can be a superposition of products $|V\rangle|J_{K_a K_c}\rangle$ with $(\Gamma_{vib}, \Gamma_{rot}) = (A_1, A_1)$ and (B_2, B_2), respectively.

For isolated vibrational states, i.e. vibrational states that are well separated from other vibrational states, it is possible to use perturbation theory to account for the effect of all of the off-diagonal matrix elements of the rotation–vibration Hamiltonian and, in this way, an analytical expression can be derived for the *effective rotational Hamiltonian* for that vibrational state. Symmetry can be used to determine which terms can be present in such an expansion. For vibrational states that are close in energy, such an approach is not appropriate and direct diagonalization of the Hamiltonian matrix, suitably truncated to cover close-lying

Table 11.6.   The possible combinations of $C_{2v}(M)$ symmetries Γ_{rv}, Γ_{vib} and Γ_{rot} [equation (11.49)] for H_2O.

Γ_{rv}	Γ_{vib}	Γ_{rot}	Γ_{rv}	Γ_{vib}	Γ_{rot}
A_1	A_1	A_1	A_2	A_1	A_2
	B_2	B_2		B_2	B_1
B_1	A_1	B_1	B_2	A_1	B_2
	B_2	A_2		B_2	A_1

vibrational states, is necessary.

The systematic development of the effective rotational Hamiltonian involves the perturbation treatment of all the terms in $\hat{H}'_{rot}$ and V_N^{anh}, using zero-order harmonic-oscillator rigid-rotor products. The development involves a consideration of both the order of magnitude of each term in $\hat{H}'_{rot} + V_N^{anh}$ together with a consideration of the order of the perturbation theory used. The effective rotational Hamiltonian is developed as a power series in the $\hat{J}_\alpha$ with 'effective' constants as coefficients. By adjusting the values of the effective constants so that the calculated term values are in optimal agreement with experiment, one can determine their values. Knowing the analytic expressions for them, as obtained from perturbation theory, one can, in principle, determine from them the values of the equilibrium rotational constants (A_e, B_e, C_e) and the force constants (k_{ij}, k_{ijk}, k_{ijkl}) given in equation (4.6).

For a diatomic molecule in an isolated singlet state, the effective rotational Hamiltonian terminated at terms quartic in J has the eigenvalues

$$\hat{H}_{eff} = B_{[V]}J(J+1) - D_{[V]}J^2(J+1)^2 \tag{11.51}$$

and the effective rotational Hamiltonian for an asymmetric top molecule terminated at quartic terms can be written as

$$\hat{H}_{rot}^{eff} = A_{[V]}\hat{J}_a^2 + B_{[V]}\hat{J}_b^2 + C_{[V]}\hat{J}_c^2 - \frac{1}{4}\sum_{\alpha\beta\gamma\delta}\tau_{\alpha\beta\gamma\delta}\hat{J}_\alpha\hat{J}_\beta\hat{J}_\gamma\hat{J}_\delta. \tag{11.52}$$

Symmetry can be used to determine which $\tau_{\alpha\beta\gamma\delta}$ must vanish. In these expressions, the effective rotational constants $A_{[V]}$, $B_{[V]}$ and $C_{[V]}$ differ for each vibrational state as do the effective *centrifugal distortion constants* $D_{[V]}$ and $\tau_{\alpha\beta\gamma\delta}$. For a symmetric top molecule, the quartic centrifugal distortion correction in the effective Hamiltonian contributes

$$-D_J J^2(J+1)^2 - D_{JK}J(J+1)K^2 - D_K K^4 \tag{11.53}$$

to the rotational eigenvalues. Higher-order perturbation theory and appropriate consideration of all the terms in $\hat{H}'_{rot}$ and V_N^{anh}, leads to centrifugal distortion

Table 11.7. The C_{3v} symmetry species of the rotation–vibration wavefunctions for CH_3F in the ν_4 fundamental state; n is a non-negative integer.

K	$\Gamma(\Phi_{rv})$	
0	E	
	$(+l)$	$(-l)$
$3n+1$	$A_1 \oplus A_2$	E
$3n+2$	E	$A_1 \oplus A_2$
$3n+3$	E	E

terms with higher even powers (sextic, octic, etc.) of the components of the angular momentum operator and such terms are often required to reproduce very precise data.

There is one circumstance in which there is a significant first-order *diagonal* correction to the energies from $\hat{H}'_{rot}$ and this involves the effect of the term $\hat{H}_2$, equation (11.43), for a degenerate vibrational state. This is called *first-order Coriolis coupling*. We use the degenerate E species ν_4 fundamental of CH_3F as an example. For the ν_4 state of CH_3F, a rotation–vibration wavefunction in the harmonic-oscillator rigid-rotor approximation is $|\nu_4, l_4\rangle |J, k, m\rangle$, where $\nu_4 = 1$ and $l_4 = \pm 1$. Each such zero-order rotation–vibration state having non-zero K is four-fold degenerate as $(k, l_4) = (+K, 1), (-K, 1), (+K, -1)$ or $(-K, -1)$. For CH_3F, the product $\hat{H}_{QP4J} = (Q_{4a}\hat{P}_{4b} - Q_{4b}\hat{P}_{4a})J_z$ is of species A_1 in C_{3v} and so $\zeta^z_{4a,4b} = -\zeta^z_{4b,4a}(= \zeta^z_4$, say) is non-vanishing. To evaluate the diagonal matrix element of the $\hat{H}_2$ term $-[\mu^e_{zz}/(hc)]\zeta^z_4\hat{H}_{QP4J}$, we use equations (4.37) and (4.39); including the effects of vibrational averaging on $A_e = \hbar^2\mu^e_{zz}/(2hc)$, the diagonal matrix element of $\hat{H}_2$ in the ν_4 state is

$$\langle\hat{H}_2\rangle = -2A_{[4]}\zeta^z_4 l_4 k. \tag{11.54}$$

Thus, each initially four-fold degenerate level is split into two; these two components are labelled as the $(+l)$ and $(-l)$ levels, respectively, and each is doubly degenerate. The $(+l)$ level is that having k and l_4 with the same sign, and the $(-l)$ level is that having k and l_4 with opposite sign. Since ζ^z_4 is positive, the $(+l)$ level is below the $(-l)$ level for each state having $K > 0$ in the ν_4 fundamental state of CH_3F. The rotation–vibration symmetry species are given in table 11.7. Vibrationally off-diagonal matrix elements of $\hat{H}_2$ split the degeneracy of the $A_1 \oplus A_2$ rotation–vibration levels within the ν_4 state, i.e. of the $(+l)$ levels having $K = 3n+1$ and the $(-l)$ levels having $K = 3n+2$. This is called *l-type doubling*.

11.6   Problems

11.1*   Prove the result quoted in equation (11.9).

11.2*   Prove the asymmetric top symmetry rule that is quoted on page 229.

11.3   The formaldehyde molecule CH_2O has C_{2v} point group symmetry. The a axis is the C_2 axis and the c axis is out of the plane. Write down the MS group of the molecule and determine the equivalent rotations of each of its elements. Determine the symmetry of the asymmetric top energy levels as a function of the evenness and oddness of K_a and K_c and contrast the results with those given for the water molecule in table 11.3.

11.4   Determine the equivalent rotations of one element from each class of $C_{3v}(M)$ for the CH_3F molecule. Confirm the symmetries of the rotational wavefunctions of CH_3F as given in table 11.2.

11.5   Use equations (4.11) and (4.12) for the translational and rotational normal coordinates to determine their symmetries for CH_3F. Determine the symmetries of the vibrational normal coordinates for CH_3F.

11.6   Determine the symmetries of the vibrational normal coordinates for the formaldehyde molecule.

11.7   Determine the non-vanishing coefficients Φ_{rst}, Φ_{rstu}, $a_r^{\alpha\beta}$ and $\zeta_{r,s}^{\alpha}$ for the molecules H_2O, CH_2O and CH_3F.

11.8   Which rotational levels of the fundamental vibrational states of the CH_3F molecule can perturb each other and what are the selection rules on K for the perturbations? In each case, determine which terms in the Hamiltonian cause the perturbations.

11.9   Determine the ΔK_a and ΔK_c selection rules for perturbations to be possible between rotation–vibration levels in the first excited states of the v_1 and v_3 fundamentals of the water molecule. Compare with the results obtained for the perturbations between the overtone $2v_2$ and each of the fundamentals v_1 and v_3. Which terms in the rotation–vibration Hamiltonian can cause each of these perturbations.

Chapter 12

Symmetry selection rules for optical transitions

12.1 Forbidden and allowed transitions

By applying the vanishing integral rule to the rovibronic transition moment integral

$$I_{\mathrm{TM}} = \int \Phi'^*_{\mathrm{rve}} \mu_A \Phi''_{\mathrm{rve}} \, d\tau \tag{12.1}$$

we determine in equation (7.86) that an electric dipole transition between the two rovibronic states Φ''_{rve} and Φ'_{rve} is rovibronically forbidden if

$$\Gamma(\Phi'^*_{\mathrm{rve}}) \otimes \Gamma(\mu_A) \otimes \Gamma(\Phi''_{\mathrm{rve}}) \not\supset \Gamma^{(s)} \tag{12.2}$$

or, equivalently, if

$$\Gamma(\Phi'^*_{\mathrm{rve}}) \otimes \Gamma(\Phi''_{\mathrm{rve}}) \not\supset \Gamma(\mu_A) \tag{12.3}$$

but now in implementing this equation, $\Gamma(\Phi'^*_{\mathrm{rve}})$, $\Gamma(\Phi''_{\mathrm{rve}})$ and $\Gamma(\mu_A)$ are MS group symmetries, rather than CNPI group symmetries, of Φ'^*_{rve}, Φ''_{rve} and μ_A, $\Gamma^{(s)}$ is the totally symmetric irreducible representation of the MS group and μ_A is the component of the molecular electric dipole moment along the space-fixed A axis [see equation (2.88)] where, in the notation of chapter 5, we now have $A = \xi, \eta$ or ζ. $\Gamma(\mu_A)$ is the one-dimensional irreducible representation of the MS group that has character $+1$ for all nuclear permutations and character -1 for all permutation–inversions.

Rovibronically forbidden transitions can gain intensity as a result of the *ortho–para* mixing effect of the nuclear hyperfine Hamiltonian. An example of such a transition is transition (b) in figure 9.3 on page 191, which is a rovibronic $A_1 \leftarrow B_1$ transition in the water molecule. For the water molecule, $\Gamma(\mu_A) = A_2$ and, since $A_1 \otimes A_2 \otimes B_1 \not\supset A_1$, this transition satisfies equation (12.2). Transitions that satisfy equation (12.2) are normally said to be *nuclear spin forbidden*, rather than rovibronically forbidden, and they are usually extremely weak.

If we set up the transition moment integral using the complete (exact) internal wavefunctions Φ_{int} [the eigenfunctions of $\hat{H}_{int}$ in equation (2.77)], then we deduce that the transition between the states Φ'_{int} and Φ''_{int} is *strictly forbidden* as an electric dipole transition if

$$\Gamma(\Phi'^*_{int}) \otimes \Gamma(\Phi''_{int}) \not\supset \Gamma(\mu_A). \tag{12.4}$$

The symmetry $\Gamma(\Phi_{int})$ is called Γ_{tot} in chapter 9, and transition (b) in figure 9.3 is between Φ_{int} states having symmetries B_1 and B_2 so that it is not forbidden by equation (12.4). A transition satisfying equation (12.4) can only gain intensity as a magnetic dipole transition or as a result of the parity violating effect discussed in section 15.2 and such strictly forbidden transitions are the weakest of all transitions.

The concept of a forbidden transition is very important.   Using equation (12.2) to determine which transitions are forbidden, rather than equation (12.4), leads to more transitions being called forbidden. For example, the *ortho–para* transition (b) in figure 9.3 is forbidden by equation (12.2) but not by equation (12.4). In general, the more approximate the formulation is, the larger are the number of forbidden transitions.   Thus, if we introduce further approximations (such as the Born–Oppenheimer, rigid-rotor and harmonic-oscillator approximations) more transitions become forbidden. The important point is that, if the approximations are appropriate, then the forbidden transitions will be weak and we can understand the main features of a spectrum without considering them.

As stated in section 7.7, and discussed in more detail in section 14.5, the classification of rovibronic states in the rotational symmetry group K(spatial) gives them the rovibronic angular momentum quantum number label J. Using the vanishing integral theorem with the group K(spatial) leads to the conclusion that transitions are nuclear spin forbidden if they are between states whose J values are both 0, or for which the change in J is larger than 1. Using K(spatial) for the complete internal wavefunctions gives them the total angular momentum quantum number label F and strictly forbidden transitions are between states whose F values are both 0, or for which the change in F is larger than 1.

Thus, at a given level of approximation, transitions of a certain type are forbidden; forbidden transitions are specified by quoting *selection rules*. In this chapter, we will determine selection rules and show how they change with the level of approximation. However, it is more appropriate to quote the selection rules in a way that specifies the transitions that are *not* forbidden; these selection rules will then be the rules that tell us which transitions are *allowed*, i.e. which are the strongest transitions in the spectrum. In equation (7.60), we have already specified the selection rules on the rovibronic angular momentum quantum number J in this way for allowed rovibronic transitions as

$$\Delta J = 0, \pm 1 \qquad \text{but} \qquad J = 0 \leftrightarrow 0 \qquad \text{is forbidden.} \tag{12.5}$$

Allowed transitions between internal states Φ_{int} satisfy

$$\Delta F = 0, \pm 1 \qquad \text{but} \qquad F = 0 \leftrightarrow 0 \qquad \text{is forbidden.} \tag{12.6}$$

From equation (12.3), the transition between the rovibronic states Φ''_{rve} and Φ'_{rve} is allowed if

$$\Gamma(\Phi'^{*}_{\text{rve}}) \otimes \Gamma(\Phi''_{\text{rve}}) \supset \Gamma(\mu_A). \tag{12.7}$$

For MS groups having only real characters (all the character tables in Appendix B have only real characters), the symmetry condition for an allowed rovibronic transition is

$$\Gamma(\Phi'_{\text{rve}}) \otimes \Gamma(\Phi''_{\text{rve}}) \supset \Gamma(\mu_A) \tag{12.8}$$

and a rovibronic transition is allowed if the product of the symmetry species of the two rovibronic states involved contains the symmetry species of μ_A.

12.2   Zero-order transition moment integrals

In the Born–Oppenheimer approximation, see equation (3.8), a rovibronic eigenfunction is the product of an electronic wavefunction $\Phi_{\text{elec},n}$ and a nuclear wavefunction $\Phi_{\text{rv},nj}$. Making the harmonic-oscillator and rigid-rotor approximations in the rotation–vibration Hamiltonian leads to the zero-order product wavefunction given in equation (5.27); simplifying the notation, the zero-order Φ''_{rve} and Φ'_{rve} are

$$\Phi''_{\text{rve}} = \Phi_{\text{elec},n''} \Phi_{\text{vib},n''v''} \Phi_{\text{rot},n''r''} \tag{12.9}$$

and

$$\Phi'_{\text{rve}} = \Phi_{\text{elec},n'} \Phi_{\text{vib},n'v'} \Phi_{\text{rot},n'r'}. \tag{12.10}$$

The electronic state label n' or n'' on the vibrational wavefunctions occurs because the harmonic force constants and, hence, the normal coordinates, change with electronic state. The electronic state label on the rotational wavefunctions occurs because the definition of the Euler angles depends in a subtle way on the geometry of the equilibrium structure and this is electronic state dependent. However, this latter dependence only has the effect of making so-called *axis-switching transitions* weakly allowed[1] and we will neglect them as part of our

[1]   Hougen J T and Watson J K G 1965 *Can. J. Phys.* **43** 298.

zero-order approximation; thus, we omit the electronic state label on the rotational wavefunctions in the following.

In terms of molecule-fixed components in the xyz axis system, a $\xi\eta\zeta$ component of the dipole moment operator can be written as

$$\mu_A = \lambda_{xA}\mu_x + \lambda_{yA}\mu_y + \lambda_{zA}\mu_z = \sum_{\alpha=x,y,z} \lambda_{\alpha A}\mu_\alpha, \quad A = \xi, \eta, \zeta \qquad (12.11)$$

where the direction cosine matrix elements $\lambda_{\alpha A}$ are given in equation (5.42) and they depend only on the Euler angles. The molecule-fixed components μ_α depend only on the vibronic coordinates, i.e. on the xyz electronic coordinates and on the vibrational displacement coordinates. We will focus on the component $A = \zeta$ in determining selection rules. In field-free space, the same results are obtained regardless of which space-fixed direction is chosen here.

Substituting equations (12.9)–(12.11) into equation (12.1) gives the following expression for the ζ component of the transition moment integral in the zero-order Born–Oppenheimer, harmonic-oscillator and rigid-rotor approximation:

$$I_{\text{TM}}^0 = \sum_{\alpha=x,y,z} \langle\Phi_{\text{rot},r'}|\langle\Phi_{\text{vib},n'v'}|\langle\Phi_{\text{elec},n'}|$$
$$\times \lambda_{\alpha\zeta}\mu_\alpha|\Phi_{\text{elec},n''}\rangle|\Phi_{\text{vib},n''v''}\rangle|\Phi_{\text{rot},r''}\rangle. \qquad (12.12)$$

Making use of the fact that only the rotational wavefunctions and the $\lambda_{\alpha\zeta}$ depend on the Euler angles and that they do not depend on the vibronic variables, we can write each of the x, y and z terms in the sum as the product of a vibronic matrix element integral of μ_α that involves integrating over vibronic variables only and a rotational matrix element integral of $\lambda_{\alpha\zeta}$ that involves integrating over Euler angles only:

$$I_{\text{TM}}^0 = \sum_{\alpha=x,y,z} \langle\Phi_{\text{vib},n'v'}|\langle\Phi_{\text{elec},n'}|\mu_\alpha|\Phi_{\text{elec},n''}\rangle|\Phi_{\text{vib},n''v''}\rangle$$
$$\times \langle\Phi_{\text{rot},r'}|\lambda_{\alpha\zeta}|\Phi_{\text{rot},r''}\rangle. \qquad (12.13)$$

The vibronic matrix element of μ_α is evaluated in a two-step procedure in which first we integrate over the electronic coordinates to obtain

$$\bar\mu_\alpha^{(n',n'')} = \mu_\alpha^{(n',n'')}(Q_1'', Q_2'', Q_3'', \ldots) = \langle\Phi_{\text{elec},n'}|\mu_\alpha|\Phi_{\text{elec},n''}\rangle_{\text{el}} \qquad (12.14)$$

where $\alpha = x, y$ or z and $\bar\mu_\alpha^{(n',n'')}$ depends on the vibrational displacement coordinates; we take these to be the normal coordinates of the electronic state n'' here. We thus obtain the zero-order transition moment integral as

$$I_{\text{TM}}^0 = \sum_{\alpha=x,y,z} \langle\Phi_{\text{vib},n'v'}|\bar\mu_\alpha^{(n',n'')}|\Phi_{\text{vib},n''v''}\rangle\langle\Phi_{\text{rot},r'}|\lambda_{\alpha\zeta}|\Phi_{\text{rot},r''}\rangle. \qquad (12.15)$$

Each of the $\alpha = x, y$ or z terms in the sum is the product of a vibrational matrix element of $\bar\mu_\alpha^{(n',n'')}$ and a rotational matrix element of $\lambda_{\alpha\zeta}$. We use this zero order expression for I_{TM} as the basis for determining selection rules.

12.3 Transitions within an electronic state

From equation (12.7), the transition between the rovibrational states $\Phi_{\text{rv},j'}$ and $\Phi_{\text{rv},j''}$ within an electronic state is allowed if

$$\Gamma(\Phi^*_{\text{rv},j'}) \otimes \Gamma(\Phi_{\text{rv},j''}) \supset \Gamma(\mu_A) \qquad (12.16)$$

where $A = \xi, \eta$ or ζ.

This rovibrational symmetry rule applies within the Born–Oppenheimer approximation. In the harmonic-oscillator and rigid-rotor approximations, the selection rules for allowed transitions are more restricted, and these approximations introduce vibrational transition moments and Hönl–London factors.

12.3.1 Vibrational transition moments and Hönl–London factors

For transitions within one electronic state, the labels (n') and (n'') on the vibrational wavefunctions are omitted since all are now associated with the one electronic state, n say, that is considered. From equation (12.15),

$$I^0_{\text{TM}} = \sum_{\alpha=x,y,z} \langle \Phi_{\text{vib},v'} | \bar{\mu}_\alpha | \Phi_{\text{vib},v''} \rangle \langle \Phi_{\text{rot},r'} | \lambda_{\alpha\zeta} | \Phi_{\text{rot},r''} \rangle \qquad (12.17)$$

where the expectation value of the dipole moment within the electronic state n as a function of the normal coordinates is given by

$$\bar{\mu}_\alpha = \mu_\alpha^{(n,n)}(Q_1, Q_2, Q_3, \ldots) = \langle \Phi_{\text{elec},n} | \mu_\alpha | \Phi_{\text{elec},n} \rangle_{\text{el}}, \qquad \alpha = x, y, z. \qquad (12.18)$$

Expressing the $\bar{\mu}_\alpha$ in terms of the vibrational displacement coordinates and comparing with equation (4.11) for the translational coordinate T_α, we find that the molecule-fixed component $\bar{\mu}_\alpha$ ($\alpha = x, y, z$) has the same symmetry in the MS group (or, equivalently, for a rigid nonlinear molecule, in the molecular point group) as the translational coordinate T_α. We can use the vanishing integral theorem for the *vibrational transition moment* integral

$$I^{(v',v'')}_{\text{vib},\alpha} = \langle \Phi_{\text{vib},v'} | \bar{\mu}_\alpha | \Phi_{\text{vib},v''} \rangle \qquad (12.19)$$

to deduce the following.

The vibrational transition moment between the vibrational states $\Phi_{\text{vib},v'}$ and $\Phi_{\text{vib},v''}$ within an electronic state can be non-zero only if

$$\Gamma(\Phi^*_{\text{vib},v'}) \otimes \Gamma(\Phi_{\text{vib},v''}) \supset \Gamma(T_\alpha) \qquad (12.20)$$

where $\Gamma(T_\alpha)$ is the species of the $\alpha(= x, y$ or $z)$ component of the translational normal coordinate.

As with the potential energy function V_N, $\bar{\mu}_\alpha$ is expanded as a power series in the normal coordinates:

$$\bar{\mu}_\alpha = \mu^e_\alpha + \sum_r \mu_{\alpha,r} Q_r + \sum_{r,s} \mu_{\alpha,rs} Q_r Q_s + \cdots \qquad (12.21)$$

which introduces the parameters μ^e_α, $\mu_{\alpha,r}$, etc. Truncating this at the linear term (this is called the *electrical harmonicity* approximation) gives

$$\bar{\mu}^0_\alpha = \mu^e_\alpha + \sum_r \mu_{\alpha,r} Q_r. \qquad (12.22)$$

The μ^e_α are the values of $\bar{\mu}_\alpha$ when the molecule is at equilibrium in the electronic state n; these are the values of the components of the equilibrium dipole moment for that electronic state. The μ^e_α are constants and must transform as the totally symmetric representation of the MS group or point group. Thus, the equilibrium dipole moment component μ^e_α can be non-vanishing only if

$$\Gamma(T_\alpha) = \Gamma^{(s)}. \qquad (12.23)$$

For example, for the water molecule, we see from table B.4 that only T_b is of symmetry A_1 and so, for the water molecule, only μ^e_b can be non-vanishing. The second term in equation (12.22), $\mu_{\alpha,r} Q_r$, must also be of the same symmetry as T_α for it to be non-vanishing, from which we deduce that the constant $\mu_{\alpha,r}$ can be non-vanishing only if

$$\Gamma(Q_r) = \Gamma(T_\alpha). \qquad (12.24)$$

The constant $\mu_{\alpha,r}$ is the value of the α component of the dipole moment gradient for the normal coordinate Q_r; thus, it is large if $\bar{\mu}_\alpha$ varies strongly with Q_r.

Substituting equation (12.22) into equation (12.19) gives the zero-order expression for the vibrational transition moment as

$$I^{(v',v'')}_{\text{vib},\alpha} = \langle \Phi^*_{\text{vib},v'} | \mu^e_\alpha + \sum_r \mu_{\alpha,r} Q_r | \Phi_{\text{vib},v''} \rangle$$

$$= \mu^e_\alpha \delta_{v'v''} + \sum_r \mu_{\alpha,r} \langle \Phi_{\text{vib},v'} | Q_r | \Phi_{\text{vib},v''} \rangle \qquad (12.25)$$

where the Kronecker delta $\delta_{v'v''}$ arises because μ_α^e is a constant and the vibrational wavefunctions are orthonormal.

The first term in equation (12.25) gives the transition moment for a purely rotational transition within one vibrational state; with the approximations made here, this is the equilibrium dipole moment component μ_α^e. Allowing for both electrical anharmonicity [i.e. the higher-order terms in equation (12.21)] and mechanical anharmonicity (i.e. cubic and quartic anharmonicity terms in the potential), this rotational transition moment becomes the vibrational expectation value

$$I_{\text{rot},\alpha}^{(v)} = \langle \Phi_{\text{vib},v} | \bar{\mu}_\alpha | \Phi_{\text{vib},v} \rangle \tag{12.26}$$

from equation (12.19). This is the dipole moment for a given vibrational state; it depends on the vibrational state and varies with isotopic substitution. From equation (12.23):

The dipole moment component $\langle \Phi_{\text{vib},v} | \bar{\mu}_\alpha | \Phi_{\text{vib},v} \rangle$ can be non-vanishing and a pure rotation spectrum allowed, if the translational coordinate T_α is totally symmetric.

The intensity of a pure rotation transition depends on the value of the dipole moment, and it can be used to determine this value.

The second term in equation (12.25) gives the transition moment for a vibrational transition and, thus, (within the electrical harmonicity approximation) only vibrational transitions, for which the matrix element of a normal coordinate Q_r is non-vanishing, are allowed. The expressions for the non-vanishing matrix elements of Q_r are given in table 4.1, from which we see that in the harmonic-oscillator approximation (and electrical harmonicity approximation) allowed vibrational transitions have the selection rule $\Delta v_r = \pm 1$, where the normal coordinate Q_r must have the symmetry of a translation T_α [from equation (12.24)]. Thus, fundamental transitions are allowed for vibrational normal coordinates that have the symmetry of a translation; such a normal coordinate is said to be *infrared active* since fundamental bands[2] occur in the infrared region.

A normal mode of a molecule is infrared active if its symmetry is equal to that of one or more of the translational coordinates T_α, where $\alpha = x, y$ or z.

---

[2] A *band* consists of all the spectral lines in a single vibrational (or vibronic) transition.

The intensity of a fundamental band depends on the value of the dipole moment gradient $\mu_{\alpha,r}$ and can be used to determine it.

In this approximation, *overtone transitions* (i.e. transitions from the vibrational ground state to vibrational states with one $v_r \geq 2$ and all other $v_{r'} = 0$) are forbidden, as are *combination tones* (i.e. transitions from the vibrational ground state to vibrational states having more than one $v_r \neq 0$). These forbidden transitions can gain intensity from both electrical anharmonicity and mechanical anharmonicity. However, they will still have to satisfy the vibrational symmetry selection rule of equation (12.20).

For the water molecule, the normal modes have symmetry $2A_1 \oplus B_2$, T_b has symmetry A_1 and T_a has symmetry B_2 (see table B.4). Thus, all three normal modes are infrared active. Similarly all six normal modes of CH_3F, having symmetry $3A_1 \oplus 3E$, are infrared active (see table B.5). For the benzene molecule (see table B.9), the translations span $A_{2u} \oplus E_{1u}$. Consequently, among the 20 normal vibrational modes of benzene [see equation (11.26)], only the one normal mode of A_{2u} symmetry and the three modes of E_{1u} symmetry are infrared active.

To obtain the expression for the line strength, and the rotational selection rules on K, of a rovibrational transition for a symmetric top or linear molecule, we rewrite the sum in equation (12.11) for $A = \zeta$ as

$$\mu_\zeta = \sum_{p=-1,0,+1} (-1)^p T_p^1(\lambda_\zeta) T_{-p}^1(\bar{\mu}) \tag{12.27}$$

where, in terms of the Cartesian components,

$$T_{+1}^1(\bar{\mu}) = \frac{1}{\sqrt{2}}(-\bar{\mu}_x - i\bar{\mu}_y) \tag{12.28}$$

$$T_{-1}^1(\bar{\mu}) = \frac{1}{\sqrt{2}}(+\bar{\mu}_x - i\bar{\mu}_y) \tag{12.29}$$

and

$$T_0^1(\bar{\mu}) = \bar{\mu}_z \tag{12.30}$$

with similar expressions for the components $T_p^1(\lambda_\zeta)$ in terms of $\lambda_{x\zeta}$, $\lambda_{y\zeta}$ and $\lambda_{z\zeta}$. These three components of a vector generate the transformation matrices of the irreducible representation $D^{(1)}$ of the group K(spatial) under the effect of overall rotation operations and they are called the *irreducible spherical tensor components*. Combining this with the fact that the wavefunctions in the integral $\langle J', k', m' | T_p^1(\lambda_\zeta) | J'', k'', m'' \rangle$ are expressed as the irreducible representation matrices $D^{(J')}$ and $D^{(J'')}$ of the group K(spatial) leads to its evaluation as an analytic expression in the angular momentum quantum numbers. The square of this expression is called a *Hönl–London factor* $A(J'', k'', J', k')$. In this way, the expression for the line strength $S(E'_{rv} \leftarrow E''_{rv})$ [see equation (2.87)] for a symmetric top or linear molecule is obtained as

$$S(E'_{rv} \leftarrow E''_{rv}) = g(2J'' + 1)|\langle \Phi'_{vib} | T_{\Delta k}^1(\bar{\mu}) | \Phi''_{vib} \rangle|^2 A(J'', k'', J', k') \tag{12.31}$$

with

$$\Delta k = k' - k'' \tag{12.32}$$

and

$$g = \begin{cases} 2g_{ns} & \text{for } (K'', K') = (0, 1) \text{ or } (1, 0) \\ g_{ns} & \text{otherwise} \end{cases} \tag{12.33}$$

where g_{ns} is the spin statistical weight factor for the transition (chapter 9), $K = |k|$, and the Hönl–London factor $A(J'', k'', J', k') = A(J'', k'', J'' + \Delta J, k'' + \Delta k)$, where $\Delta J = J' - J''$, is given in table 12.1; $A(J'', k'', J'' + \Delta J, k'' + \Delta k)$ vanishes for $|\Delta J| > 1$ and for $|\Delta k| > 1$. The selection rules on k are

$$\Delta k = 0 \tag{12.34}$$

if the vibrational transition moment $\langle \Phi'_{vib} | T_0^1(\bar{\mu}) | \Phi''_{vib} \rangle = \langle \Phi'_{vib} | \bar{\mu}_z | \Phi''_{vib} \rangle$ is non-vanishing, and

$$\Delta k = \pm 1 \tag{12.35}$$

if the vibrational transition moment $\langle \Phi'_{vib} | T_{\pm 1}^1(\bar{\mu}) | \Phi''_{vib} \rangle$ is non-vanishing (so that $\langle \Phi'_{vib} | \bar{\mu}_x | \Phi''_{vib} \rangle$ and/or $\langle \Phi'_{vib} | \bar{\mu}_y | \Phi''_{vib} \rangle$ are non-vanishing). Equation (12.31) gives the line strength for the transition from the energy level E''_{rv} to the level E'_{rv}. The expression accounts for the m-degeneracy (see section 2.7) and for the nuclear spin degeneracy (chapter 9).

There is no agreement as to whether one should label energy levels, and express selection rules, using signed or unsigned angular momentum quantum numbers. That is, should one use the *signed* quantum number k and the signed quantum numbers l_r in the list $L = (l_{t+1}, l_{t+2}, \ldots, l_f)$, as we do here, or should one use the *unsigned* quantum numbers K and $|\sum_r l_r|$? Theoreticians, like us, are more likely to use signed angular momentum quantum numbers, whereas experimentalists (see, for example, the books by Herzberg) are more likely to use unsigned angular momentum quantum numbers. The theoretical development of the selection rules in terms of signed angular momentum quantum numbers is straightforward. However, it means that we label a zero-order energy level using the quantum numbers (J, k, L) [see equation (11.47)], so that (J, K, L) and $(J, -K, -L)$ [where $-L = (-l_{t+1}, -l_{t+2}, \ldots, -l_f)$] describe the same energy level. This comes about because of time reversal symmetry (see section 14.8); the molecular energy is invariant to the reversal of all momenta (linear and angular). Thus, in the harmonic-oscillator rigid-rotor approximation, a transition $(J, k, L) = (J', K', L') \leftarrow (J'', K'', L'')$ can alternatively be assigned as $(J, k, L) = (J', -K', -L') \leftarrow (J'', -K'', -L'')$. In what follows, we first derive the selection rules using rotation–vibration wavefunctions and signed angular momentum quantum numbers and then also express the results in terms of unsigned quantum numbers.

The rotational selection rules of equations (12.34) and (12.35) give rise to parallel and perpendicular bands, respectively, as we will see in section 12.3.3. Mechanical and electrical anharmonicity do not change these rotational selection

Table 12.1. Hönl–London factors $A(J, k, J + \Delta J, k + \Delta k)$.

	$\Delta k = 0$	$\Delta k = +1^{\mathrm{a}}$
$\Delta J = -1$	$\frac{J^2 - k^2}{J(2J+1)}$	$\frac{(J-1-k)(J-k)}{2J(2J+1)}$
$\Delta J = 0$	$\frac{k^2}{J(J+1)}$	$\frac{(J+1+k)(J-k)}{2J(J+1)}$
$\Delta J = +1$	$\frac{(J+1)^2 - k^2}{(J+1)(2J+1)}$	$\frac{(J+2+k)(J+1+k)}{2(J+1)(2J+1)}$

^{a} Hönl–London factors with $\Delta k = -1$ are obtained using the relation $A(J, k, J + \Delta J, k - 1) = A(J, -k, J + \Delta J, -k + 1)$.

rules, so that equation (12.31) is valid for symmetric tops and linear molecules in the approximation of neglecting both rotation–vibration interaction and the breakdown of the Born–Oppenheimer approximation.

To obtain selection rules on K_a and K_c for an asymmetric top molecule, we write equation (12.17) as

$$I_{\mathrm{TM}}^0 = \sum_{\alpha=a,b,c} I_{\mathrm{vib},\alpha}^{(v',v'')} \langle \Phi_{\mathrm{rot},r'} | \lambda_{\alpha\zeta} | \Phi_{\mathrm{rot},r''} \rangle \qquad (12.36)$$

where $I_{\mathrm{vib},\alpha}^{(v',v'')}$ is given in equation (12.25), and apply the vanishing integral rule to the integral involving $\lambda_{\alpha\zeta}$ that occurs here. We know that $\bar{\mu}_\alpha^0$ [see equation (12.22)] in the integral $I_{\mathrm{vib},\alpha}^{(v',v'')}$ transforms as T_α in the MS group and it can be shown that $\lambda_{\alpha\zeta}$ transforms as $\hat{J}_\alpha$ (or R_α). We can use the asymmetric top symmetry rule on page 229 to determine the MS group symmetry of the $\Phi_{\mathrm{rot},r}$. Using the MS group symmetries of the $\Phi_{\mathrm{rot},r}$ and $\lambda_{\alpha\zeta}$, the application of the vanishing integral rule to $\langle \Phi_{\mathrm{rot},r'} | \lambda_{\alpha\zeta} | \Phi_{\mathrm{rot},r''} \rangle$ leads to the selection rules

$$
\begin{aligned}
\Delta K_a &= \text{even} & \Delta K_c &= \text{odd} & \text{if } \alpha &= a \\
\Delta K_a &= \text{odd} & \Delta K_c &= \text{odd} & \text{if } \alpha &= b \\
\Delta K_a &= \text{odd} & \Delta K_c &= \text{even} & \text{if } \alpha &= c.
\end{aligned}
\qquad (12.37)
$$

These three different rotational selection rules give rise to a-type, b-type and c-type bands, respectively, each of which has a characteristic appearance. For the water molecule, the pure rotation spectrum will be a b-type band since only T_b is of symmetry A_1. If the molecule is a near prolate rotor, then $\Delta K_a = \text{even(odd)}$ can be replaced by $\Delta K_a = 0(\pm 1)$ for the strong transitions; if the molecule is a near oblate rotor, then $\Delta K_c = \text{even(odd)}$ can be replaced by $\Delta K_c = 0(\pm 1)$ for the strong transitions.

12.3.2 The rotational spectrum of the CO molecule

A simple application of the above is to the rotational absorption spectrum of the CO molecule, for which a part is shown in figure 1.2 on page 4. From equation (11.51), with neglect of centrifugal distortion, the rotational term values for CO in its $v = 0$ vibrational ground state are

$$F_{\text{rot}} = B_{[0]}J(J+1). \tag{12.38}$$

In an absorption transition, the energy of the final state is necessarily higher than that of the initial state, and so, of the three possibilities for ΔJ, only $\Delta J = 1$ is possible. Thus, rotational transitions take place at the wavenumbers

$$\tilde{\nu}_J = B_{[0]}(J+1)(J+2) - B_{[0]}J(J+1) = 2B_{[0]}(J+1). \tag{12.39}$$

From equation (12.31), with $k' = k'' = 0$, the line strength is

$$S(J+1 \leftarrow J) = g_{\text{ns}}(2J+1)|I_{\text{rot},z}^{(0)}|^2 A(J, 0, J+1, 0) \tag{12.40}$$

where $g_{\text{ns}} = 1$ for $^{12}\text{C}^{16}\text{O}$, $I_{\text{rot},z}^{(0)}$ is the dipole moment of CO in the vibrational ground state [see equation (12.26)] and $A(J, 0, J+1, 0)$ is given in table 12.1. This gives

$$S(J+1 \leftarrow J) = |I_{\text{rot},z}^{(0)}|^2(J+1). \tag{12.41}$$

With $B_{[0]} = 1.922\,53$ cm^{-1} and[3] $|I_{\text{rot},z}^{(0)}| = 0.122$ D, we substitute equations (12.38), (12.39) and (12.41) into equation (1.9) in order to compute the intensity of individual lines in the rotational spectrum of CO. Figure 12.1 shows the result as a stick diagram for the region of the CO rotational spectrum plotted in figure 1.2. To calculate the transmittance τ [see equation (1.3)], in order to compare quantitatively with figure 1.2, we would need the path length l, concentration c and the instrument lineshape function [in order to generate the function $\epsilon(\tilde{\nu})$].

12.3.3 Parallel and perpendicular bands of CH$_3$F

The normal coordinates of the CH$_3$F molecule span the representation

$$\Gamma_Q = 3A_1 \oplus 3E \tag{12.42}$$

of $C_{3v}(\text{M})$, the translational coordinate T_z is of species A_1 and the pair of translational coordinates T_x and T_y are of species E (see table B.5). We simulate the $\nu_1[\Gamma(Q_1) = A_1]$ and $\nu_4[\Gamma(Q_{4a}, Q_{4b}) = E]$ fundamental absorption bands to show how the different rotational selection rules in equations (12.34) and (12.35), $\Delta k = 0$ for the ν_1 band and $\Delta k = \pm 1$ for the ν_4 band, affect the appearance of the bands.

---

[3] 1 D = 1 debye $\approx 3.335\,64 \times 10^{-30}$ C m.

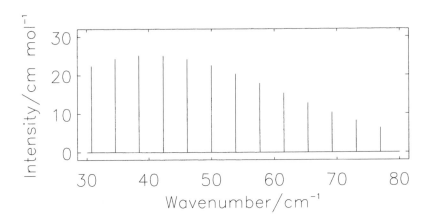

Figure 12.1. Part of the absorption spectrum of carbon monoxide simulated at 300 K. This simulation should be compared to the experimental spectrum in figure 1.2.

Neglecting centrifugal distortion the term values in the ground state ($V = 0$) and in the v_1 state ($V = 1$) are given by

$$F = G_{[V]} + B_{[V]}J(J + 1) + (A_{[V]} - B_{[V]})K^2 \qquad (12.43)$$

where $G_{[V]}$ is the vibrational term value, and $A_{[V]}$ and $B_{[V]}$ are the vibrationally averaged rotational constants. For the v_4 state, we must take into account the first-order Coriolis coupling correction given in equation (11.54); the term values, neglecting centrifugal distortion, are

$$F = G_{[4]} + B_{[4]}J(J + 1) + (A_{[4]} - B_{[4]})K^2 - 2A_{[4]}\zeta_4^z l_4 k$$
$$= G_{[4]} + B_{[4]}J(J + 1) + (A_{[4]} - B_{[4]})K^2 \pm 2A_{[4]}\zeta_4^z K. \qquad (12.44)$$

For each value of $K > 0$ the term value splits into two and the components are labelled the $(+l)$ and $(-l)$ states as explained after equation (11.54). In table 12.2, we give the experimentally determined values of the parameters, where, in the notation $v_4^{\pm 1}$, the superscript is the vibrational angular momentum quantum number l_4.

All vibrational transitions have the same selection rules on J and spectral lines satisfying $\Delta J = -1, 0$ and $+1$ are called P, Q and R lines, respectively. For the v_1 band, it is straightforward to use equation (12.43) for the ground state and v_1 state, with the parameters from table 12.2, to determine the positions of the lines using the selection rules $\Delta k = 0$ and $\Delta J = 0, \pm 1$. In figure 12.2, we show a simulation of the v_1 absorption band of CH_3F. This is called a parallel band since its intensity is determined by the vibrational matrix element of $\bar{\mu}_z$, the dipole moment component parallel to the C_3 symmetry axis of the molecule. We draw stick spectra where the height of the stick is the relative intensity determined using

Table 12.2. Molecular parameters (in cm^{-1}) for CH$_3$F. Parameter values from Champion J P *et al* 1982 *J. Mol. Spectrosc.* **96** 422.

	$|0\rangle$	$|v_1\rangle$	$|v_4^{\pm 1}\rangle$
$G_{[V]}$	0.0	2916.643	2998.438
$A_{[V]}$	5.182 009	5.131 72	5.146 15
$B_{[V]}$	0.851 794 25	0.851 804	0.852 428
$A_{[4]}\zeta_4^z$			0.243 16

equation (1.9). We cannot calculate absolute intensities since we do not know the value of the factor $|\langle v_1|T_0^1(\mu)|0\rangle|^2/Q$, where Q is the partition function, but this factor is the same for all transitions in the band and so relative intensities can be computed. In the figure, we draw separately the *sub-bands* for $K = 0, 1, 2, 3$ and 4; in the bottom display, we combine these, together with sub-bands for $K = 5, 6, \ldots, 50$, to get the complete parallel band. Within each sub-band, the P-lines form a series (the P-*branch*) on the low-frequency side of the sub-band centre and the R-branch series is on the high-frequency side of the sub-band centre. In the region of the sub-band centre, the compact Q-branch occurs. In a parallel band, the sub-band Q-branches all fall in the same region at the centre of the band. In calculating the intensities, we use the spin statistical weight factors of CH$_3$F given in table 9.4 and we further use the fact that, for states with $K = 3, 6, 9, 12, \ldots$ in the vibrational ground state, the rovibronic symmetry Γ_{rve} is $A_1 \oplus A_2$ (this follows from the results in table 11.2 in conjunction with the fact that the electronic and vibrational wavefunctions have A_1 symmetry). In our approximation, the A_1 and A_2 energies are coincident and the intensities for the transitions involving them add together.

For the v_4 band, equation (12.35) applies since (Q_{4a}, Q_{4b}) have the same symmetry as (T_x, T_y). This is called a perpendicular band since its intensity is determined by the vibrational matrix elements of $T_{\pm 1}^1(\bar{\mu})$ or, equivalently, $(\bar{\mu}_x, \bar{\mu}_y)$ which are components perpendicular to the C_3 symmetry axis of the molecule. Detailed symmetry analysis (see problem 12.5) shows that for the vibrational transition moments to be non-vanishing, the more restrictive selection rule

$$\Delta(k - l_4) = 0 \tag{12.45}$$

must be satisfied. In the present case, this condition can be rewritten as

$$\Delta k = l_4 = \pm 1. \tag{12.46}$$

This equation implies that among the four vibrational transition moments $\langle v_4^{\pm 1}|T_{\pm 1}^1|0\rangle$ (where the signs are uncorrelated), only $\langle v_4^{+1}|T_{+1}^1|0\rangle$ and $\langle v_4^{-1}|T_{-1}^1|0\rangle$ are non-vanishing. We can use symmetry to derive a relation

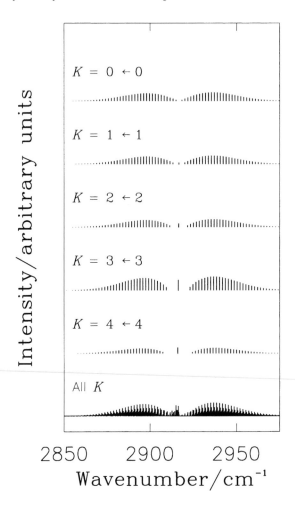

Figure 12.2. A simulation of the ν_1 absorption band of CH_3F at the absolute temperature $T = 300$ K. We draw separately the sub-bands for $K = 0, 1, 2, 3$ and 4; in the bottom display, we combine these and sub-bands with $K = 5, 6, \ldots, 50$, to get the complete parallel band.

between these two matrix elements. With an appropriate choice of phase factors for the two vibrational functions $|\nu_4^{+1}\rangle$ and $|\nu_4^{-1}\rangle$, it can be shown that (see problem 12.3)

$$(23)^*|\nu_4^{+1}\rangle = |\nu_4^{-1}\rangle \qquad (12.47)$$

where $(23)^*$ is an element of the MS group and the ground-state vibrational wavefunction $|0\rangle$ is totally symmetric so that $(23)^*|0\rangle = |0\rangle$. The operation $(23)^*$

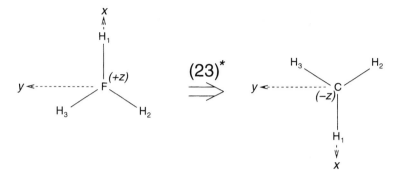

Figure 12.3. The molecule-fixed xyz axes for the CH$_3$F molecule and the effect of the MS group operation (23)* on these axes. In the initial situation, the z axis points out of the plane of the page but it is reversed by (23)*.

reverses the direction of the molecular dipole moment vector in space but it also changes the xyz axes as shown in figure 12.3. Thus[4],

$$(23)^*(\bar{\mu}_x, \bar{\mu}_y, \bar{\mu}_z) = (\bar{\mu}_x, -\bar{\mu}_y, \bar{\mu}_z) \tag{12.48}$$

which, in conjunction with equations (12.28) and (12.29), gives

$$(23)^* T_{+1}^1 = -T_{-1}^1. \tag{12.49}$$

In section 7.5.1, we proved the vanishing integral rule for H$_2$O and, as part of the proof, we showed that the value of an integral of a function depending on the coordinates chosen for H$_2$O is unchanged when a symmetry operation is applied to the integrand. This is true for all molecules and all symmetry operations. We make use of this result by applying the symmetry operation (23)* to the integrand of $\langle v_4^{+1} | T_{+1}^1 | 0 \rangle$. The fact that the value of the integral is unchanged by the application of the symmetry operation leads to the relation

$$\langle v_4^{+1} | T_{+1}^1 | 0 \rangle = -\langle v_4^{-1} | T_{-1}^1 | 0 \rangle \tag{12.50}$$

so that the line strengths in the perpendicular band in equation (12.31) are determined by a single parameter, $|\langle v_4^{+1} | T_{+1}^1 | 0 \rangle|^2 = |\langle v_4^{-1} | T_{-1}^1 | 0 \rangle|^2$.

Equation (12.46) implies that $\Delta K = +1$ transitions are to $(+l)$ levels and $\Delta K = -1$ transitions are to $(-l)$ levels in the v_4 band. These selection rules satisfy equation (12.16) that the rovibrational symmetries of allowed transitions be connected by the symmetry of μ_A ($= A_2$ for CH$_3$F) as one can see from the symmetries as given in tables 11.2 and 11.7. In figure 12.4, we show a simulation of the v_4 band drawn in a manner analogous to figure 12.2. We label the sub-bands $K'(\pm l) \leftarrow K''$. In figure 12.4, we draw five sub-bands separately and, in

---

[4] This is just as for the point group operation σ_{xz} onto which (23)* maps; see figure 8.8, but note the different convention used for locating the xyz axes in an H$_3^+$ molecule.

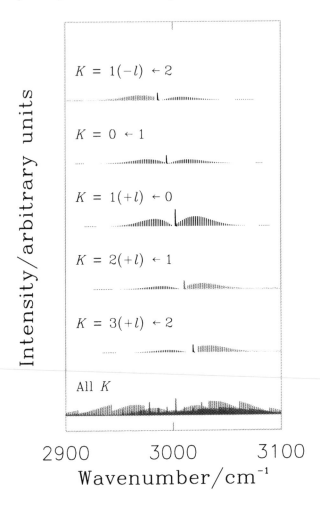

Figure 12.4. A simulation of the ν_4 absorption band of CH_3F at the absolute temperature $T = 300$ K. We draw the five sub-bands separately, and, in the bottom display, we combine these sub-bands, and all other sub-bands with $K'' \leq 50$, to produce the complete perpendicular band.

the bottom display, we combine these sub-bands, and all other sub-bands with $K'' \leq 50$, to produce the complete perpendicular band.

In a perpendicular band, the sub-band Q-branches do not all fall in the same region at the centre of the band but instead they have an approximately equidistant spacing of $2[A(1 - \zeta^z) - B]$. The spectra of H_3^+ shown in figures 1.1 and 1.4 are parts of the ν_2 fundamental band of this molecule. The ν_2 normal mode has E' symmetry in $D_{3h}(M)$ (see table B.8) and the ν_2 band is a perpendicular

Table 12.3. Observed transition wavenumbers[a] in the ν_2 band of H_3^+.

(J', k', l'_2)[b]	(J'', k'')	$\tilde{\nu}_{obs}/cm^{-1}$
$(0, 0, \mp 1)$	$(1, \pm 1)$	2457.277
$(3, \pm 1, \pm 1)$	$(3, 0)$	2509.063
$[(2, \pm 2, \pm 1) - (2, 0, \mp 1)]_{II}$	$(2, \pm 1)$	2518.198
$(1, \pm 1, \pm 1)$	$(1, 0)$	2529.711
$(1, 0, \mp 1)$	$(1, \pm 1)$	2545.412
$(2, \pm 1, \mp 1)$	$(2, \pm 2)$	2554.655
$(3, \pm 2, \mp 1)$	$(3, \pm 3)$	2561.486
$[(2, \pm 2, \pm 1) - (2, 0, \mp 1)]_{II}$	$(1, \pm 1)$	2691.430
$(2, \pm 1, \pm 1)$	$(1, 0)$	2725.885
$[(2, \pm 2, \pm 1) - (2, 0, \mp 1)]_{I}$	$(1, \pm 1)$	2726.208
$[(3, \pm 3, \pm 1) - (3, \pm 1, \mp 1)]_{II}$	$(2, \pm 2)$	2762.057
$[(3, \pm 3, \pm 1) - (3, \pm 1, \mp 1)]_{I}$	$(2, \pm 2)$	2823.125
$[(3, \pm 2, \pm 1) - (3, 0, \mp 1)]_{I}$	$(2, \pm 1)$	2826.022
$[(4, \pm 4, \pm 1) - (4, \pm 2, \mp 1)]_{II}$	$(3, \pm 3)$	2829.911
$[(4, \pm 4, \pm 1) - (4, \pm 2, \mp 1)]_{I}$	$(3, \pm 3)$	2918.013

[a]From Oka T 1980 *Phys. Rev. Lett.* **45** 531.

[b]States mixed by *l*-resonance are labelled as $[(J', k^{(1)}, l_2^{(1)}) - (J', k^{(2)}, l_2^{(2)})]_{I/II}$ (see the text).

band with, in principle, the same structure as the ν_4 fundamental band of CH_3F. However, the simple model for the molecular energies that we can use to make a realistic simulation of the bands for CH_3F, is not satisfactory for H_3^+ and its ν_2 band does not have the relatively simple appearance of figure 12.4. In table 12.3, we list the assignments of the strongest transitions in figures 1.1 and 1.4. For the upper states of some of these lines, the quantum numbers (J', k', l'_2) cannot be unambiguously assigned since there is a so-called (2,2) *l*-resonance; this is an interaction [caused by the Hamiltonian term $\hat{H}_2$ in equation (11.43), which has off-diagonal matrix elements between the rovibrational levels of the ν_2 state and the rovibrational levels of non-degenerate vibrational states, including the ground state, having $l_2 = 0$] that mixes the zero-order states $(J', k', l'_2) = (J', k^{(1)}, l_2^{(1)})$ and $(J', k^{(2)}, l_2^{(2)})$, where $k^{(1)} - k^{(2)} = l_2^{(1)} - l_2^{(2)} = \pm 2$. The corresponding upper states in table 12.3 are labelled $[(J', k^{(1)}, l_2^{(1)}) - (J', k^{(2)}, l_2^{(2)})]_{I/II}$ where the indices I and II distinguish the two mixed states.

12.3.4   Rotation–vibration interaction

Rotation–vibration interaction relaxes the selection rules given in equations (12.34), (12.35), (12.37) and (12.46) but the rovibration symmetry rule of equation (12.16) still applies if *ortho–para* mixing and the breakdown of the Born–Oppenheimer approximation are not included. Thus, for CH_3F (see table B.5), when rotation–vibration interaction is included, all rovibrational transitions satisfying

$$A_1 \leftrightarrow A_2 \quad \text{or} \quad E \leftrightarrow E \tag{12.51}$$

become possible. The symmetries given in table 11.2 apply to the rovibrational levels of CH_3F in its ground state and in its ν_1 state; the symmetries given in table 11.7 apply in the ν_4 state. Using the details of the symmetry analyses that lead to the results in these tables, particularly the effect of the MS group operation (123) and the rovibrational symmetry selection rules of equation (12.51), one can derive the general selection rule

$$\Delta k - \sum_j \Delta l_j = 0, \pm 3, \pm 6, \pm 9, \ldots = 3t, \tag{12.52}$$

where $t = 0, \pm 1, \pm 2, \ldots$, and $\Delta l_j = l'_j - l''_j$. For the ν_1 band, this reduces to

$$\Delta k = 3t \tag{12.53}$$

since all $l_j = 0$ in the ν_1 state, whereas, for the ν_4 band, we obtain (see problem 12.5)

$$\Delta k = l_4 + 3t. \tag{12.54}$$

For the ν_1 band of CH_3F, the allowed transitions have $\Delta k = 0$ and, for the ν_4 band, they have $\Delta k = l_4$. For the ν_1 band, the forbidden transitions have $\Delta k = \pm 3, \pm 6, \pm 9, \ldots$, so that transitions such as $K = 3 \leftarrow 0$, $K = 2 \leftarrow 1$ or $K = 4 \leftarrow 2$ can occur. For the ν_4 band, the forbidden transitions satisfy $\Delta k = l_4 \pm 3, l_4 \pm 6, l_4 \pm 9, \ldots$, so that transitions such as $K = 2(-l) \leftarrow 0$, $K = 3(-l) \leftarrow 1$ or $K = 2(+l) \leftarrow 2$ can occur. These forbidden transitions are, generally, very much weaker than the transitions allowed by equations (12.34) and (12.46).

Rotation–vibration interaction makes rotational transitions in the H_3^+ and CH_4 molecules allowed. For both of these molecules, one might think that their point group symmetry rigorously precludes the possibility of them having a pure rotational spectrum; neither has a dipole moment. For the H_3^+ molecule, the T_α have symmetry $A_2'' \oplus E'$ (see table B.8) and so none is totally symmetric; in the ground vibrational state, the dipole moment components $\langle \Phi_{\text{vib},0} | \bar{\mu}_\alpha | \Phi_{\text{vib},0} \rangle$ vanish and pure rotation transitions are forbidden according to equation (12.20). However, they are not forbidden according to the rovibrational symmetry selection rule of equation (12.16) and transitions having $\Delta k = \pm 3, \pm 6, \ldots$ are possible as a result of rotation–vibration interaction. In classical terms, the centrifugal distortion caused by rotation of the H_3^+ molecule about a C_2 axis (passing through one H atom and through the molecular centre of mass) gives the molecule C_{2v} symmetry and, in such a geometry, it has a dipole moment along that C_2 axis. Quantum mechanically, the rotation–vibration interaction mixes the rovibrational wavefunctions of the ground vibrational state with the rovibrational wavefunctions of the v_2 excited state and the pure rotation spectrum steals intensity from the allowed v_2 fundamental of E' symmetry.

For centrosymmetric rigid molecules such as the hydrogen molecule or the benzene molecule, the point group contains the vibronic inversion operation i. This is not the same as the CNPI group operation E^* and its effect does not define the parity ($\pm$) of the state. Its effect defines the 'g' or 'u' symmetry of the vibronic state. For such a molecule, the translational coordinates can only be of u symmetry and vibrational transitions between two u vibrational states or two g vibrational states are forbidden. For example, in benzene, the translational coordinates have symmetry $A_{2u} \oplus E_{1u}$ in $D_{6h}(M)$, see table B.9. The MS group[5] symmetry operation onto which the point group operation i is mapped is called $\hat{O}_i$ [for benzene using the labelling of table B.9, $\hat{O}_i = (14)(25)(36)(ad)(be)(cf)^*$] and it is *always* an operation having R^0 as its equivalent rotation. This means that the effect of the point group operation i on the rovibronic variables (i.e. all variables except nuclear spin) is the same as the effect of the MS group operation $\hat{O}_i$. $\hat{O}_i$ is a symmetry operation for the rovibronic Hamiltonian[6] and so i is also a symmetry operation for the rovibronic Hamiltonian; it is not just a symmetry operation of the vibronic Hamiltonian like most point group operations. This means that the labels g and u are good symmetry labels for the rovibronic states of a centrosymmetric rigid molecule. All the rovibronic states formed from a vibronic state of g(u) symmetry have g(u) symmetry, as a result of the fact that all rotational wavefunctions must be g since R^0 is the equivalent rotation of $\hat{O}_i$. There can be no rotation–vibration interaction between the levels of u and g vibrational states and, hence, rovibrational transitions between two u vibrational states or two g vibrational states are forbidden even with inclusion of rotation–vibration interaction. This means that pure rotation transitions cannot steal intensity from allowed rovibrational transitions and that pure rotation transitions in a centrosymmetric rigid molecule cannot be induced by rotation–vibration interaction. They cannot be induced by a breakdown of the Born–Oppenheimer approximation either (since i is a symmetry operation for the rovibronic Hamiltonian) but they can be induced by the nuclear spin hyperfine Hamiltonian (even in H_2) or by a non-rigidity that produces a new MS group as we discuss in chapter 13. Such transitions must satisfy the strict selection rules given in equations (12.4) and (12.6).

---

[5] For a linear centrosymmetric rigid molecule, this would be the EMS group $D_{\infty h}(EM)$ given in table B.17.

[6] $\hat{O}_i$, like all operations of the MS group, is also a symmetry operation for the full Hamiltonian including $\hat{H}_{hfs}$ but the point group operation i is *not* a symmetry operation of the full Hamiltonian.

12.4   Transitions between electronic states

Within the Born–Oppenheimer, harmonic-oscillator and rigid-rotor approximations, the transition moment I_{TM}^0 is given by equation (12.13) for transitions between electronic states n' and n''. To develop symmetry selection rules for allowed
transitions, we need an MS group that accommodates all accessible versions of
the molecule in both electronic states. If the two electronic states have the same
MS group, then this one MS group can be used for expressing the symmetry
selection rules for transitions between the states. If the electronic states have different MS groups, then the MS group required will be more complicated to set
up. Such a situation for the N and V electronic states of ethylene is discussed
in section 13.6, where it is explained that for the N (electronic ground) state, the
MS group is $D_{2h}(M)$, for the excited V state the MS group is $D_{2d}(M)$ but if both
electronic states are to be considered together, the appropriate MS group is G_{16}.
Let us say that the electronic wavefunctions $\Phi_{elec,n'}^*$ and $\Phi_{elec,n''}$ have the symmetries Γ'^*_{elec} and Γ''_{elec} and the vibrational wavefunctions $\Phi_{vib,n'v'}^*$ and $\Phi_{vib,n''v''}$ have
the symmetries Γ'^*_{vib} and Γ''_{vib}, in the appropriate MS group. The dipole moment
component μ_α in equation (12.13) has the symmetry $\Gamma(\mu_\alpha)$. With this notation,
I_{TM}^0 is non-zero and the transition is allowed if the symmetries satisfy

$$\Gamma'^*_{elec} \otimes \Gamma'^*_{vib} \otimes \Gamma''_{elec} \otimes \Gamma''_{vib} \supset \Gamma(\mu_\alpha). \tag{12.55}$$

If we define $\Gamma_{ve} = \Gamma_{vib} \otimes \Gamma_{elec}$, we can reformulate equation (12.55) as an
equation involving vibronic symmetries.

We say that the transition is *vibronically allowed,* if the vibronic
symmetries satisfy

$$\Gamma'^*_{ve} \otimes \Gamma''_{ve} \supset \Gamma(\mu_\alpha). \tag{12.56}$$

The rotational selection rules depend on α and are given as in equations (12.34), (12.35) and (12.37). The vibronic symmetry selection rule will
not be changed by anharmonicity or by vibronic mixing caused by the breakdown
of the Born–Oppenheimer approximation since these effects do not change the
vibronic symmetry.

More restrictive selection rules can be obtained.   We begin with
equation (12.15) for the zero-order transition moment integral I_{TM}^0.   In
equation (12.14),

$$\bar{\mu}_\alpha^{(n',n'')} = \mu_\alpha^{(n',n'')}(Q_1, Q_2, Q_3, \ldots) = \langle \Phi_{elec,n'} | \mu_\alpha | \Phi_{elec,n''} \rangle_{el} \tag{12.57}$$

where $\alpha = x, y$ or z. As in equation (12.21), we make the normal coordinate expansion

$$\bar{\mu}_\alpha^{(n',n'')} = \mu_{\alpha,0}^{(n',n'')} + \sum_r \mu_{\alpha,r}^{(n',n'')} Q_r + \sum_{r,s} \mu_{\alpha,rs}^{(n',n'')} Q_r Q_s + \cdots \quad (12.58)$$

which introduces the parameters $\mu_{\alpha,0}^{(n',n'')}$, $\mu_{\alpha,r}^{(n',n'')}$, etc. In this expansion, the normal coordinates are those of one of the electronic states n' or n'' involved in the transition.

If the symmetries are such that

$$\Gamma_{\text{elec}}'^* \otimes \Gamma_{\text{elec}}'' \supset \Gamma(\mu_\alpha) \quad (12.59)$$

we say that the transition is *electronically allowed*.

The transition must simultaneously satisfy equation (12.55) in order to be vibronically allowed and so, in an electronically allowed transition,

$$\Gamma_{\text{vib}}' = \Gamma_{\text{vib}}''. \quad (12.60)$$

In this case, the constant term $\mu_{\alpha,0}^{(n',n'')}$ in equation (12.58) can be non-vanishing and, since this term normally contributes much more to the vibronic transition moments $\langle \Phi_{\text{vib},v'j'} | \bar{\mu}_\alpha^{(n',n'')} | \Phi_{\text{vib},v''j''} \rangle$ than the subsequent higher-order terms in the normal coordinates, electronically allowed transitions are generally strong.

In estimating the intensities of electronically allowed transitions, we truncate the expansion in equation (12.58) after the constant term $\mu_{\alpha,0}^{(n',n'')}$. In this approximation, the square of the vibronic transition moment is given by

$$|\langle \Phi_{\text{vib},v'j'} | \bar{\mu}_\alpha^{(n',n'')} | \Phi_{\text{vib},v''j''} \rangle|^2 = |\mu_{\alpha,0}^{(n',n'')}|^2 |\langle \Phi_{\text{vib},v'j'} | \Phi_{\text{vib},v''j''} \rangle|^2 \quad (12.61)$$

which involves the *Franck–Condon factor* $|\langle \Phi_{\text{vib},v'j'} | \Phi_{\text{vib},v''j''} \rangle|^2$; this is the square of the overlap integral between the initial and final vibrational wavefunctions. In this approximation the Franck–Condon factors give the relative intensities of the vibronic bands within an electronic band system and this is referred to as the *Franck–Condon principle*.

If the equilibrium geometries of the two electronic states involved in the transition are very different, then the Franck–Condon principle results in the occurrence of long *progressions* of vibrational bands with significant intensity. Such a long progression in a typical diatomic molecule is shown in figure 12.5.

In this figure, we have drawn two model potential energy curves for the ^{12}CH molecule; each of the potential energy curves is given as a *Morse potential*

$$V(r)/(hc) = T_e + D_e[1 - e^{-a(r-r_e)}]^2 \tag{12.62}$$

where r is the internuclear distance, r_e is its equilibrium value, T_e is the value of the potential energy at $r = r_e$, D_e is the dissociation energy and a is a parameter that determines the curvature of the potential energy at $r = r_e$. For the lower electronic state in figure 12.5, we set $T_e'' = 0$, $r_e'' = 1$ Å, $D_e'' = 30\,000$ cm^{-1} and $a'' = 1.5$ Å^{-1}, whereas for the upper electronic state, we have chosen $T_e' = D_e' = 20\,000$ cm^{-1}, $r_e' = 1.35$ Å and $a' = 1.5$ Å^{-1}. We calculate the $v'' = 0$ vibrational wavefunction for the lower electronic state and the vibrational wavefunctions with $v' \leq 12$ for the upper state; these wavefunctions are plotted in figure 12.5 at their respective energy values. The Franck–Condon factors for the $v = v' \leftarrow 0$ transitions are plotted as a histogram to the right of the upper potential energy curve. The largest Franck–Condon factor of 0.11 is obtained for the transition $v = 3 \leftarrow 0$ but the Franck–Condon factors vary slowly with v'. This is because the steep inner wall of the upper-state potential energy, where the upper-state wavefunctions have their inner classical turning points, is situated near $r = 1$ Å, the equilibrium r-value for the lower-state potential. At equilibrium, the $v'' = 0$ lower-state vibrational wavefunction has its maximum amplitude and the excited upper-state wavefunctions have significant amplitude at their inner classical turning points.

An electronically forbidden transition is a transition that does not satisfy equation (12.59). Vibronically allowed transitions satisfy equation (12.56) and such transitions could be electronically forbidden. Such an electronically forbidden but vibronically allowed transition can have appreciable intensity if the electronic transition moment $\bar{\mu}_\alpha^{(n',n'')}$ depends strongly on the vibrational coordinates, in which case the terms of first and higher order in the normal coordinates in equation (12.58) have significant magnitude. This situation is referred to as the *Herzberg–Teller effect*.

12.5 Raman transitions

In section 1.6, we introduced the Raman effect and we now discuss the selection rules for Raman transitions. We consider rovibrational Raman transitions within the electronic ground state (which is assumed to be non-degenerate) of a molecule. Thus, the levels with energies E_i and E_f in figure 1.8 are different rotation–vibration levels in the electronic ground state. In the *polarizability approximation*, the Raman line strength $S_{\text{Raman}}(E_{\text{rv}}' \leftarrow E_{\text{rv}}'')$ (see section 1.6) is expressed in

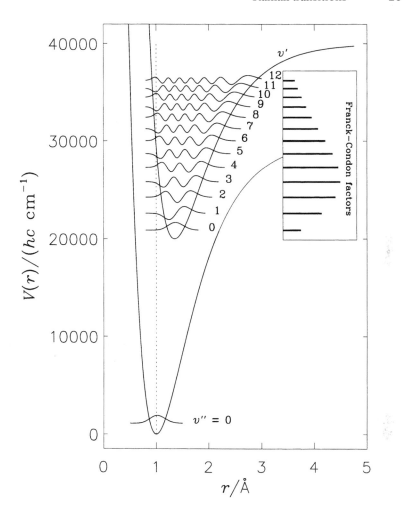

Figure 12.5. The Franck–Condon factors for $v = v' \leftarrow 0$ transitions in absorption for a typical diatomic molecule. The vibrational wavefunction with $v'' = 0$ for the lower electronic state and those with $v' \leq 12$ for the upper electronic state are drawn at the respective energies. A vertical dotted line indicates $r = r_e'' = 1$ Å and the Franck–Condon factors for the $v = v' \leftarrow 0$ transitions are plotted as a histogram. The maximum Franck–Condon factor for the $v = 3 \leftarrow 0$ transition has the value 0.11.

terms of matrix elements of the *static electric polarizability tensor* with elements

$$\bar{\alpha}_{AB}(\theta, \phi, \chi, Q_1, Q_2, Q_3, \ldots) = \sum_{n' \neq 0} \left[\frac{\bar{\mu}_B^{(0,n')} \bar{\mu}_A^{(n',0)}}{V_{\text{elec}}^{(n')} - V_{\text{elec}}^{(0)}} + \frac{\bar{\mu}_A^{(0,n')} \bar{\mu}_B^{(n',0)}}{V_{\text{elec}}^{(n')} - V_{\text{elec}}^{(0)}} \right]$$

$$(12.63)$$

where $A, B = \xi, \eta$, or ζ label the space-fixed axes. In equation (12.63), the electronic integrals

$$\bar{\mu}_A^{(n',n'')}(\theta, \phi, \chi, Q_1, Q_2, Q_3, \ldots) = \langle \Phi_{\text{elec},n'} | \mu_A | \Phi_{\text{elec},n''} \rangle_{\text{el}} \qquad A = \xi, \eta, \zeta, \tag{12.64}$$

are defined by analogy with equation (12.57) and $V_{\text{elec}}^{(n')}$ (which depends on the normal coordinates Q_r) is the value of the Born–Oppenheimer potential energy function for the electronic state n'. The electronic ground state has the electronic wavefunction $\Phi_{\text{elec},0}$ and the Born–Oppenheimer potential energy function $V_{\text{elec}}^{(0)}$. The electronic integral $\bar{\mu}_A^{(n',n'')}$ depends not only on the normal coordinates Q_r but also on the Euler angles θ, ϕ, χ because μ_A depends on these angles [see equation (12.11)]. The sum in equation (12.63) extends over all electronic states other than the ground state.

The Raman line strength $S_{\text{Raman}}(E'_{\text{rv}} \leftarrow E''_{\text{rv}})$ is given by an expression analogous to equation (2.87):

$$S_{\text{Raman}}(E'_{\text{rv}} \leftarrow E''_{\text{rv}}) = \sum_{\Phi'_{\text{rv}}, \Phi''_{\text{rv}}} \sum_{A,B=\xi,\eta,\zeta} |\langle \Phi'_{\text{rv}} | \bar{\alpha}_{AB} | \Phi''_{\text{rv}} \rangle|^2 \tag{12.65}$$

where Φ''_{rv} and Φ'_{rv} are the rovibrational wavefunctions of the initial and final states, respectively, in the Raman transition. Thus, for Raman transitions, we derive, by analogy with equation (12.3), that the Raman transition is forbidden in the polarizability approximation when

$$\Gamma(\Phi'^*_{\text{rv}}) \otimes \Gamma(\Phi''_{\text{rv}}) \not\supset \Gamma(\bar{\alpha}_{AB}) \tag{12.66}$$

where $\Gamma(\bar{\alpha}_{AB})$ is the MS group symmetry of $\bar{\alpha}_{AB}$. For a non-degenerate electronic ground state, we always have $\Gamma(\bar{\alpha}_{AB}) = \Gamma^{(s)}$, the totally symmetric species of the MS group in question and so the Raman transition is allowed if $\Gamma(\Phi'_{\text{rv}}) = \Gamma(\Phi''_{\text{rv}})$.

In the Born–Oppenheimer, harmonic-oscillator and rigid-rotor approximation for a symmetric top or linear molecule, $S_{\text{Raman}}(E'_{\text{rv}} \leftarrow E''_{\text{rv}})$ is given by an expression similar to equation (12.31):

$$S_{\text{Raman}}(E'_{\text{rv}} \leftarrow E''_{\text{rv}}) = g(2J'' + 1)|\langle \Phi'_{\text{vib}} | T^\omega_{\Delta k}(\bar{\alpha}) | \Phi''_{\text{vib}} \rangle|^2$$
$$\times A^{(\omega)}(J'', k'', J', k'). \tag{12.67}$$

Here, the quantum numbers are defined as in equation (12.31) whereas

$$g = \begin{cases} 2g_{\text{ns}} & \text{for } (K'', K') = (0, 1), (1, 0), (0, 2) \text{ or } (2, 0) \\ g_{\text{ns}} & \text{otherwise} \end{cases} \tag{12.68}$$

where g_{ns} is the spin statistical weight factor. There are two types of Raman scattering: *Isotropic Raman scattering* with $\omega = 0$ in equation (12.67) and *anisotropic Raman scattering* with $\omega = 2$. The irreducible tensor components $T^\omega_{\Delta k}(\bar\alpha)$ are given by

$$T^0_0(\bar\alpha) = -\frac{1}{\sqrt{3}}[\bar\alpha_{xx} + \bar\alpha_{yy} + \bar\alpha_{zz}] \tag{12.69}$$

$$T^2_0(\bar\alpha) = \frac{1}{\sqrt{6}}[2\bar\alpha_{zz} - \bar\alpha_{xx} - \bar\alpha_{yy}] \tag{12.70}$$

$$T^2_{\pm 1}(\bar\alpha) = \mp\bar\alpha_{xz} - i\bar\alpha_{yz} \tag{12.71}$$

$$T^2_{\pm 2}(\bar\alpha) = \tfrac{1}{2}[\bar\alpha_{xx} - \bar\alpha_{yy}] \pm i\bar\alpha_{xy} \tag{12.72}$$

where all signs are correlated and $T^\omega_{\Delta k}(\bar\alpha) = 0$ for $|\Delta k| > \omega$. In equations (12.69)–(12.72), the $\bar\alpha_{\beta\gamma}(\beta, \gamma = x, y, z)$ are the components along the molecule-fixed xyz axes of the polarizability tensor elements. Expressions for these components are obtained by replacing $\xi\eta\zeta$ in equation (12.63) by xyz; they depend solely on the normal coordinates Q_r. The polarizability tensor is symmetric and there are six distinct molecule-fixed elements $\bar\alpha_{xx}, \bar\alpha_{yy}, \bar\alpha_{zz}, \bar\alpha_{xy} = \bar\alpha_{yx}, \bar\alpha_{xz} = \bar\alpha_{zx}$ and $\bar\alpha_{yz} = \bar\alpha_{zy}$.

The six polarizability tensor elements $\bar\alpha_{xx}, \bar\alpha_{yy}, \bar\alpha_{zz}, \bar\alpha_{xy}, \bar\alpha_{xz}$ and $\bar\alpha_{yz}$ generate the same representation of the MS group as the six products $T^2_x, T^2_y, T^2_z, T_xT_y, T_xT_z$ and T_yT_z of translational coordinates.

The symmetries of the $\bar\alpha_{\beta\gamma}$ components are normally indicated in group character tables (where they are called $\alpha_{\beta\gamma}$).

For isotropic Raman scattering ($\omega = 0$), the rotational factor in equation (12.67) is given by

$$A^{(0)}(J'', k'', J', k') = \delta_{k'k''}\delta_{J'J''} \tag{12.73}$$

where $\delta_{k'k''}$ are $\delta_{J'J''}$ are Kronecker deltas. Isotropic Raman scattering takes place between states with $J' = J''$ and $k' = k''$ so it gives rise to Q-branches only. The rotational factors for anisotropic Raman scattering are given in table 12.4; these are non-vanishing[7] for $|\Delta J| = |J' - J''| \leq 2$, $J'' + J' \geq 2$ and $|\Delta k| = |k' - k''| \leq 2$. Thus, in addition to the P, Q and R branches found in electric dipole spectra, anisotropic Raman spectra also have O-branches with $\Delta J = -2$ and S-branches with $\Delta J = 2$.

We derive further selection rules by applying the vanishing integral rule to the matrix element $\langle\Phi'_{\text{vib}}|T^\omega_{\Delta k}(\bar\alpha)|\Phi''_{\text{vib}}\rangle$ in equation (12.67). For CH$_3$F, $T^0_0(\bar\alpha)$

---

[7] The selection rules $|\Delta J| \leq 2$ and $J'' + J' \geq 2$ for anisotropic Raman scattering and $\Delta J = 0$ for isotropic scattering result from rotational symmetry described by K(spatial), and they are completely general in the absence of *ortho–para* mixing.

Table 12.4. Rotational factors $A^{(2)}(J, k, J + \Delta J, k + \Delta k)$ in the intensities of anisotropic Raman scattering transitions. Rotational factors with $\Delta k < 0$ are obtained using the relation $A^{(2)}(J, k, J + \Delta J, k - \Delta k) = A^{(2)}(J, -k, J + \Delta J, -k + \Delta k)$.

ΔJ	Δk	$A^{(2)}(J, k, J + \Delta J, k + \Delta k)$
-2	0	$\dfrac{3(J-k)(J-k-1)(J+k)(J+k-1)}{2(J-1)(2J-1)J(2J+1)}$
	1	$\dfrac{(J-k)(J-k-1)(J-k-2)(J+k)}{(J-1)(2J-1)J(2J+1)}$
	2	$\dfrac{(J-k-3)(J-k-2)(J-k-1)(J-k)}{4(J-1)(2J-1)J(2J+1)}$
-1	0	$3k^2 \dfrac{J^2-k^2}{(J-1)J(J+1)(2J+1)}$
	1	$(J + 2k + 1)^2 \dfrac{(J-k)(J-k-1)}{2(J-1)J(J+1)(2J+1)}$
	2	$\dfrac{(J+k+1)(J-k-2)(J-k-1)(J-k)}{2(J-1)J(J+1)(2J+1)}$
0	0	$\dfrac{[3k^2-J(J+1)]^2}{(2J-1)J(J+1)(2J+3)}$
	1	$(2k + 1)^2 \dfrac{3(J-k)(J+k+1)}{2(2J-1)J(J+1)(2J+3)}$
	2	$\dfrac{3(J+k+1)(J+k+2)(J-k-1)(J-k)}{2(2J-1)J(J+1)(2J+3)}$
$+1$	0	$3k^2 \dfrac{(J+1)^2-k^2}{J(J+1)(J+2)(2J+1)}$
	1	$(2k - J)^2 \dfrac{(J+k+2)(J+k+1)}{2J(J+1)(J+2)(2J+1)}$
	2	$\dfrac{(J+k+1)(J+k+2)(J+k+3)(J-k)}{2J(J+1)(J+2)(2J+1)}$
$+2$	0	$\dfrac{3(J-k+2)(J-k+1)(J+k+2)(J+k+1)}{2(J+1)(2J+3)(J+2)(2J+1)}$
	1	$\dfrac{(J-k+1)(J+k+3)(J+k+2)(J+k+1)}{(2J+1)(J+1)(2J+3)(J+2)}$
	2	$\dfrac{(J+k+1)(J+k+2)(J+k+3)(J+k+4)}{4(J+1)(2J+3)(J+2)(2J+1)}$

and $T_0^2(\bar{\alpha})$ each have A_1 symmetry, whereas the pairs $[T_{-2}^2(\bar{\alpha}), T_{+2}^2(\bar{\alpha})]$ and $[T_{-1}^2(\bar{\alpha}), T_{+1}^2(\bar{\alpha})]$ each have E symmetry. Transitions from the vibrational ground state to the A_1 fundamental levels v_1, v_2 and v_3 each have an isotropic component and an anisotropic component. Both components contain transitions with $\Delta k = 0$ but the isotropic component consists of Q-branches only, whereas the anisotropic component consists of O, P, Q, R and S branches. The fundamental Raman transitions to the E fundamental levels $v_4^{\pm 1}$, $v_5^{\pm 1}$ and $v_6^{\pm 1}$ are due entirely to anisotropic Raman scattering. They contain O, P, Q, R and S branches with $\Delta k = \pm 1, \pm 2$.

For vibrational Raman bands with isotropic and anisotropic components, the fraction of isotropically-scattered intensity to anisotropically-scattered intensity detected in a given experiment depends on the angle between the electric field vector of the exciting light and that of the scattered light. This angle is determined by the direction of observation relative to the direction of the exciting light.

In the Born–Oppenheimer approximation and neglecting rotation–vibration interaction, Raman transitions of CH_3F are subject to the selection rule

$$\Delta \left(k - \sum_j l_j \right) = 0, \pm 3 \tag{12.74}$$

which is analogous to, but less restrictive than, equation (12.45). Consequently, rovibrational Raman transitions to fundamental-level states with $l_j = +1$ have $\Delta k = +1$ or -2, whereas, for states with $l_j = -1$, the transitions have $\Delta k = -1$ or $+2$.

For all of the fundamental levels of CH_3F, Raman transitions from the vibrational ground state are allowed. We say that all the fundamental levels of CH_3F are *Raman active*.

In general, a normal mode is Raman active if its symmetry is equal to the symmetry as one or more of the translational coordinate products T_x^2, T_y^2, T_z^2, $T_x T_y$, $T_x T_z$ and $T_y T_z$.

For benzene (table B.9), $T_0^0(\bar{\alpha})$ and $T_0^2(\bar{\alpha})$ both have A_{1g} symmetry in $D_{6h}(M)$, $[T_{-1}^2(\bar{\alpha}), T_{+1}^2(\bar{\alpha})]$ have E_{1g} symmetry and $[T_{-2}^2(\bar{\alpha}), T_{+2}^2(\bar{\alpha})]$ have E_{2g} symmetry. Of the 20 fundamental normal modes of benzene [equation (11.26)], the two of A_{1g} symmetry are Raman active and the fundamental Raman bands have isotropic and anisotropic components with $\Delta k = 0$. The one normal mode of E_{1g} symmetry gives rise to an anisotropic-scattering fundamental Raman band with $\Delta k = \pm 1$ and each of the four E_{2g} normal modes gives rise to an anisotropic-scattering fundamental Raman band with $\Delta k = \pm 2$.

In CH_3F, all the normal modes are infrared active and Raman active and so obviously a normal mode can be both infrared active and Raman active. However, in a centrosymmetric molecule like benzene, the dipole moment components $\bar{\mu}_x$, $\bar{\mu}_y$ and $\bar{\mu}_z$ have u symmetry, whereas the polarizability tensor components $\bar{\alpha}_{xx}, \bar{\alpha}_{yy}, \bar{\alpha}_{zz}, \bar{\alpha}_{xy}, \bar{\alpha}_{xz}$ and $\bar{\alpha}_{yz}$ have g symmetry. Allowed vibrational transitions are u↔g in an infrared spectrum but are g↔g or u↔u in a Raman spectrum. Consequently,

---

in a centrosymmetric molecule there is an exclusion rule: No normal mode can be both infrared and Raman active.

As explained at the end of section 12.3.4, the point group inversion operation i is spoilt as a symmetry operation by *ortho–para* mixing and such mixing would, therefore, spoil the exclusion rule. This rule can also be spoilt by non-rigidity effects as observed for ethylene; see section 13.6.

12.6   Problems

12.1   Use the results given in section 12.3.2 to assign the lines (i.e. label them using upper and lower state J values) in the experimental CO spectrum given in figure 1.2.

12.2   Use equations (12.43) and (12.44), together with the zero-order-approximation selection rules discussed in section 12.3.3, to derive expressions for the wavenumbers of P, Q and R branch transitions in the v_1 and v_4 fundamental absorption bands of CH_3F. Sketch the qualitative appearance of the sub-bands.

12.3*   The v_4 fundamental level of CH_3F is described by two vibrational wavefunctions $|v_4^{-1}\rangle = |\psi_0\rangle |1^{-1}\rangle$ and $|v_4^{+1}\rangle = |\psi_0\rangle |1^{+1}\rangle$, where $|\psi_0\rangle$ is totally symmetric in the molecular symmetry group $C_{3v}(M)$ and given by

$$|\psi_0\rangle = |v_1 = 0\rangle|v_2 = 0\rangle|v_3 = 0\rangle|v_5^{l_5} = 0^0\rangle|v_6^{l_6} = 0^0\rangle. \qquad (12.75)$$

The functions $|v_4^{l_4}\rangle = |1^{-1}\rangle$ and $|1^{+1}\rangle$ are given by equation (4.38) and depend on the coordinates Q_4 and α_4. It can be shown that

$$(123)Q_4 = (23)^*Q_4 = Q_4 \qquad (12.76)$$
$$(123)\alpha_4 = \alpha_4 - 2\pi/3 \qquad (12.77)$$
$$(23)^*\alpha_4 = -\alpha_4. \qquad (12.78)$$

Use these relations to determine the transformation properties of the functions $|1^{-1}\rangle$ and $|1^{+1}\rangle$ under (123) and (23)*.

12.4*   Use table 7.4 to show that all operations in $C_{3v}(M)$ can be written as products involving (123) and (23)* so that the transformation properties under all operations in the group can be derived from those under (123) and (23)*. Verify that $|\psi_0\rangle|1^{-1}\rangle$ and $|\psi_0\rangle|1^{+1}\rangle$ (problem 12.3) have E symmetry in $C_{3v}(M)$.

12.5*   Use the answers to problems 12.3 and 12.4, together with equations (11.8) and (12.1), to prove equation (12.54).

12.6   What would the rotational *Raman* spectrum of CO look like?

12.7   Use equations (12.43) and (12.44), together with the zero-order-approximation selection rules discussed in section 12.5, to derive expressions for the wavenumbers of O, P, Q, R and S branch transitions in the ν_1 and ν_4 fundamental Raman bands of CH_3F. Sketch the qualitative appearance of the sub-bands.

12.8 The molecule F_2O has $\boldsymbol{C}_{2v}$ symmetry at equilibrium, where the OF distance is 1.405 Å and $\angle(FOF) = 103.0°$. Calculate the rotational constants A_e, B_e, C_e for F_2O. Which types of transition (i.e. a-type, b-type or c-type) occur in the rotational spectrum of F_2O? Determine, in the rigid-rotor approximation, the frequency of the $J = 1 \leftarrow 0$ transition in the rotational spectrum.

Chapter 13

The symmetry groups of non-rigid molecules

13.1 The MS group of a non-rigid molecule

A rigid molecule is defined as being such that the barriers between its versions are insuperable and, as a result, there are no observable tunnelling splittings. In section 8.2, the MS group of a rigid molecule was defined as the subgroup of the complete nuclear permutation inversion (CNPI) group obtained by deleting unfeasible operations. For a rigid molecule an unfeasible operation is one that causes a coordinate change that moves the molecule from one version to another. To generalize the definition of the MS group to non-rigid molecules, we generalize the definition of an unfeasible operation:

> The MS group of a molecule is the subgroup of the CNPI group obtained by removing unfeasible operations, where an unfeasible operation causes a coordinate change that moves the molecule from one version to another across an insuperable energy barrier.

With this generalized definition, feasible operations either do not change the numbered version or they change it to another version into which there is observable tunnelling.

For a non-rigid molecule, there are one or more contortional[1] *large amplitude* vibrations that give rise to tunnelling splittings. In developing the zero-order harmonic-oscillator rigid-rotor Hamiltonian for a rigid molecule, in chapters 4 and 5, the Taylor's series expansions of the potential function and μ

---

[1] We use the general word *contortion* for any large amplitude tunnelling motion such as inversion or internal rotation.

tensor elements were truncated at their leading terms; see equations (4.6) and (5.29). For a non-rigid molecule, these Taylor's series expansions are made in the small-amplitude vibrational coordinates only, with coefficients that are analytical functions of the (large amplitude) contortional coordinate (or coordinates). Using the leading terms in these Taylor's series expansions gives the zero-order rotation–contortion–vibration Hamiltonian and this can usually be separated into rotational, contortional and (small-amplitude) vibrational parts. The MS group of the non-rigid molecule can be used to classify the *roconvibrational* states of the molecule, to determine selection rules for transitions and to determine which states can perturb each other.

13.2   The ammonia molecule

The point group symmetry of the ground electronic state equilibrium structure of the NH_3 molecule is C_{3v} and, just as for CH_3F (see figure 8.1), there are two versions of this structure as shown in figure 13.1. However, unlike the situation for CH_3F, for NH_3 in its vibrational ground state, there is an observable tunnelling splitting of about 0.8 cm^{-1}. Thus, the MS group of ammonia is not the same as that for CH_3F since operations such as (12) and E^* that interconvert the versions are now feasible using the extended definition. All operations of the CNPI group are feasible for NH_3 and the MS group is the group $D_{3h}(M)$ as for H_3^+.

A very important use of the MS group for a non-rigid molecule can be illustrated using the NH_3 molecule and this concerns the determination of the relative intensities of spectral lines that are split as a result of tunnelling. If one used a very low resolution spectrometer that did not reveal the inversion splitting of the lines in ammonia, one could understand the relative intensities of the lines using the statistical weights determined using the $C_{3v}(M)$ group. As for CH_3F, see equation (9.27), the proton spin species is $4A_1 \oplus 2E$ in the $C_{3v}(M)$ group. The ^{14}N nucleus has a spin $I = 1$ so its spin functions span $3A_1$ and the total nuclear spin species for $^{14}NH_3$ is

$$\Gamma_{ns} = 12A_1 \oplus 6E. \tag{13.1}$$

As for CH_3F, the overall species Γ_{tot} is restricted by the Pauli exclusion principle to be either A_1 or A_2 in $C_{3v}(M)$ and, as a result, from equation (13.1) we can determine that each level has a nuclear spin statistical weight of 12 using the $C_{3v}(M)$ group. With inversion tunnelling splittings resolved, the MS group becomes $D_{3h}(M)$, which has six different symmetry species rather than the three in $C_{3v}(M)$. The statistical weights of tunnelling $^{14}NH_3$ can be determined as for H_3^+ in section 9.3.2 using the $D_{3h}(M)$ group but we must multiply the g_{ns} given in table 9.3 for H_3^+ by 3 to allow for the ^{14}N nuclear spin degeneracy.

Using the reverse correlation table for $C_{3v}(M) \rightarrow D_{3h}(M)$ as given in table 7.7 we can see what happens when the levels of $^{14}NH_3$ as classified in $C_{3v}(M)$ are split by inversion tunnelling. With the nuclear spin statistical weights added, the result is given in table 13.1. From this table, we see that levels

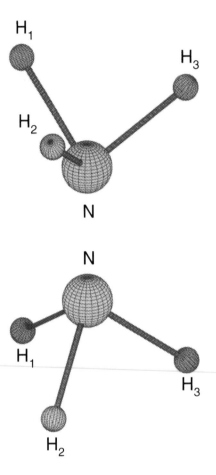

Figure 13.1. The two versions of the NH$_3$ molecule.

Table 13.1. The reverse correlation table for C_{3v}(M) to D_{3h}(M) for the ^{14}NH$_3$ ammonia molecule with nuclear spin statistical weights added.

C_{3v}(M)	D_{3h}(M)
$A_1(12)$	$A_1'(0) \oplus A_2''(12)$
$A_2(12)$	$A_2'(12) \oplus A_1''(0)$
$E(12)$	$E'(6) \oplus E''(6)$

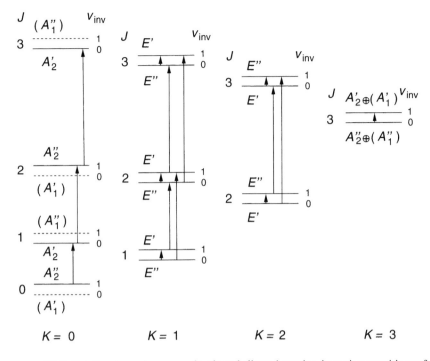

Figure 13.2. Rotation-inversion energy levels and allowed rotation-inversion transitions of NH_3, with $J \leq 3$ and $v_{inv} = 0$ or 1. Missing levels (i.e. levels with zero spin statistical weight) are indicated by broken lines and their symmetry species are bracketed; allowed electric dipole transitions are indicated by arrows.

of species A_1 or A_2 in $C_{3v}(M)$ will not be split into two observable levels by inversion tunnelling since one component will have zero statistical weight and be missing. The E levels will split into two observable levels and each of the levels will have a statistical weight of 6.

The derivation of symmetry selection rules for electric dipole transitions for NH_3 is identical to the derivation for H_3^+ since these two molecules have the same MS group. Consequently, it follows from the discussion of equations (7.87) and (7.88) in conjunction with table 13.1 that the nuclear spin allowed transitions for NH_3 satisfy the conditions $A_2' \leftrightarrow A_2''$ or $E' \leftrightarrow E''$. In figure 13.2, we show the rotation-inversion transitions of NH_3 involving levels with $J \leq 3$ that are allowed in the Born–Oppenheimer approximation if there is no rotation–vibration coupling. The quantum number v_{inv} labels the inversion-split levels. In principle, as a result of inversion tunnelling, each rotational level should give rise to a pair of levels having $v_{inv} = 0$ or 1. In practice, one of the components of such a pair can sometimes be missing as discussed earlier; the missing levels are indicated by broken lines in the figure.

13.3   Torsionally tunnelling ethylene

As discussed in section 8.3.5, the ethylene molecule is a rigid molecule in its ground electronic state and there are 12 versions of its equilibrium structure. These versions are shown in figure 8.9. The MS group of the version shown in figure 8.9(a) is the group of eight operations

$$\{E, (12)(34), (13)(24)(56), (14)(23)(56),$$
$$E^*, (12)(34)^*, (13)(24)(56)^*, (14)(23)(56)^*\}. \qquad (13.2)$$

The group is $D_{2h}(M)$ and its character table is given in table B.7.

The barrier between the forms (a) and (e) in figure 8.9 is the lowest of the barriers separating (a) from the other versions. If there was observed torsional tunnelling in the spectrum and one wished to symmetry label torsionally split levels, it would be necessary to apply the generalized definition of feasibility given earlier in order to obtain the MS group. With this definition, the permutations (12) or (34) become feasible, since the barrier separating versions (a) and (e) is now considered not to be insuperable. Also all products of these operations with the operations of $D_{2h}(M)$ given in equation (13.2) would also become feasible. However, operations such as (243) which interconverts versions (a) and (b), (23) which interconverts versions (a) and (c), (24) which interconverts versions (a) and (d), (56) which interconverts versions (a) and (g) and (234) which interconverts versions (a) and (f), all interconvert versions that are separated by an insuperable energy barrier and they remain unfeasible.

There are six equivalent sets of versions: [(a),(e)], [(b),(d)], [(c),(f)], [(g),(k)], [(h),(j)] and [(i),(l)] that are separated from each other by insuperable energy barriers. We can take any one of the six sets for the purpose of understanding the MS group and symmetry labelling the energy levels. Each would give the same energy level pattern and symmetry labels so that, in the event that torsional tunnelling was observed, there would be a six-fold structural degeneracy on each energy level.

For the set [(a),(e)], the operations in the MS group of torsionally tunnelling ethylene are obtained by adding (12) and (34) to the operations of the MS group of the rigid version, given in equation (13.2), and by including all the distinct products of the operations. They are:

$$\{E, (12)(34), (13)(24)(56), (14)(23)(56)$$
$$E^*, (12)(34)^*, (13)(24)(56)^*, (14)(23)(56)^*$$
$$(12), (34), (1324)(56), (1423)(56)$$
$$(12)^*, (34)^*, (1324)(56)^*, (1423)(56)^*\}. \qquad (13.3)$$

This is a group of 16 nuclear permutation and permutation-inversion operations that is called G_{16}. The character table of this group is given in table B.21. Regrettably there are two different conventions in the literature for labelling

Table 13.2. The correlation table for G_{16} to $D_{2h}(M)$.

G_{16}	$D_{2h}(M)$	G_{16}	$D_{2h}(M)$
A_1^+	A_g	A_1^-	A_u
A_2^+	B_{1u}	A_2^-	B_{1g}
B_1^+	A_g	B_1^-	A_u
B_2^+	B_{1u}	B_2^-	B_{1g}
E^+	$B_{3g} \oplus B_{2u}$	E^-	$B_{2g} \oplus B_{3u}$

Table 13.3. The reverse correlation table for $D_{2h}(M)$ to G_{16} with statistical weights added for $^{12}C_2H_4$.

$D_{2h}(M)$	G_{16}	$D_{2h}(M)$	G_{16}
$A_g(7)$	$A_1^+(1) \oplus B_1^+(6)$	$B_{2g}(3)$	$E^-(3)$
$A_u(7)$	$A_1^-(6) \oplus B_1^-(1)$	$B_{2u}(3)$	$E^+(3)$
$B_{1g}(3)$	$A_2^-(0) \oplus B_2^-(3)$	$B_{3g}(3)$	$E^+(3)$
$B_{1u}(3)$	$A_2^+(3) \oplus B_2^+(0)$	$B_{3u}(3)$	$E^-(3)$

the irreducible representations. In the remainder of this section, in order to be consistent with the use in the ethylene literature, we use the MW convention for labelling the irreducible representations.

In the paragraph containing equations (7.111) and (7.112), it is shown how to use character tables to determine the correlation of the symmetry species of a group to those of one of its subgroups. We can use this theory to determine the correlation of the species of G_{16} to those of its subgroup $D_{2h}(M)$ and these results are given in table 13.2.

Making use of equation (7.113), it is possible to determine the reverse correlation table $D_{2h}(M) \rightarrow G_{16}$ which gives the species in G_{16} induced by each symmetry species of $D_{2h}(M)$; these results are given in table 13.3. The numbers in brackets for each symmetry species are the nuclear spin statistical weights for $^{12}C_2H_4$, determined for the groups G_{16} and $D_{2h}(M)$ using the methods discussed in chapter 9.

The results in table 13.3 show the effect of torsional tunnelling on each of the eight symmetry types of rovibrational energy level of $^{12}C_2H_4$ in the group $D_{2h}(M)$. That is, once we have symmetry classified each level of the rigid molecule using $D_{2h}(M)$, we can simply look in table 13.3 to see what the effect of torsional tunnelling will be. Rovibrational levels of species A_g and A_u in $D_{2h}(M)$ have a statistical weight of seven; torsional tunnelling splits each into two, one with a statistical weight of six and one with a statistical weight of one, so that transitions originating in this pair of levels will have an intensity ratio 6:1.

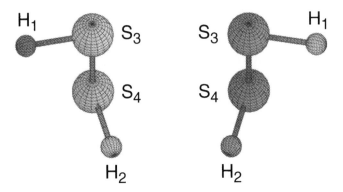

Figure 13.3. The two conformers of an HSSH molecule in the electronic ground state.

Levels of species B_{1g} and B_{1u} each have a statistical weight of three in $D_{2h}(M)$; torsional tunnelling produces one level with statistical weight of three and one with a statistical weight of zero so it is missing, and there is no splitting of such lines regardless of the torsional barrier height. Levels of species B_{2g}, B_{2u}, B_{3g} and B_{3u} each correlate with an E species of G_{16} and are not split by torsional tunnelling. This example shows that a group theory analysis is of great help in predicting how levels will split and in determining the intensity ratios of the split levels, without doing any numerical calculation of the tunnelling splittings.

For torsional tunnelling C_2H_4, the group G_{16} can be used to classify rotational, torsional and vibrational wavefunctions, to determine selection rules and to determine which levels can perturb each other. We will not go into these applications here but the use of G_{16} for understanding the Raman spectrum is discussed in section 13.6.

13.4   Intensity alternations for HSSH and DSSD

The spin statistical weights for H_2O and D_2O (table 9.1) give rise to intensity alternations of 1:3 and 6:3 ($= 2$:1), respectively, in their spectra. However, H_2O is a strongly asymmetric top with a κ value [see equation (5.84) and figure 5.6] of -0.42. As a result, the rotational energy level pattern and, hence, the line structure of its spectra are irregular and it is difficult to recognize the intensity alternation. The disulphane molecule $H^{32}S^{32}SH$ has the same 1:3 alternation but with $\kappa = -0.99996$, it is an almost perfect prolate symmetric top with regular line patterns in its spectra and an easily recognized intensity alternation.

Figure 13.3 shows the two conformers of an HSSH molecule in the electronic ground state; the two protons are labelled 1 and 2 and the two ^{32}S nuclei 3 and 4. The two conformers have dihedral angles (the angles between the $H_1S_3S_4$

and $H_2S_4S_3$ planes) close to $90°$. In high-resolution spectroscopic experiments, splittings due to tunnelling between the conformers are observed and so the appropriate MS group for disulfane is

$$\{E, (12)(34), E^*, (12)(34)^*\}. \tag{13.4}$$

This group is called G_4, see table B.18, and it is isomorphic to the MS group for the water molecule, with the obvious mapping:

$$
\begin{array}{ccccc}
H_2O: & E & (12) & E^* & (12)^* \\
HSSH: & E & (12)(34) & E^* & (12)(34)^*.
\end{array}
\tag{13.5}
$$

We label the rovibronic states of HSSH by the irreducible representations A_1, A_2, B_1 and B_2, using the Γ_3 notation of table B.18. The CNPI group of HSSH contains, in addition to the feasible elements in equation (13.4), the unfeasible elements (12), (34), (12)* and (34)*. If torsional tunnelling were not observed in HSSH, then its MS group would simply be the group $\{E,(12)\}$ and, since this MS group contains neither E^* nor any permutation–inversion element, HSSH would be chiral (see section 14.7).

A ^{32}S nucleus has spin $I = 0$ and so the possible spin functions for the nuclei 3 and 4 in $H^{32}S^{32}SH$ are

$$\delta_3 = |I_3, \sigma_3\rangle = |0, 0\rangle \quad \text{and} \quad \delta_4 = |I_4, \sigma_4\rangle = |0, 0\rangle. \tag{13.6}$$

By analogy with equations (9.8)–(9.11), we form the following four possible spin functions for $H^{32}S^{32}SH$:

$$\Phi_{ns,1} = \alpha\alpha\delta_3\delta_4 \qquad \Phi_{ns,2} = \beta\beta\delta_3\delta_4 \qquad \Phi_{ns,3} = \frac{1}{\sqrt{2}}[\alpha\beta + \beta\alpha]\delta_3\delta_4 \tag{13.7}$$

and

$$\Phi_{ns,4} = \frac{1}{\sqrt{2}}[\alpha\beta - \beta\alpha]\delta_3\delta_4. \tag{13.8}$$

The functions $\Phi_{ns,1}$, $\Phi_{ns,2}$ and $\Phi_{ns,3}$ each have symmetry A_1, whereas $\Phi_{ns,4}$ has symmetry B_2.

The operation (12)(34) in equation (13.4) simultaneously interchanges the two protons, which are fermions, and the two ^{32}S nuclei, which are bosons. The complete internal wavefunction for HSSH is required by Bose–Einstein statistics to remain unchanged under the interchange (34) of the two ^{32}S nuclei whereas, because of Fermi–Dirac statistics, it must change sign under the interchange (12) of the two protons. Thus, it must change sign under the combined interchange operation (12)(34), which means that, exactly as for H_2O, the complete internal wavefunctions for HSSH have $\Gamma_{tot} = B_1$ or B_2.

The determination of the spin statistical weights for HSSH proceeds in a manner completely analogous to that for H_2O. The four nuclear spin functions

Table 13.4. Spin statistical weights for HSSH and DSSD.

$H^{32}S^{32}SH$				$D^{32}S^{32}SD$			
Γ_{rve}	$\Gamma_{ns,t}$	Γ_{tot}	g_{ns}	Γ_{rve}	$\Gamma_{ns,t}$	Γ_{tot}	g_{ns}
A_1	B_2	B_2	1	A_1	$6A_1$	A_1	6
A_2	B_2	B_1	1	A_2	$6A_1$	A_2	6
B_1	$3A_1$	B_1	3	B_1	$3B_2$	A_2	3
B_2	$3A_1$	B_2	3	B_2	$3B_2$	A_1	3

for HSSH in equations (13.7) and (13.8) transform according to the reducible representation $3A_1 \oplus B_2$ and, with $\Gamma_{tot} = B_1$ or B_2, the spin statistical weights for HSSH are identical to those of H_2O given in table 9.1. Similarly, the spin statistical weights for DSSD are identical to those of D_2O. We give, in table 13.4, the spin statistical weights for HSSH and DSSD—this table is essentially a copy of table 9.1.

The symmetry selection rules for allowed electric dipole transitions in HSSH are identical to those of H_2O given in equation (7.59). The series of rotational transitions

$$J_{K_a K_c} = J_{1J} \leftarrow J_{0J} \tag{13.9}$$

in the ground vibronic state, for $J = 1, 2, 3, \ldots$ have been obtained experimentally and they are shown in figure 13.4. The experimental technique used to obtain the spectrum does not yield the transmittance τ (see section 1.1) directly, instead a signal proportional to the second derivative $d^2\tau/d\nu^2$ is obtained as a function of the frequency. An absorption line, which corresponds to a dip in the transmittance, produces a lineshape function with one central positive lobe flanked by two smaller negative lobes. In figure 13.4, we label the transitions in equation (13.9) by J and the 1:3 intensity alternation for J even:odd is clearly visible.

13.5   The water dimer and the water trimer

13.5.1   Water dimer

The equilibrium structure of the water dimer, as obtained from both *ab initio* calculation and experimental spectroscopic study, is shown in figure 13.5 and one wants to understand the nature and quantitative strength of the interaction between the monomer units as a function of their separation and relative orientation. The thin bond between an H atom on one monomer (the donor monomer) and the O atom of the other (the acceptor monomer) is called a *hydrogen bond* and it is understood as an intermonomer electrostatic interaction between an incompletely

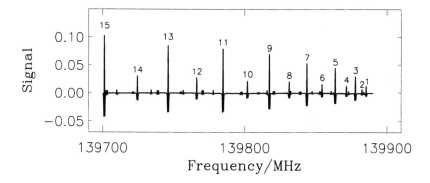

Figure 13.4. The sequence of $H^{32}S^{32}SH$ transitions defined in equation (13.9), recorded at the University of Cologne by G Winnewisser and F Lewen using the Cologne Terahertz Spectrometer. The transitions are labelled by J and the 1:3 intensity alternation for J even:odd is clearly visible. The weak unlabelled transitions are 'hot transitions' (see section 1.4) or transitions in HSSH isotopomers containing ^{34}S.

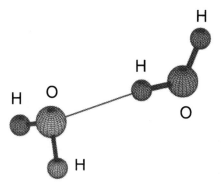

Figure 13.5. The equilibrium structure of the water dimer. The hydrogen bond between the water monomers is drawn as a thin line from the H atom of the donor monomer at the right to the O atom of the acceptor monomer at the left.

screened positively charged H nucleus on the donor monomer and the pair of electrons in one of the 'lone-pair' sp^3 orbitals on the O atom of the acceptor monomer. This weak bond holds the monomer units together. The study of the spectrum of this molecule, with the aim of characterizing the intermonomer potential, is of great importance for unraveling the hydrogen bond interactions in liquid water, which are continually forming and breaking as the monomer units move about.

The CNPI group has order $h = 4! \times 2! \times 2 = 96$ and the point group of the

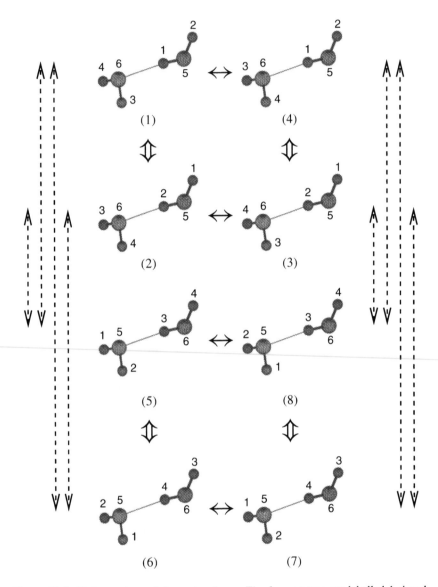

Figure 13.6. Eight versions of the water dimer. The four protons are labelled 1–4 and the two oxygen nuclei are labelled 5 and 6, respectively (see also figure 13.5). Tunnelling pathways (see the text) are indicated by double-headed arrows: Single arrows (↔) indicate acceptor tunnelling, broken arrows (◄-►) indicate donor–acceptor interchange tunnelling and double arrows (⇔) indicate donor tunnelling.

equilibrium structure[2] is C_s of order two. So there are $96/2 = 48$ versions of the equilibrium structure. Eight versions of the water dimer are shown in figure 13.6 and these versions do not have very high potential energy barriers separating them. There are five sets of eight equivalent versions obtained from these by breaking the OH bonds, such as would involve dimers like $H_1O_6H_4–H_2O_5H_3$ but we ignore them since they are behind insuperable energy barriers and only cause an unresolved six-fold structural degeneracy. The intermonomer potential energy surface for the water dimer has three distinct barriers between the eight versions in figure 13.6 and they are indicated by the differently drawn double-headed arrows. We will consider the MS groups appropriate for the water dimer depending on the various tunnelling possibilities among these versions and deduce the relative intensities of the lines that are split by these tunnelling motions.

If the water dimer were rigid with C_s point group symmetry then the MS group of version (1) would be

$$C_s(M) = \{E, (34)^*\}. \tag{13.10}$$

In this circumstance, each energy level would have a hidden eight-fold degeneracy from these versions. This degeneracy becomes revealed as the various tunnelling splittings resulting from the tunnellings indicated in figure 13.6 are resolved.

The lowest of the three barriers between the versions in figure 13.6 is that between versions (1) and (4), for example. This is called *acceptor tunnelling* or *acceptor switching* and, by *ab initio* calculation, it has a barrier of about 200 cm^{-1}. This is essentially an internal rotation of the acceptor monomer about its C_2 axis. If splittings in the spectrum from acceptor tunnelling are observed then, for versions (1) and (4), the permutation (34) becomes feasible and the MS group becomes

$$G_4 = \{E, (34), E^*, (34)^*\}. \tag{13.11}$$

Each level would now have a hidden four-fold degeneracy from the other versions depicted in figure 13.6.

The next higher barrier is that between versions (1) and (5), for example, and in this tunnelling motion the donor and acceptor monomers change roles; thus, this is called *donor–acceptor interchange* tunnelling. In this motion, the O_6H_3 bond of the acceptor monomer in version (1) rotates in to form the H-bond, while at the same time in a concerted manner the O_5H_1 bond of the donor monomer rotates out of the H-bond[3]. By *ab initio* calculation, this has a barrier of about 300 cm^{-1}. If donor–acceptor tunnelling and acceptor tunnelling both occur, then all eight of the versions in figure 13.6 become accessible to each other and the MS group becomes the group G_{16} given in equation (13.3). The same MS group as for torsionally tunnelling ethylene is obtained. We see that the MS groups of

---

[2] The nuclei of the donor monomer and the oxygen nucleus of the acceptor monomer define the plane of reflection symmetry.

[3] Making this tunnelling motion from version (1), it would then be necessary to rotate the molecule in space in order to obtain version (5) aligned as it is in figure 13.6.

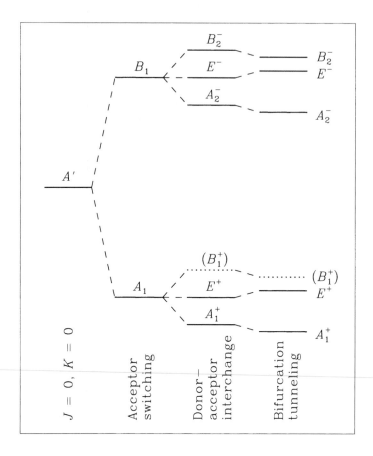

Figure 13.7. The splitting pattern of the $J = K = 0$ level of $(H_2O)_2$. From Fellers R S *et al* 1999 *J. Chem. Phys.* **110** 6306. The level of B_1^+ symmetry has zero nuclear spin statistical weight and is missing.

non-rigid molecules having different equilibrium structures can be the same. For historical reasons, the symmetry species for the water dimer are named using the LH convention.

The highest of the three tunnelling barriers between the version in figure 13.6 is the *donor tunnelling* or *bifurcation* path that, for example, connects versions (1) and (2). This has an *ab initio* barrier of about 650 cm^{-1}. However, the molecule can tunnel from (1) to (2) by moving through version (5) using two donor–acceptor interchanges and this means than no new splittings are introduced by including the donor-tunnelling. However, the barrier to donor-tunnelling will quantitatively affect the energy level splittings.

To recognize the fingerprint of these splittings in the spectrum, it is necessary to determine the nuclear spin statistical weights for $(H_2{}^{16}O)_2$ in the groups $C_s(M)$, G_4 and G_{16} and to determine the reverse correlation of the species $C_s(M) \rightarrow G_4 \rightarrow G_{16}$. The result is [where we use the $\Gamma(LH)$ notation for the irreducible representations of the G_{16} group in table B.21 and the Γ_3 notation for the G_4 group in table B.18]

$$
A'(16) \rightarrow \begin{cases} B_1(12) & \rightarrow & A_2{}^-(3) & \oplus & B_2{}^-(6) & \oplus & E^-(3) \\ A_1(4) & \rightarrow & A_1{}^+(1) & \oplus & B_1{}^+(0) & \oplus & E^+(3) \end{cases}
$$

$$
A''(16) \rightarrow \begin{cases} B_2(12) & \rightarrow & A_2{}^+(3) & \oplus & B_2{}^+(6) & \oplus & E^+(3) \\ A_2(4) & \rightarrow & A_1{}^-(1) & \oplus & B_1{}^-(0) & \oplus & E^-(3). \end{cases}
$$

$$(13.12)$$

The splitting pattern of the $J = K = 0$ level is shown in figure 13.7, and the experimentally derived term values of the $J = 0$ and 1, $K = 0$, levels are shown in figure 13.8; we have indicated the allowed transitions in the latter figure. From this experimental determination, we see that tunnelling between all eight versions is observed.

13.5.2 Water trimer

The equilibrium structure of the water trimer, again as obtained from both *ab initio* calculation and experimental spectroscopic study, is shown in figure 13.9. This molecule is important because its energy levels will reveal the effect on the intermonomer potential of the approach of a third water molecule. This is crucial to developing an accurate model of the potential for liquid water. The equilibrium structure has no symmetry (point group C_1 containing only the identity E). The CNPI group has order $h = 6! \times 3! \times 2 = 8640$. Thus, there are 8640 versions of this structure. Including all possible tunnellings except those that involve the breaking of OH bonds (as done previously for the water dimer) leads to 96 versions of such a numbered equilibrium structure. There are 90 equivalent sets of these 96 versions. Experimental studies have been made of the tunnelling splittings and their relative intensities, in order to determine the appropriate MS group and hence the feasible tunnelling routes. For the fully protonated species, the spectral splittings are too large to allow reliable relative intensities to be obtained and the fully deuterated molecule (which has much smaller splittings due to the heavier mass of the deuteron) is used. The quartet tunnelling splittings observed on two rotational transitions of $(D_2O)_3$ are shown in figure 13.10. The quartet structure and the 76:108:54:11 statistical weight pattern agrees with the predictions[4] obtained using an MS group of 48 elements. This means that there are two subsets of 48 versions within which tunnelling is observed to occur.

---

[4] Wales D J 1993 *J. Am. Chem. Soc.* **115** 11 180.

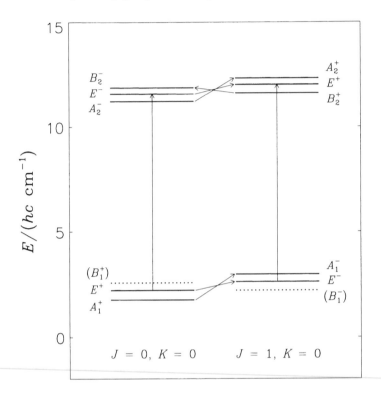

Figure 13.8. Experimentally derived term values for $(H_2O)_2$. Adapted from Fellers R S *et al* 1999 *J. Chem. Phys.* **110** 6306. We have added, using arrows, the allowed electric dipole transitions between the levels. The levels of B_1^+ and B_1^- symmetry have zero nuclear spin statistical weight and are missing.

13.6 Ethylene and its Raman spectrum

In section 13.3, we discussed the MS group of ethylene in the event that torsional tunnelling splittings were resolved. In its ground electronic state, the torsional barrier is about 25 000 cm^{-1} and the reader may have wondered why we chose such an unrealistic example. The reason concerns the interpretation of the resonance Raman scattering spectrum of ethylene. If this spectrum is excited by illuminating a gas phase sample with radiation that is resonant with the V-N electronic transition[5], it is necessary to use the group G_{16} in order to interpret the results. The N state of ethylene is the electronic ground state and the V state is an excited electronic state.

The wavenumber values of the lines in the Raman scattering spectrum are

---

[5] Hence, the designation *resonance* Raman scattering.

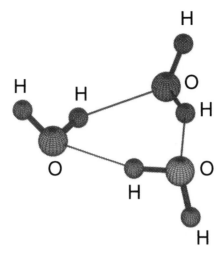

Figure 13.9. The equilibrium structure of the water trimer.

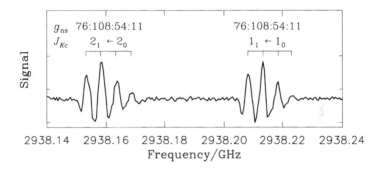

Figure 13.10. An experimental spectrum of the fully deuterated water trimer $(D_2O)_3$. Adapted from Liu K *et al* 1994 *J. Am. Chem. Soc.* **116** 3507.

shifted from the wavenumber of the exciting line by an amount that corresponds to the rotation–vibration term values; see figure 1.8. The selection rules in the resonance Raman effect can be obtained by applying those of the electric dipole transitions twice: Once for the transition induced by the exciting radiation up from the ground electronic state rotation–vibration level to the excited electronic state level; and, second for the transition back down from the excited electronic state level to the final ground electronic state rotation–vibration level that is characterized by the scattered radiation. For ethylene in its V excited electronic state, the equilibrium structure is twisted away from the $\boldsymbol{D}_{2h}$ symmetry of the ground electronic state to have a torsional angle of 90° so that the point group symmetry is D_{2d} like ground-state allene. As a result, internal rotation deformation becomes possible in this Raman process.

> We, thus, have to broaden the definition of feasibility to include tunnelling made possible by electronic transition when we wish to set up the MS group for situations in which a change in electronic state occurs.

With this generalization of the definition of feasibility, the MS group that one uses to interpret the Raman spectrum of ethylene is the group G_{16}.

A selection rule for the Raman spectrum of centro-symmetric molecules like ethylene is that g↔u transitions are forbidden (see section 12.5). Yet, in the resonance Raman spectrum of ethylene odd overtones $v = 0 \rightarrow 1, 0 \rightarrow 3, 0 \rightarrow 5$, etc. of the torsional vibration, of u symmetry, weakly occur. This is explained because such transitions are no longer forbidden in the G_{16} group appropriate when torsion is feasible. These vibrational transitions are accompanied by rotational transitions[6] involving $\Delta K_a = \pm 1$.

Thus, electronic transition increases the number of feasible operations and gives rise to an enlarged MS group if the transition makes deformations possible that carry the molecule over insuperable potential energy barriers. In the same way, the breakdown of the Born–Oppenheimer approximation can also lead to an enlarged MS group.

It can be laborious to determine the character table of a 'new' MS group, and to use it to obtain nuclear spin statistical weights. However, a free software package called GAP[7] is available for these tasks.

13.7 Problems

13.1 Determine the MS groups for hydrogen peroxide H_2O_2, ethane C_2H_6 and methanol CH_3OH, in the event (as is the case) that torsional tunnelling splittings are observed.

13.2 Construct the reverse correlation table like table 13.1 with statistical weights added that are appropriate for $^{14}ND_3$.

13.3 Construct the reverse correlation table like table 13.3 with statistical weights added that are appropriate for $^{12}C_2D_4$.

---

[6] See Watson J K G *et al* 1996 *J. Chem. Phys.* **105** 1348.
[7] Schmied R and Lehmann K K 2004 *J. Mol. Spectrosc.* **26** 201.

PART 4

OTHER SYMMETRIES AND SYMMETRY VIOLATION

Chapter 14

Other symmetries

14.1 The fourth postulate of quantum mechanics

So far in this book, apart from a brief discussion in section 1.5 on lineshapes, we have ignored the possibility of time dependence.

To allow for time dependence, it is postulated that state functions depending on coordinates q and time t satisfy the time-dependent Schrödinger equation

$$i\hbar\frac{\partial \Psi_n(q,t)}{\partial t} = \hat{H}\Psi_n(q,t) \qquad (14.1)$$

where $\hat{H}$ is the Hamiltonian set up according to the first postulate.

This is the fourth postulate of quantum mechanics and it is valid whether or not the Hamiltonian contains t.

For a Hamiltonian that does not depend on time (such as that for an isolated molecule in free space), we substitute

$$\Psi_n(q,t) = \Phi_n(q)f_n(t) \qquad (14.2)$$

into equation (14.1) to give

$$i\hbar\Phi_n(q)f'_n(t) = f_n(t)[\hat{H}\Phi_n(q)] \qquad (14.3)$$

where $f'_n(t) = \partial f_n(t)/\partial t$. Dividing both sides by $\Phi_n(q)f_n(t)$, we obtain

$$i\hbar\frac{f'_n(t)}{f_n(t)} = \frac{\hat{H}\Phi_n(q)}{\Phi_n(q)}. \qquad (14.4)$$

293

The two sides of this equation can only be equal if both of them are equal to the same constant and, from the form of the right-hand side, the constant is a stationary state eigenvalue E_n of $\hat{H}$ with $\Phi(q) = \Phi_n(q)$ being the related eigenfunction. Putting the left-hand side of equation (14.4) equal to E_n gives

$$\frac{\partial}{\partial t} f_n(t) = -i \frac{E_n}{\hbar} f_n(t) \tag{14.5}$$

which [see equation (2.1)] we can solve to obtain

$$f_n(t) = e^{(-iE_n/\hbar)t} \tag{14.6}$$

so that, for stationary states, we have

$$\Psi_n(q, t) = \Phi_n(q) e^{(-iE_n/\hbar)t} \tag{14.7}$$

where

$$\hat{H} \Phi_n(q) = E_n \Phi_n(q). \tag{14.8}$$

In situations where the time dependence of the Hamiltonian can be separated as a small perturbation, V' say, one can use the stationary-state eigenfunctions of the time-independent part of the Hamiltonian given in equation (14.7) as a basis set for treating the effect of V'. This is done in treating the effect of weak electromagnetic radiation on an isolated gas phase molecule and it leads to the expression for the absorption line strength in terms of matrix elements of the dipole moment operator between time-independent eigenfunctions $\Phi_n(q)$ of the molecular Hamiltonian [see equation (2.87)].

14.2   Conservation laws

The time-dependent Schrödinger equation can be used to show how symmetry operations lead to conservation laws. Let $\hat{O}$ be a symmetry operation of a molecular Hamiltonian $\hat{H}$; in the general case, $\hat{H}$ can depend on time. Since $\hat{O}$ is a symmetry operation,

$$[\hat{H}, \hat{O}] = 0. \tag{14.9}$$

To determine how the expectation value of $\hat{O}$ for a time-dependent state function $\psi_n(q, t)$ say, changes with time, we evaluate

$$\frac{\partial}{\partial t} \langle \psi_n(q, t) | \hat{O} | \psi_n(q, t) \rangle$$

$$= \left\langle \frac{\partial \psi_n(q, t)}{\partial t} | \hat{O} | \psi_n(q, t) \right\rangle + \left\langle \psi_n(q, t) | \frac{\partial}{\partial t} (\hat{O} \psi_n(q, t)) \right\rangle$$

$$= \left\langle \frac{\partial \psi_n(q, t)}{\partial t} | \hat{O} | \psi_n(q, t) \right\rangle + \left\langle \psi_n(q, t) | \hat{O} | \frac{\partial \psi_n(q, t)}{\partial t} \right\rangle \tag{14.10}$$

where, in the last equality, we have used the fact that $\hat{O}$ does not depend explicitly on t. We obtain $\partial \psi_n(\boldsymbol{q}, t)/\partial t$ from equation (14.1) and insert the resulting expression in equation (14.10); this gives

$$
\frac{\partial}{\partial t} \langle \psi_n(\boldsymbol{q}, t) | \hat{O} | \psi_n(\boldsymbol{q}, t) \rangle
$$

$$
= \frac{\mathrm{i}}{\hbar} (\langle \hat{H} \psi_n(\boldsymbol{q}, t) | \hat{O} | \psi_n(\boldsymbol{q}, t) \rangle - \langle \psi_n(\boldsymbol{q}, t) | \hat{O} | \hat{H} \psi_n(\boldsymbol{q}, t) \rangle)
$$

$$
= \frac{\mathrm{i}}{\hbar} \langle \psi_n(\boldsymbol{q}, t) | [\hat{H}, \hat{O}] | \psi_n(\boldsymbol{q}, t) \rangle = 0 \qquad (14.11)
$$

where we have used equation (14.9) in conjunction with the fact that $\hat{H}$ is Hermitian [see the first equality in equation (2.12)]. Equation (14.11) shows that the expectation value of an operation that commutes with the Hamiltonian (i.e. a symmetry operation) does not change with the passage of time; it is said that the symmetry is *conserved*.

One interesting conservation law is the conservation of fermion and boson permutation symmetry; this is an aspect of the fifth postulate of quantum mechanics that is stated on page 179. Any permutation of identical particles is a symmetry operation since it commutes with the Hamiltonian of the Universe. Hence, once created, a particle that obeys Fermi–Dirac statistics or Bose–Einstein statistics (see section 9.1) will always continue to obey such statistics.

14.3 Electron permutation symmetry

Any permutation of the positions and spins of the electrons in a molecule is a symmetry operation. The use of this symmetry is implied when we use Slater determinant basis functions. In this way, all electrons are treated equivalently and the complete (orbital and spin) electronic wavefunctions are anti-symmetric with respect to any odd permutation of the electrons (obeying the Pauli exclusion principle).

For an n-electron molecule, the complete electron permutation group (called the *symmetric group* S_n, with order $n!$) finds specialized use in the *symmetric group approach* to performing CI electronic wavefunction calculations. In this approach, one does not use Slater determinants. Instead one formally sets up orbital and spin functions that transform irreducibly in S_n and then one combines them in such a way that the product spin-orbital functions transform as the one-dimensional irreducible representation $\Gamma^{(\mathrm{e})}(\mathrm{A})$ having character $+1$ under all even electron permutations and character -1 under all odd electron permutations. Such symmetry behaviour is required to assure compliance with the Pauli exclusion principle. Two representations of S_n whose product contains $\Gamma^{(\mathrm{e})}(\mathrm{A})$ are said to be *dual* or *associate* and they must be of the same dimension. In actual practice, the spin functions are never actually constructed but rather the orbital functions are constructed in what is called a *spin-adapted* way. Using this technique in a many-electron problem can simplify what would otherwise be a rather laborious procedure.

The restriction that the symmetry of the electronic spin-orbital functions has to be $\Gamma^{(e)}(A)$, i.e. the Pauli exclusion principle, leads to the periodic system of the elements. It was pointed out in section 3.3.3 that for a one-electron atom the AO energies depend only on the quantum number n. However, in a many-electron atom, the AO energies depend also on the quantum number l because an outer electron moves in the effective field of the inner electrons. The nuclear charge is 'screened' by the inner electrons to an extent that depends both on the l and n quantum numbers for the outer electron. With some simplifying approximations, we can say that, for a many-electron atom, the orbitals lie in the order 1s, 2s, 2p, 3s, 3p, 4s, 3d, 4p, 5s, 4d, 5p, 6s, ... As we add electrons one at a time under the restriction of the Pauli exclusion principle, which means we can only place two electrons (having opposite spins) in an orbital, there will be periodically recurring outer orbital structures. As an example, for the rare earth elements, we have:

Be $(1s)^2(2s)^2$

Mg $(1s)^2(2s)^2(2p)^6(3s)^2$

Ca $(1s)^2(2s)^2(2p)^6(3s)^2(3p)^6(4s)^2$

Sr $(1s)^2(2s)^2(2p)^6(3s)^2(3p)^6(4s)^2(3d)^{10}(4p)^6(5s)^2$

Ba $(1s)^2(2s)^2(2p)^6(3s)^2(3p)^6(4s)^2(3d)^{10}(4p)^6(5s)^2 (4d)^{10}(5p)^6(6s)^2$.

Without the Pauli exclusion principle, there would be no such periodically recurring outermost electronic orbital structure. The chemical properties of an atom, like those of a molecule, are very largely determined by the orbital configuration of the outermost (highest-energy) electrons. This means that there are periodically recurring chemical properties and, hence, the periodic table. This is a consequence of electron permutation symmetry and the Pauli exclusion principle.

14.4   Translational symmetry

In section 2.5, we discussed space-fixed XYZ axes and the separation of the translation and internal (rovibronic) energies. This separation was achieved by referring the coordinates of the l particles in the molecule (N nuclei and $l - N$ electrons) to XYZ axes that are parallel to the XYZ axes but with origin at the molecular centre of mass (X_0, Y_0, Z_0). In this way, the molecule is described by the $3l$ coordinates

$$X_0, Y_0, Z_0, X_2, Y_2, Z_2, X_3, Y_3, Z_3, \ldots, X_l, Y_l, Z_l.$$

The coordinates (X_1, Y_1, Z_1) are redundant since they can be determined from the condition that the XYZ axes have their origin at the molecular centre of mass [see equation (2.51)].

In describing translational operations, we use the so-called *active* picture in which the XYZ axes remain fixed in space and the molecule is moved. The alternative and equivalent *passive* picture involves keeping the molecule fixed and moving the XYZ axes in the opposite direction. A translational operation changes the XYZ coordinates of all nuclei and electrons in the molecule by constant amounts, $(\Delta X, \Delta Y, \Delta Z)$ say, so that

$$(X_i, Y_i, Z_i) \rightarrow (X_i + \Delta X, Y_i + \Delta Y, Z_i + \Delta Z). \tag{14.12}$$

Such an operation has the effect of changing the centre-of-mass coordinates

$$(X_0, Y_0, Z_0) \rightarrow (X_0 + \Delta X, Y_0 + \Delta Y, Z_0 + \Delta Z) \tag{14.13}$$

whereas the coordinates $X_2, Y_2, Z_2, X_3, Y_3, Z_3, \ldots, X_l, Y_l, Z_l$ are unchanged. The spins of the l particles are unaffected by such a translational operation and, in this section, we ignore spin. This translational operation is denoted $R_T^{(\Delta X, \Delta Y, \Delta Z)}$ and it is a symmetry operation for the molecular Hamiltonian of a single isolated molecule in uniform field-free space; the classical expression for the spin-free molecular energy E, see equation (2.54), is unaffected by $R_T^{(\Delta X, \Delta Y, \Delta Z)}$. Equivalently, all $R_T^{(\Delta X, \Delta Y, \Delta Z)}$ commute with the total molecular Hamiltonian[1] $\hat{H}$,

$$[R_T^{(\Delta X, \Delta Y, \Delta Z)}, \hat{H}] = 0. \tag{14.14}$$

The infinite set of all translational symmetry operations $R_T^{(\Delta X, \Delta Y, \Delta Z)}$ constitute a symmetry group that is called the translational symmetry group G_T.

It is necessary to define the effect of a translational symmetry operation on a function. Figure 14.1 shows how an NH_3 molecule, for example, is displaced a distance ΔX along the X axis by the translational symmetry operation that changes X_0 to $X'_0 = X_0 + \Delta X$. Together with the molecule, we have drawn a sine wave symbolizing the molecular wavefunction, Φ_j say. We have marked one wavecrest to keep track of the way the function is displaced by the symmetry operation. For the physical situation to be unchanged by the symmetry operation, the marked wavecrest and, thus, the entire wavefunction, is displaced by ΔX along the X axis as shown in figure 14.1. We define the effect of $R_T^{(\Delta X, \Delta Y, \Delta Z)}$ on a wavefunction in the usual way [see the first equality in equation (7.7)] by writing

$$R_T^{(\Delta X, \Delta Y, \Delta Z)} \Phi_j(X_0, Y_0, Z_0, X_2, Y_2, Z_2, X_3, Y_3, Z_3, \ldots, X_l, Y_l, Z_l)$$
$$= \Phi_j^{R_T^{(\Delta X, \Delta Y, \Delta Z)}}(X_0, Y_0, Z_0, X_2, Y_2, Z_2, X_3, Y_3, Z_3, \ldots, X_l, Y_l, Z_l). \tag{14.15}$$

---

[1] The total molecular Hamiltonian $\hat{H}$ is the sum of the translational Hamiltonian $\hat{H}_{trans}$ from equation (2.60) and the complete internal Hamiltonian $\hat{H}_{int}$ from equation (2.77).

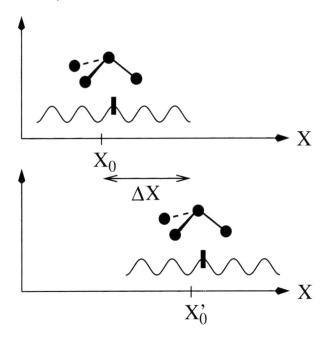

Figure 14.1. An NH_3 molecule and its wavefunction, symbolized by a sine wave, before (top) and after (bottom) a translational symmetry operation R_T. The wavefunction in the upper figure is Φ_j and that in the lower is $\Phi_j^{R_T}$.

where $\Phi_j^{R_T}$ is a new function of the coordinates. As we can appreciate from figure 14.1, in the active picture, the translated wavefunction is such that

$$\Phi_j^{R_T^{(\Delta X, \Delta Y, \Delta Z)}} (X_0 + \Delta X, Y_0 + \Delta Y, Z_0 + \Delta Z,$$
$$X_2, Y_2, Z_2, X_3, Y_3, Z_3, \ldots, X_l, Y_l, Z_l)$$
$$= \Phi_j(X_0, Y_0, Z_0, X_2, Y_2, Z_2, X_3, Y_3, Z_3, \ldots, X_l, Y_l, Z_l). \quad (14.16)$$

To follow the notation used in equation (7.7), we would write this equation as

$$\Phi^R(X_0', Y_0', Z_0', X_2', Y_2', Z_2', X_3', Y_3', Z_3', \ldots X_l, Y_l, Z_l)$$
$$= \Phi(X_0, Y_0, Z_0, X_2, Y_2, Z_2, X_3, Y_3, Z_3, \ldots, X_l, Y_l, Z_l). \quad (14.17)$$

This definition, which is the 'opposite' of that which has to be used for permutations as we define them [see the second equality in equation (7.7)], causes the wavefunction to 'move with the molecule' as shown for the X direction in figure 14.1. We can rewrite these equations as

$$R_T^{(\Delta X, \Delta Y, \Delta Z)} \Phi_j(X_0, Y_0, Z_0, X_2, Y_2, Z_2, X_3, Y_3, Z_3, \ldots, X_l, Y_l, Z_l)$$

$$= \Phi_j^{R_T^{(\Delta X, \Delta Y, \Delta Z)}}(X_0, Y_0, Z_0, X_2, Y_2, Z_2, X_3, Y_3, Z_3, \ldots, X_l, Y_l, Z_l)$$

$$= \Phi_j(X_0 - \Delta X, Y_0 - \Delta Y, Z_0 - \Delta Z,$$

$$X_2, Y_2, Z_2, X_3, Y_3, Z_3, \ldots, X_l, Y_l, Z_l). \tag{14.18}$$

We are now in a position to relate translational symmetry and the conservation of linear momentum. The quantum mechanical operators representing the translational linear momentum of a molecule are

$$(\hat{P}_X, \hat{P}_Y, \hat{P}_Z) = \left(-i\hbar \frac{\partial}{\partial X_0}, -i\hbar \frac{\partial}{\partial Y_0}, -i\hbar \frac{\partial}{\partial Z_0} \right) \tag{14.19}$$

and, in order to determine the relationship between $R_T^{(\Delta X, \Delta Y, \Delta Z)}$ and the $(\hat{P}_X, \hat{P}_Y, \hat{P}_Z)$ operators, we consider a translation $R_T^{(\delta X, 0, 0)}$ where δX is infinitesimally small. In this case, we can approximate the right-hand side of the second equality in equation (14.18) by a first-order Taylor's series expansion:

$$R_T^{(\delta X, 0, 0)} \Phi_j(X_0, Y_0, Z_0, X_2, Y_2, Z_2, \ldots, X_l, Y_l, Z_l)$$

$$= \Phi_j(X_0 - \delta X, Y_0, Z_0, X_2, Y_2, Z_2, \ldots, X_l, Y_l, Z_l)$$

$$= \Phi_j(X_0, Y_0, Z_0, X_2, Y_2, Z_2, \ldots, X_l, Y_l, Z_l) - \frac{\partial \Phi_j}{\partial X_0} \delta X. \tag{14.20}$$

From the definition of the translational linear momentum operator $\hat{P}_X$ [in equation (14.19)], we see that

$$\frac{\partial \Phi_j}{\partial X_0} = \frac{i}{\hbar} \hat{P}_X \Phi_j \tag{14.21}$$

and, by introducing this identity in equation (14.20), we obtain

$$R_T^{(\delta X, 0, 0)} \Phi_j = \Phi_j - \frac{i}{\hbar} \delta X \hat{P}_X \Phi_j \tag{14.22}$$

where we have omitted the coordinate arguments for brevity. Since the function Φ_j in equation (14.22) is arbitrary, it follows that we can write the symmetry operation as

$$R_T^{(\delta X, 0, 0)} = 1 - \frac{i}{\hbar} \delta X \hat{P}_X. \tag{14.23}$$

The operation $R_T^{(\Delta X, 0, 0)}$, for which ΔX is an arbitrary finite length, obviously has the same effect on a wavefunction as the operation $R_T^{(\delta X, 0, 0)}$ applied to the wavefunction $\Delta X / \delta X$ times. We simply divide the translation by ΔX into $\Delta X / \delta X$ steps, each step of length δX. This remains true in the limit of $\delta X \to 0$. Thus,

$$R_T^{(\Delta X, 0, 0)} = \lim_{\delta X \to 0} (R_T^{(\delta X, 0, 0)})^{\frac{\Delta X}{\delta X}} = \lim_{\delta X \to 0} \left(1 - \frac{i}{\hbar} \delta X \hat{P}_X \right)^{\frac{\Delta X}{\delta X}}$$

$$= \exp\left(-\frac{i}{\hbar} \Delta X \hat{P}_X \right) \tag{14.24}$$

where we have used the general identity

$$\lim_{x \to 0} (1 + ax)^{y/x} = \exp(ay). \tag{14.25}$$

We can derive expressions analogous to equation (14.24) for $R_T^{(0,\Delta Y,0)}$ and $R_T^{(0,0,\Delta Z)}$ and we can resolve a general translation $R_T^{(\Delta X,\Delta Y,\Delta Z)}$ as

$$R_T^{(\Delta X,\Delta Y,\Delta Z)} = R_T^{(\Delta X,0,0)} R_T^{(0,\Delta Y,0)} R_T^{(0,0,\Delta Z)}. \tag{14.26}$$

Consequently,

$$R_T^{(\Delta X,\Delta Y,\Delta Z)} = \exp\left[-\frac{i}{\hbar}\left(\Delta X \hat{P}_X + \Delta Y \hat{P}_Y + \Delta Z \hat{P}_Z\right)\right]. \tag{14.27}$$

In equation (14.27), the argument of the exponential involves operators and we deal with such an exponential by using the Taylor's series expansion

$$\exp(i\hat{O}) = 1 + i\hat{O} + \frac{1}{2!}(i\hat{O})^2 + \cdots \tag{14.28}$$

where $\hat{O}$ is a Hermitian operator. $R_T^{(\Delta X,\Delta Y,\Delta Z)}$ is a symmetry operation, i.e. it commutes with the total Hamiltonian for arbitrary $(\Delta X, \Delta Y, \Delta Z)$. The equations

$$[\hat{H}, \hat{P}_X] = [\hat{H}, \hat{P}_Y] = [\hat{H}, \hat{P}_Z] = 0 \tag{14.29}$$

imply

$$[\hat{H}, \hat{P}_X^n] = [\hat{H}, \hat{P}_Y^n] = [\hat{H}, \hat{P}_Z^n] = 0. \tag{14.30}$$

Substituting equation (14.28) into equation (14.27), we see that $R_T^{(\Delta X,\Delta Y,\Delta Z)}$ is a symmetry operation if and only if the $\hat{P}_A$ commute with $\hat{H}$. As a result, using equation (14.11), we see that the expectation value of each linear momentum operator is conserved in time. The conservation of linear momentum and the translational invariance of the molecular Hamiltonian [equation (14.14)] follow from each other.

The translational wavefunctions [see equation (2.69)] are eigenfunctions of the linear momentum operators $\hat{P}_A$ or linear combinations of such eigenfunctions and from the relation between the translational symmetry operations and the linear momentum operators, equation (14.27), we see that the effect of a translational symmetry operation is determined solely by the $\boldsymbol{k}$ vector, with components (k_X, k_Y, k_Z), which defines the linear momentum. In practice, translational states are discussed by using the $\boldsymbol{k}$ vector and the law of conservation of linear momentum and, although this implies translational symmetry and its conservation, the translational symmetry group is rarely invoked explicitly.

14.5   Rotational symmetry

Using the active picture, a rotational operation is an overall rotation of a molecule in space about an axis that passes through the centre of mass of the molecule. We have already introduced the XYZ axes that are parallel to the space-fixed XYZ axes but which have their origin at the molecular centre of mass (X_0, Y_0, Z_0). We now introduce the $X'Y'Z'$ axes which are initially coincident with the XYZ axes but which are attached to the molecule. If we rotate the molecule in space about an axis passing though its centre of mass, the $X'Y'Z'$ axes are rotated and the Euler angles specifying its orientation with respect to the XYZ axes will describe the rotation that the molecule has undergone. We call these Euler angles β, α and γ; these are defined in figure 14.2 and, using them, the components of the molecular orbital angular momentum $\hat{\boldsymbol{J}}$ (see section 2.7) are given by

$$\hat{J}_X = -i\hbar \left(-\sin\alpha \frac{\partial}{\partial \beta} + \csc\beta \cos\alpha \frac{\partial}{\partial \gamma} - \cot\beta \cos\alpha \frac{\partial}{\partial \alpha} \right) \quad (14.31)$$

$$\hat{J}_Y = -i\hbar \left(\cos\alpha \frac{\partial}{\partial \beta} + \csc\beta \sin\alpha \frac{\partial}{\partial \gamma} - \cot\beta \sin\alpha \frac{\partial}{\partial \alpha} \right) \quad (14.32)$$

and

$$\hat{J}_Z = -i\hbar \frac{\partial}{\partial \alpha}. \quad (14.33)$$

We define the rotational symmetry operation $R_R^{(\Delta\beta, \Delta\alpha, \Delta\gamma)}$ as being that rotation of the molecule in space that is described by the Euler angle changes

$$(\beta, \alpha, \gamma) \rightarrow (\beta + \Delta\beta, \alpha + \Delta\alpha, \gamma + \Delta\gamma). \quad (14.34)$$

A rotational operation such as this is a symmetry operation for a molecule in free space since space is *isotropic*; translational invariance follows from the fact that space is *uniform*.

Just as with the above treatment of translational symmetry, we want to derive an expression for $R_R^{(\Delta\beta, \Delta\alpha, \Delta\gamma)}$ so that, when applied to a wavefunction, it causes the change in the Euler angles given in equation (14.34). Because of the analogy between equation (14.33) and the definition of $\hat{P}_X$ in equation (14.19), we can repeat the arguments expressed in equations (14.20)–(14.24) with X replaced by α to show that

$$R_R^{(0, \Delta\alpha, 0)} = \exp\left(-\frac{i}{\hbar} \Delta\alpha \hat{J}_Z \right). \quad (14.35)$$

A more complicated derivation leads to the following expression for a general rotation operation:

$$R_R^{(\Delta\beta, \Delta\alpha, \Delta\gamma)} = \exp\left(-\frac{i}{\hbar} \Delta\alpha \hat{J}_Z \right) \exp\left(-\frac{i}{\hbar} \Delta\beta \hat{J}_Y \right) \exp\left(-\frac{i}{\hbar} \Delta\gamma \hat{J}_Z \right) \quad (14.36)$$

where, because $\hat{J}_Y$ and $\hat{J}_Z$ do not commute (see section 2.7), the exponential factors must be applied in the correct order. The infinite set of all rotation

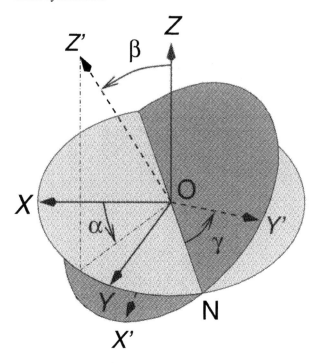

Figure 14.2. The definition of the Euler angles β, α and γ that relate the rotated axes $X'Y'Z'$ to the space-fixed axes XYZ. The origin of both axis systems is at the molecular centre of mass O and the node line ON is directed so that a right-handed screw is driven along ON in its positive direction by twisting it from Z to Z' through β where $0 \leq \beta \leq \pi$. α and γ have the ranges 0 to 2π. γ is measured from the node line.

operations $R_R^{(\Delta\beta, \Delta\alpha, \Delta\gamma)}$ form a group which we call the rotational group K(spatial); it is a symmetry group of the molecular rovibronic Hamiltonian $\hat{H}_{rve}$ because the operations in it commute with $\hat{H}_{rve}$:

$$[R_R^{(\Delta\beta, \Delta\alpha, \Delta\gamma)}, \hat{H}_{rve}] = 0. \tag{14.37}$$

This follows since $\hat{H}_{rve}$ commutes with $\hat{J}_Y$ and $\hat{J}_Z$ (see section 2.7). $\hat{H}_{rve}$ also commutes with $\hat{J}_X$ and we can write an equation like equation (14.11) in which we replace the operator $\hat{O}$ by the angular momentum $\hat{J}_A$. Thus, in the same way that translational symmetry is related to the conservation of linear momentum, so rotational symmetry is related to the conservation of angular momentum. These are two examples of the more general *Noether's theorem* first proved by Emmy Noether[2]; which relates conservation laws to symmetry.

[2] Noether E 1918 *Nachr. d. König. Gesellsch. d. Wiss. zu Göttingen, Math. Phys. Klasse* 235; English translation, Tavel M A 1971 *Transport Theory and Statistical Physics* **1** 183.

As discussed in section 2.7, eigenfunctions of $\hat{H}_{\text{rve}}$ can be set up that are simultaneously eigenfunctions of $\hat{J}^2$ [with eigenvalues $J(J + 1)\hbar^2, J = 0, 1, 2 \ldots$] and $\hat{J}_Z$ [with eigenvalues $m\hbar, m = -J, -J + 1, \ldots, J$] and they simplify the diagonalization of the matrix representation of $\hat{H}_{\text{rve}}$. Such a function will transform according to the mth row of the irreducible representation $D^{(J)}$ of K(spatial). With these basis functions, the matrix representation of $\hat{H}_{\text{rve}}$ will be block diagonal in J and m. The matrices in the irreducible representations $D^{(J)}$ of K(spatial) are the so-called D-matrices given in equation (5.52). Labelling states using the group K(spatial) is equivalent to labelling them using the rotational angular momentum quantum numbers J and m.

In using the group K(spatial), one needs to know that the product of two irreducible representations can be reduced as follows:

$$D^{(J_1)} \otimes D^{(J_2)} = D^{(J_1+J_2)} \oplus D^{(J_1+J_2-1)} \oplus \cdots \oplus D^{(|J_1-J_2|)}. \qquad (14.38)$$

This is a statement of the rule of vector addition. The totally symmetric irreducible representation of K(spatial) is the representation $D^{(0)}$. Thus, from equation (7.83), the vanishing integral rule, matrix elements between rovibronic states having J values of J' and J'' (transforming as $D^{(J')}$ and $D^{(J'')}$, respectively) of the field-free molecular rovibronic Hamiltonian (which transforms as $D^{(0)}$) will vanish if

$$D^{(J'')} \otimes D^{(J')} \not\supset D^{(0)} \qquad (14.39)$$

i.e. if $J' \neq J''$. The dipole moment operator [see equation (2.88)] transforms as $D^{(1)}$. which means that electric dipole rovibronic transitions are forbidden between rovibronic states states having J values of J' and J'' if

$$D^{(J'')} \otimes D^{(1)} \otimes D^{(J')} \not\supset D^{(0)}. \qquad (14.40)$$

Thus, electric dipole transitions are forbidden between states whose angular momentum quantum numbers differ by more than one unit or if they are both zero. This has already been mentioned on page 141 in connection with determining forbidden electric dipole transitions for the water molecule, and in equation (12.5).

In this discussion of rotational symmetry, we have neglected electron and nuclear spin angular momentum and have only been concerned with the rovibronic orbital angular momentum of the molecule. For a singlet electronic state, this angular momentum is denoted $\hat{J}$. For a non-singlet electronic state, we use the symbol $\hat{J}$ for the sum of the rovibronic orbital angular momentum (now called $\hat{N}$) and the electron spin angular momentum $\hat{S}$. If nuclear spin hyperfine structure is resolved, we add the total nuclear spin angular momentum $\hat{I}$ to $\hat{J}$ to form $\hat{F}$ [see equations (2.85) and (2.86)], and we classify the states of the internal Hamiltonian $\hat{H}_{\text{int}}$ in K(spatial). This introduces the total angular momentum quantum numbers F and m_F instead of J and m. If we allow for all possible interactions in the molecule, including those involving the nuclear spins, then it is only the total angular momentum $\hat{F}$ that is conserved, and we classify the internal states Φ_{int} in the symmetry group K(spatial). The internal Hamiltonian $\hat{H}_{int}$ rigorously commutes with the operations of K(spatial). The total angular momentum quantum number F is the vector sum of the rovibronic and total nuclear spin angular momentum quantum numbers J and I, so that $F = J + I, J + I - 1, \ldots, |J - I|$ [see equation (14.38)]. Thus, in the event that nuclear hyperfine interactions are resolved, equations (14.39) and (14.40) are replaced by equations involving F' and F'' rather than J' and J''; see equation (12.6).

The symmetry of the rovibronic and internal states in K(spatial) gives new symmetry information over and above that obtained by using the molecular symmetry group G_{MS} for any molecule. Thus, to label by symmetry the states of a molecule, we use both of these groups. The symmetry labelling obtained consists of the angular momentum quantum numbers J and m (or F and m_F if there is resolved nuclear hyperfine structure) from K(spatial), together with an irreducible representation label from G_{MS}.

14.6   Charge conjugation

The electrostatic potential energy of a molecule, given in equation (2.45), is unchanged if we change the sign of the charge of each particle. Thus, the complete internal electromagnetic Hamiltonian given in equation (2.77) is invariant to the operation of 'changing the sign of the charge of every particle'. However, in a complete theoretical treatment allowing for relativistic effects, it is necessary to generalize this symmetry operation to that of changing every particle into its antiparticle. This symmetry operation is called *charge conjugation* C. When C is applied to a charged particle, the resultant antiparticle has the opposite charge but the operation C also applies to a neutral particle such as a neutron (changing it to an antineutron).

The symmetry operation C by itself is of some use in elementary particle physics and it is involved by implication when the symmetry of a molecule such as dipositronium Ps_2 is considered; the dipositronium molecule consists of two positrons (antielectrons that have the mass of an electron and charge $+e$) and two electrons (each with charge $-e$). Because of positron–electron annihilation, this short-lived species has not yet been directly observed. For Ps_2, which is rather like the hydrogen molecule, the positron 'nuclei' each have the same mass as the electron and, as a result, the Born–Oppenheimer approximation is not appropriate. If we number the electrons 1 and 2 and the positrons 3 and 4, the following group of permutations is the complete permutation symmetry group for the molecule:[3]

$$\{E, (12), (34), (12)(34), [13][24], [14][23], [1423], [1324]\}.$$

[3] Kinghorn D B and Poshusta R D 1993 *Phys. Rev.* A **47** 3671.

Table 14.1. The character table for the complete permutation symmetry group of dipositronium Ps_2. The electrons are numbered 1 and 2 and the positrons are numbered 3 and 4. Permutations in parentheses are of identical particles. Permutations in square brackets involve the exchange of both electrons with both positrons and they are symmetry operations because of charge conjugation symmetry.

:	E	[1423] [1324]	(12)(34)	[13][24] [14][23]	(12) (34)
A_1 :	1	1	1	1	1
A_2 :	1	1	1	-1	-1
B_1 :	1	-1	1	1	-1
B_2 :	1	-1	1	-1	1
E :	2	0	-2	0	0

The permutations in parentheses are of identical particles (electrons and/or positrons). The permutations in square brackets involve exchange of the pair of electrons with the pair of positrons and these are symmetry operations because of charge conjugation symmetry; the simultaneous exchange of both electrons with both positrons leaves the Hamiltonian invariant. The group is isomorphic to the D_{2d} point group and its character table is given in table 14.1. Applying the Pauli exclusion principle for Ps_2, we find that the overall states (i.e. with the inclusion of spin) can only have symmetries A_2 or B_1. Parity and angular momentum can also be used as symmetry labels on the states.

14.7 Parity

The complete internal electromagnetic Hamiltonian $\hat{H}_{int}$ for any molecule commutes with the parity operation E^* and E^* is a symmetry operation. Using the MS group, there are three situations that have to be distinguished in considering the parity label of the Φ_{int} states:

- Molecules for which E^* is in the MS group.
- Chiral molecules for which neither E^* nor any permutation inversion operation are in the MS group.
- Molecules for which E^* is not in the MS group but for which there is at least one permutation–inversion operation in the MS group.

For molecules like ethylene, in which the MS group contains the operation E^*, there are two allowed MS group symmetry species for Φ_{int} and it is easy to determine the parity from the character under E^* ($+1$ or -1) of these two allowed species. For such a molecule, the Φ_{int} are eigenfunctions of E^* with eigenvalue $+1$ or -1 as the state has parity $+$ or $-$. The internal states Φ_{int} are non-degenerate, so that equation (7.25) applies (with $R = E^*$ and $\psi_n = \Phi_{int}$) and equation (7.29) is true. The transformation properties of Φ_{int} under the effect of any permutation of identical nuclei is fixed by the statistics and the only symmetry distinction that the MS group can make is that of parity. Thus, for molecules having an MS group that contains E^*, we could as well label the symmetries of the complete internal states by parity as by the irreducible representations of the MS group. Only internal states of the same parity can perturb each other and electric dipole transitions between internal states of the same parity are forbidden by the vanishing integral rule since the electric dipole moment operator [see equation (2.88)] has negative parity. However, in building up Φ_{int} from its component parts, we use the MS group symmetry labels and we do not lose anything (from the point of view of determining which states can perturb each other or which transitions are forbidden) by labelling the internal states in the same way. For all planar molecules, E^* is in the MS group and so the Φ_{int} states are non-degenerate and have definite parity.

If the MS group of a molecule contains neither E^* nor any permutation–inversion operation, then it will have two forms of its equilibrium structure that are the mirror images of each other and there will be an insuperable energy barrier between these forms on the potential energy surface. Such a molecule is called *chiral* or *optically active* and the two forms, called *enantiomers*, are physically distinguishable. The CHFClBr molecule is an example of such a molecule and its two enantiomers are drawn in figure 14.3. For a chiral molecule, each of its enantiomers rotate the plane of polarization of linearly polarized light in opposite directions. For each enantiomer, its wavefunctions do not observably tunnel into the potential minimum that supports the other enantiomer. The internal energies of one enantiomer are degenerate with those of the other enantiomer so that each internal energy level is doubly degenerate (one state being for one enantiomer and the other being for the other enantiomer); each has the same non-degenerate symmetry species in the MS group determined by the nuclear spin statistics. The E^* operation interconverts enantiomers and, hence, the Φ_{int} of either enantiomer is not an eigenfunction of E^*. These functions cannot be assigned a definite parity.

Molecules for which the MS group does not contain E^* but for which it does contain permutation–inversion symmetry operations are not chiral even though the E^* inversion operation interconverts versions of the molecule that are separated by an insuperable energy barrier. Versions differ only in the sense of the numbering of identical particles and they cannot be physically distinguished. The CH$_3$F molecule is an example of such a molecule and its two versions are drawn in figure 8.1. For molecules of this type, there are always two allowed MS group symmetry labels for the Φ_{int}. These labels can be used to determine which states perturb each other and which transitions are symmetry forbidden. In this picture, just as for an optically active molecule, the Φ_{int} are not eigenfunctions of E^* and they do not have a definite parity. However, if one allows for inversion tunnelling and/or for nuclear hyperfine interactions, non-degenerate states of definite parity

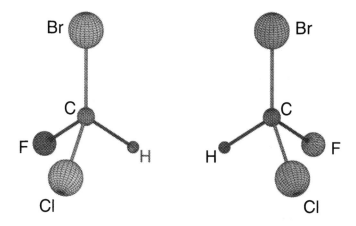

Figure 14.3. The two enantiomers of the molecule CHFClBr.

result. The two possible MS group symmetry species obtained for Φ_{int} can be viewed as the *incipient* parity labels for the, in principle, resolvable states.

We use the CH_3F molecule as an example of this type of molecule and we have already discussed some aspects of its symmetry using its MS group $C_{3v}(M)$ in chapter 9. The rules of statistics (see section 9.4) dictate that the complete internal states of CH_3F can be of symmetry $\Gamma_{tot} = A_1$ or A_2 (they cannot be of symmetry E) in $C_{3v}(M)$. The proton nuclear spin states are of symmetry $^4A_1$ and ^2E. Rovibronic states of A_1 or A_2 symmetry can only be in the $^4A_1$ proton spin state, to give Φ_{int} states of symmetry A_1 or A_2, respectively. Rovibronic states of E symmetry can only be in the ^2E proton spin state to give Φ_{int} states of both A_1 and A_2 symmetry. This pair of internal states arising from an E rovibronic state are 'accidentally' degenerate[4] in the absence of nuclear hyperfine splitting effects and in the absence of inversion tunnelling.

What happens to the symmetries of the Φ_{int} states of CH_3F if there is observable inversion tunnelling so that the MS group becomes the CNPI group $D_{3h}(M)$? From the discussion in chapter 8 and the reverse correlation table 7.7 for $C_{3v}(M) \rightarrow D_{3h}(M)$, one might initially think that internal states of symmetry A_1 split into states of symmetry $A_1' \oplus A_2''$ and that internal states of symmetry A_2 split into states of symmetry $A_2' \oplus A_1''$. But internal states of symmetry A_1' or A_1'' in $D_{3h}(M)$ are forbidden by the rules of Fermi–Dirac statistics (as is the case for H_3^+, discussed in chapter 9). Thus, internal states of symmetry A_1 in $C_{3v}(M)$ become internal states of symmetry A_2'' in $D_{3h}(M)$ if there is inversion tunnelling. Similarly, internal states of symmetry A_2 in $C_{3v}(M)$ become internal states of symmetry A_2' in $D_{3h}(M)$ if there is inversion tunnelling. These C_{3v} states are not

---

[4] Being of symmetry A_1 and A_2, they are not forced to be degenerate by symmetry.

subject to any further splitting from the tunnelling. The group $D_{3h}(M)$ contains E^* and the eigenstates of the tunnelling CH_3F molecule have a definite parity: A_2'' states have $-$ parity and A_2' states have $+$ parity. Thus, we could say that Φ_{rve} states of non-tunnelling CH_3F of A_1 symmetry in $C_{3v}(M)$ have incipient $-$ parity, since this is the unique parity of the state that arises if there is tunnelling and when $D_{3h}(M)$, and a tunnelling wavefunction, is used. Similarly, the Φ_{rve} states of non-tunnelling CH_3F of A_2 symmetry in $C_{3v}(M)$ have incipient $+$ parity since this is the unique parity of the Φ_{int} state that arises if there is tunnelling. For Φ_{rve} states of non-tunnelling CH_3F of E symmetry, this approach tells us that these states are of incipient double parity (i.e. $+$ and $-$); these initially degenerate states will be split by inversion tunnelling and also by nuclear hyperfine effects.

For molecules of this type, E^* is not feasible since there is an insuperable barrier to inversion. However, it is possible to label the Φ_{rve} states using incipient parity. Some rovibronic states, such as those of A_1 and A_2 symmetry for CH_3F, will have a unique parity and some, such as those of E symmetry for CH_3F, will have double parity. If nuclear hyperfine splittings are important, then the parity label would be the appropriate symmetry label to use, in conjunction with the angular momentum label F from K(spatial). One way to approach such a problem when there is significant hyperfine splitting is to convert from MS symmetry to the parity label at the end of the symmetry labelling procedure, when one finally needs the symmetry labels on the complete internal states.

14.8   Time reversal

The complete internal electrodynamic Hamiltonian is invariant to the operation of reversing all linear and angular momenta, including spin angular momenta. We could call it the 'motion reversal' symmetry operation but it is called, somewhat enigmatically, the 'time reversal' symmetry operation (T). It has an enigmatic name and it is rather special since it is *anti-unitary* which, as far as it is necessary to understand here, means that we cannot combine it with the operations of the groups K(spatial) and G_{MS} and add a symmetry label under time-reversal to the angular momentum labels and MS group irreducible representation label. As a result, the time-reversal symmetry operation is not used to provide another symmetry label on the energy levels. However, it does have some applications in the study of molecular symmetry.

There are some molecular symmetry groups for which there are one or more pairs of irreducible representations, Γ and Γ^* say, that are the complex conjugates of each other. Time-reversal symmetry means that, for every level of symmetry Γ, there will be a level of symmetry Γ^* that is degenerate with it. As an example, we give in table 14.2 the character table of the MS group $C_3(M)$, and the irreducible representations E_+ and E_- form such a pair. Because of time-reversal symmetry, every energy level of E_+ symmetry in $C_3(M)$ will coincide with a level of E_- symmetry. Such a pair of levels are said to be *separably*

Table 14.2. The character table of the $C_3(M)$ group. $\omega = \exp(2\pi i/3)$.

$C_3(M)$:	E	(123)	(132)
A_1 :	1	1	1
E_+ :	1	ω	ω^2
E_- :	1	ω^2	ω

Table 14.3. The condensed character table of the $C_3(M)$ group.

$C_3(M)$:	E	(123), (132)	
A:	1	1	
E:	2	-1	sep

degenerate and, because of this, the character table is usually condensed by adding the characters of the separably degenerate representations. Such a condensed character table for $C_3(M)$ is given in table 14.3, where the fact that E is the sum of separably degenerate irreducible representations is indicated by writing 'sep' for it as shown.

Time-reversal symmetry is of use in providing extra information as to whether matrix elements of operators between the components of a degenerate state vanish[5]. Also, T can be used to help us appreciate which terms can and cannot be present when setting up a Hamiltonian for a special case. In particular, in the development of the (Hermitian) effective rotational Hamiltonian (see page 241), the time-reversal symmetry requirement that the Hamiltonian be invariant to the reversal of all momenta means that such a Hamiltonian can only involve even powers of the angular momentum operators.

---

[5] See section 4 of Watson J K G 1974 *J. Mol. Spectrosc.* **50** 281.

Chapter 15

Symmetry violation

15.1 The electroweak Hamiltonian

In the theory of the symmetry of fundamental particles, the inversion symmetry operation, or parity operation, is called P. However, in the MS group it is called E^* and that is the notation we have used up to now. In this chapter, we use the P notation for the parity operation since we wish to make contact with results obtained in studies of fundamental particles.

We consider here the possible violation of four types of symmetry operation: The parity operation P, the charge-conjugation operation C, the time-reversal operation T and the 'identical' particle permutation operation. We also consider the violation, or possible violation, of the product symmetry operations CP and CPT. Such symmetry violations are normally thought of as being studied in fundamental particle physics but they can lead to spectroscopically measurable consequences for atoms and molecules.

In the *standard model* of fundamental particle physics, matter is built up using 12 fundamental particles (six quarks and six leptons – the electron is a lepton) which interact via three types of force (the strong, the electromagnetic and the weak). The standard model does not account for gravity and it does not explain or predict the masses of the 12 fundamental particles. Thus, it cannot be the whole story and there are several alternate theories that go beyond the standard model. Experimental tests of the predictions of such theories are important and particle physics experiments can provide such tests. However, it is also the case that by making atomic and molecular spectroscopy measurements at high precision and sensitivity, one will also obtain results that provide useful tests.

So far in our discussion of the atomic and molecular Hamiltonian, the only interaction force that we have considered to act between nuclei and electrons is the electromagnetic force. The gravitational force between an electron and a proton is about 10^{-39} of the electrostatic force between them, and it can be safely neglected. The strong force is responsible for the binding that holds the quarks within protons and neutrons together, and the residual strong force binds protons and neutrons together to form atomic nuclei. Electrons are not susceptible to the strong force and so an atomic or molecular Hamiltonian does not involve it. The weak force has a very short range and it is indeed 'weak'; however, it assumes an importance that is belied by its weakness and short-range nature since it allows processes to occur that would otherwise be forbidden. This is the case when we include the effect of the weak force in an atomic or molecular Hamiltonian.

The theory of electromagnetic interactions came about as a result of Maxwell's unification of the theories of electric and magnetic interactions. In a similar way, the standard model unifies the weak force with the electromagnetic force and this means that we should consider both forces on the same footing. A consequence of electroweak unification is the prediction of the weak neutral current interaction; this short-range force acts between nuclei and electrons in an atom or molecule. Including it gives the *electroweak* Hamiltonian rather than the electromagnetic Hamiltonian that we have used so far.

15.2 Parity (P) violation

The strong and electromagnetic interactions conserve parity but the weak interaction does not. As a result, the commutator of P with the atomic or molecular electroweak Hamiltonian is not zero and states of opposite parity are very slightly mixed. Thus, within the standard model, parity violation in an atom or molecule occurs because of the weak neutral current interaction, and the extent of this parity violation can be calculated.

An example of parity violation for an atom is the parity-forbidden electric dipole transition between the (nominally parity $+$) 6S ground state and the (also nominally parity $+$) 7S excited state of caesium at 540 nm[1]. The standard model predicts that these S states are, in fact, slightly contaminated (by about 10^{-11}) with (parity $-$) P state character. As a result, the parity-forbidden 7S $\leftarrow$ 6S transition has a non-vanishing electric dipole transition amplitude of about 10^{-10} D. Experiment and theory are compared by determining the value of the *weak charge* Q_W, that characterizes the strength of the weak neutral current interaction; currently, there is no statistically meaningful difference between the experimental and theoretical values of Q_W for the caesium nucleus[2]. This is good evidence for the correctness of standard model of electroweak theory and its computational implementation; the latter demands, among other things, a very high precision *ab initio* calculation of the electronic wavefunction including relativistic effects. The effects of parity violation have also been observed in atoms other than caesium.

The interest in parity violation in molecules does not stem from the fact that certain (electromagnetically strictly forbidden) electric dipole transitions become extremely weakly allowed but from an interesting effect that the parity-violating weak neutral current interaction has on the electronic and rovibronic energies

---

[1] For details, see Wood C S *et al* 1999 *Can. J. Phys.* **77** 7.

[2] Koslov M G *et al* 2001 *Phys. Rev. Lett.* **86** 3260.

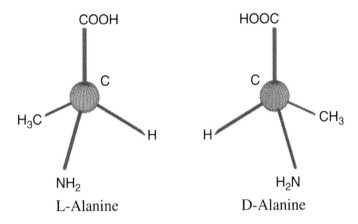

Figure 15.1. The two enantiomers of alanine.

of chiral (or optically active) molecules. The symmetry properties of chiral molecules are briefly discussed on page 306. Using the electroweak Hamiltonian for a chiral molecule leads to the result that the two enantiomers do not have exactly the same electronic energies at their respective equilibrium geometries. Also the shapes of the potential energy surfaces around these two minima are slightly different. As a result, the separations of the rovibronic energies of the two enantiomers of a chiral molecule are not identical to each other. No experiment has yet been able to measure such differences[3]. For electroweak interactions involving atoms having nuclear charge number $Z \sim 10$, the parity-violating energy difference ΔE_{PV} is calculated to be of the order 10^{-36} J, which corresponds to a frequency difference of order 10^{-3} Hz or a wavenumber difference of order 10^{-13} cm^{-1}. However, ΔE_{PV} depends strongly on the charges of the nuclei involved[4].

The more precise measurement and calculation of parity-violating transitions in atoms is an active area of research because it is a comparatively inexpensive method for testing one aspect of the standard model of theoretical physics. Theoretical and experimental studies of parity violation for chiral molecules is also an active area of spectroscopic research because of its possible involvement in the evolution of living organisms. How this involvement might have come about we now explain.

In figure 15.1, we show the D and L enantiomers of the chiral amino acid alanine $H_2N(H)C(CH_3)COOH$. Replacing the CH_3 group in the alanine molecule by other side chain groups produces other amino acids, the enantiomers of which

---

[3] Daussy Ch *et al* 1999 *Phys. Rev. Lett.* **83** 1554 find transition frequencies of the two enantiomers of CHFClBr at around 9.3 μm to be equal to within 13 Hz.

[4] See, for example, Laerdahl J K and Schwerdtfeger P 1999 *Phys. Rev.* A **60** 4439.

can also be labelled D and L in the same way[5]. Two alanine molecules can form the dimer $H_2N(H)C(CH_3)$-CO-NH-$(H)C(CH_3)COOH$ by elimination of H_2O and the formation of the -CO-NH- amide (or peptide) link. Protein molecules are polymers (polypeptides) of amino acids connected by amide links. In living organisms on earth, protein molecules only involve the L enantiomers of the chiral amino acids, except in very rare cases. In a similar way, the nucleic acid polymer molecules RNA and DNA in earthly living organisms only involve D-sugars. A crucial result of this *homochirality* of bio-organic molecules is that these two types of polymer are able to form helices, and this helical structure is a basic requirement for the replication processes of life as we know it. The origin of homochirality is unknown even though it seems to be a prerequisite for the origin of life.

There have been many suggestions for how the homochirality of the bio-organic molecules of life has come about[6]. One suggestion is that the parity violating electroweak interaction is responsible for giving L-amino acids, and D-sugars, slightly lower energies than their enantiomeric counterparts, and that this energy advantage over the aeons of evolutionary time is responsible for the observed homochirality. There are many unanswered questions. How were such complex bio-molecules formed? How did they polymerize? How were they protected from degradation? Were key reactions in the gas phase, in solution or on surfaces? Was the infall of extraterrestrial prebiotic material significant? In the absence of answers to these questions, the role of the parity-violating electroweak interaction, which slightly lifts the degeneracy of the rovibronic energies of the enantiomers of chiral molecules, is unclear. Nonetheless, the measurement and calculation of the parity-violating energy differences between enantiomers are very interesting problems in the fields of high-precision molecular spectroscopy and *ab initio* quantum chemistry.

15.3   CP violation

The operation of charge conjugation C was introduced in section 14.6 and it is the operation of changing every particle in the system under study to its antiparticle. The weak interaction and the electroweak Hamiltonian are not invariant to C. As with parity, the extent of C violation can be calculated using the standard model. The direct application of C symmetry is rather limited but CP symmetry, the combined operation of C with P, is important to consider.

The observation of P violation in weak interactions[7] made many theoretical physicists despondent. But they could cheer themselves up by thinking that if only they had defined parity as being what we now call the combined operation

---

[5] The glycine molecule is obtained by replacing the CH_3 group by H and this is the only amino acid that is not chiral.

[6] See, for example, Bailey J *et al* 1988 *Science* **281** 672 and references therein.

[7] Lee T D and Yang C N 1956 *Phys. Rev.* **104** 254, and Wu C S *et al* 1957 *Phys. Rev.* **105** 1413.

CP, everything would be alright; CP would then be a symmetry operation of all interactions including the weak interaction. This view resulted from a deeply held conviction that somehow the Universe had to have such symmetry to be beautiful. That hope proved unfounded. In 1964, CP violation was discovered to occur in the decay of K mesons[8] and, since then, CP violation has also been found to occur in the decay of B mesons. Some CP violation can be accommodated in the standard model but extensions to the standard model generally imply stronger CP violation and, consequently, it is necessary to measure the extent of CP violation in as many situations as possible.

CP violation is important because its operation in the early Universe can be invoked to solve the matter–antimatter puzzle. The puzzle is that matter and antimatter would have been produced in equal quantities by pair production as the early Universe cooled and matter–antimatter annihilation would then produce a Universe eventually consisting only of photons; exact matter/antimatter symmetry in the Universe is clearly not the case. By observations of the cosmic radiation background, it is found the the Universe today consists of matter and photons in a ratio of about 10^9 photons for every particle of matter. From this, we infer that, as the Universe cooled, there were $(10^9 + 1)$ particles of matter for every 10^9 particles of antimatter (a very small asymmetry) so that their mutual annihilation removed antimatter and produced one particle of matter for every 10^9 photons. It is observed that CP violation allows nuclear decays very slightly unsymmetrical in the production of particles and antiparticles; such nuclear decays could have competed with matter–antimatter annihilation as the Universe cooled to produce the required very small asymmetry. Thus, in the same way as P violation can be invoked as being vital to explain the formation of life (because it could be the cause of homochirality in key bio-organic molecules), so CP violation can be invoked to explain the existence of matter in preponderance to antimatter in the Universe. The lack of P and CP symmetry produces a far more interesting Universe than one which is completely symmetric under these operations, even if it is not so 'beautiful' from a symmetry point of view.

The measurement of the precise extent of CP violation in atomic and molecular spectroscopy experiments actually involves the assumption of CPT invariance, where CPT invariance is invariance under the triple product of charge conjugation, parity and time reversal. Presuming CPT invariance (which we discuss in section 15.5), a measurement of the extent of T violation will also give the extent of CP violation, since the CP violation has to cancel out exactly the effect of the T violation. Current neutron, atomic and molecular measurements seek to determine the extent of T violation as a way of determining the extent of CP violation.

---

[8] Christenson J H *et al* 1964 *Phys. Rev. Lett.* **13** 138.

15.4   T violation

The time-reversal symmetry operation was introduced in section 14.8.   T symmetry applies to strong and electromagnetic interactions but not to weak interactions. The measurement of T violation involves determining a non-zero value for a quantity that would be precisely zero under T symmetry. The classic example is d_n, the electric dipole moment (EDM) of the neutron[9]. The EDM of the electron d_e is another example. A non-vanishing particle EDM d can only be parallel or antiparallel to the direction of the spin s, since spin is the only defined alignment direction in a particle. The dipole moment is invariant to T but T reverses s, which means that d must vanish under T symmetry since that is the only way it can equal its negative. Thus, a non-zero d implies T violation; it also implies P violation (P reverses d but not s) but that is not so interesting since P violation is understood (or so we think). The measurement of d_n has been a longstanding goal and, nowadays, attempts to measure d_e are also ongoing.

In the standard model, the EDM of the neutron

$$d_n(\text{standard model}) \approx 10^{-34}e \text{ cm} \tag{15.1}$$

which is far too small ever to be measured[10] but in extensions to the standard model values are obtained that could be detected in practical experiments. In a recent sensitive neutron experiment[11], an upper limit of

$$|d_n(\text{experiment})| \le (6 \pm 1) \times 10^{-26}e \text{ cm} \tag{15.2}$$

could be set, which is close to the predicted order of magnitude for d_n obtained in some theories.

Atomic or molecular eigenstates can be labelled $|F, m_F\rangle$ using the total (orbital plus spin) angular momentum quantum number F and the projection quantum number m_F. T reverses angular momenta converting the state $|F, m_F\rangle$ to the state $|F, -m_F\rangle$. In an electric field, the two states $|F, m_F\rangle$ and $|F, -m_F\rangle$ remain degenerate if there is T symmetry but, with T violation, a non-vanishing d_e will cause a splitting between them. In an experiment[12] on atomic thallium, the lack of observation of such a splitting between the $|F = 1, m_F = \pm 1\rangle$ hyperfine sublevels of the ground state leads to an upper limit for the EDM of the electron of

$$d_e(\text{experiment}) \le (7 \pm 8) \times 10^{-28}e \text{ cm.} \tag{15.3}$$

An experiment[13] on ^{174}YbF involves an attempt to measure the splitting in an electric field between the $m_F = \pm 1$ sublevels of the $F = 1, N = 0$ level in the

[9] Purcell E M and Ramsey N F 1950 *Phys. Rev.* **78** 807.
[10] 1 e cm $\approx$ 1.602 18 $\times$ 10^{-21} C m.
[11] Harris P G *et al* 1999 *Phys. Rev. Lett.* **82** 904.
[12] Regan B C *et al* 2002 *Phys. Rev. Lett.* **88** 071805.
[13] Hudson J J *et al* 2002 *Phys. Rev. Lett.* **89** 023003.

$X^2\Sigma^+$ ground vibronic state. The null result of this experiment leads to an upper limit of $d_e \leq (-0.2\pm3.2) \times 10^{-26}\, e$ cm but improved precision would be achieved by doing the experiment on trapped YbF molecules; progress in trapping YbF molecules has recently been made using an Alternating Gradient Decelerator[14]. The standard model predicts the EDM of the electron as

$$d_e(\text{standard model}) \leq 10^{-38}e \text{ cm} \qquad (15.4)$$

but several theories going beyond the standard model have predictions in the range close to the experimental upper limit[15].

15.5   Testing for CPT violation

The standard model has the symmetry of the combined operation CPT, and CPT violation has so far defied experimental observation. However, string theory (which is one extension of the standard model) could break CPT symmetry[16] and there have been many suggestions for possible ways of experimentally testing for CPT violation. Two involve atoms.

The positronium atom Ps is a bound state of an electron and the positively charged antielectron (the positron). The self-annihilating decay of ortho-Ps to three gamma particles has been studied[17] to look for asymmetry indicating CPT violation. The amplitude of the CPT-violating asymmetry is found to be 0.0026 ± 0.0031. Of interest to spectroscopists is the possibility of studying the spectrum of antihydrogen. An antihydrogen atom consists of a negatively charged antiproton and a positron. To test for CPT violation, one would measure the 1S-2S 'positronic' transition frequency of antihydrogen and compare it with that of the 1S-2S electronic transition in normal hydrogen. A difference would indicate CPT violation. Two recent experiments, the so-called ATHENA[18] and ATRAP[19] collaborations, both using the antiproton factory at CERN, have managed to make cold antihydrogen atoms. The next step is to trap and cool the antihydrogen atoms into the 1S ground state for long enough to make the required precision measurement of the 1S–2S transition frequency.

15.6   Testing for permutation symmetry violation

The fifth postulate of quantum mechanics is introduced in section 9.1, it states that the complete wavefunction (including spin) of a system of particles is changed in sign by an interchange of two identical fermions in the system, but is unchanged

[14] Tarbut M R *et al* 2004 *Phys. Rev. Lett.* **92** 173002.

[15] Commins E D 1999 *Adv. At. Mol. Opt. Phys.* **40** 1.

[16] See, for example, Kostelecký V A and Potting R 1995 *Phys. Rev.* D **51** 3923.

[17] Vetter P A and Freedman S J 2003 *Phys. Rev. Lett.* **91** 263401.

[18] Amoretti M *et al* 2002 *Nature* **419** 456.

[19] Gabrielse G *et al* 2002 *Phys. Rev. Lett.* **89** 213401.

by the interchange of two identical bosons. Electron permutation symmetry and the Pauli exclusion principle are introduced in section 3.2, and the way that the Pauli exclusion principle leads to a periodic structure in the chemical properties of the elements is discussed in section 14.3. The application of the fifth postulate to nuclear permutation symmetry in order to obtain nuclear spin statistical weights is discussed in chapter 9. Because of the overwhelming experimental evidence in favour of the fifth postulate, most people consider it to be a fact and consider it a kind of bureaucratic oversight that there is no proof written down in a way that is simple and acceptable to all. There have been complicated 'proofs' advanced, but it is fair to say that none of these has been found completely acceptable; why should such a simple postulate require anything complicated in its proof? Maybe because it is not true.

The theoretical possibility of its violation can be sought in several ways. One would be the possibility that for electrons, say, there is a property (as yet unidentified) for which the vast majority of electrons share the same value but for which a tiny number have a different value; in this way, not all permutations of electrons would lead to an indistinguishable situation. It is said that Fermi considered this possibility but predicted that it would have had drastic effects on the properties of the elements over the billions of years of their existence. Another possibility is that particles are not pure bosons or pure fermions. An algebra with such 'intermediate' statistics can be constructed[20] and such studies may lead to a greater understanding of how, why and in what circumstances, the violation of the fifth postulate might occur. Finally, string theory does not have particles as its fundamental entities and so must require something more general than the fifth postulate. In this situation, the fifth postulate would have to be considered as being incomplete and the effect of its completion might lead to greater insights as to whether it can be violated by particles. Perhaps this is all grasping at straws in the hope that something as sensational as its violation might occur. However, in the light of the discovery of P violation and of CP violation, it would seem prudent, just as in the case of CPT symmetry, to subject the fifth postulate to experimental test. At least one should experimentally determine a quantitative upper limit for its possible violation.

Upper time limits for Pauli exclusion principle (PEP) violating decays of ^{12}C nuclei to nuclei having three nucleons (protons or neutrons, each of which are fermions) in a 1s shell have been obtained using the NEMO-2 detector. The signature of such decays would be emitted γ, β^- or β^+ rays at the appropriate energies. Three such decays have been looked for with a null result and they are shown in figure 15.2. This leads to a lower limit of more than 10^{24} years for these PEP violating processes to occur. A very similar type of experiment for the electron[21] involves the search for X-rays emitted from a current-carrying copper strip that signify the cascade of a third electron into the 1s level. The null result of

[20] Greenberg O W and Delgado J D 2001 *Phys. Lett.* A **288** 139.
[21] Ramberg E and Snow G A 1960 *Phys. Lett.* B **238** 438.

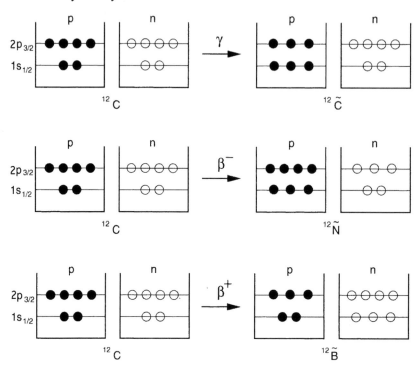

Figure 15.2. Three possible nuclear decays that lead to states with three fermions in a 1s shell; these are forbidden by the Pauli exclusion principle. From Arnold R *et al* 1999 *Eur. Phys. J.* A **6** 361.

this experiment is interpreted to mean that any PEP violation is at the level of less than 1.7×10^{-26}. An experiment that has a more straightforward interpretation involves looking for a transition between *paronic* states of helium[22]. Paronic states of helium are states that are symmetric with respect to electron exchange and that are, therefore, forbidden by PEP. Accurate calculations of the energies of paronic states of helium have been made[23]. The non-observation of such a transition in helium leads to an upper limit of 5×10^{-6} on the PEP violation. A further study involving electrons is concerned with the contribution of the PEP to the excess noise in a conductor[24]. All these experiments are concerned with fermions. An experiment on photons[25], which are bosons, leads to an upper limit for finding photons in exchange-antisymmetric states of 1.2×10^{-7}.

The use of a highly sensitive molecular spectroscopy experiment to look for a violation of the fifth postulate involves trying to observe a transition between

[22] Deilamian K *et al* 1995 *Phys. Rev. Lett.* **74** 4787.

[23] Drake G W F 1989 *Phys. Rev.* A **39** 897.

[24] Kurdak C *et al* 1996 *Surf. Sci.* **361/362** 705.

[25] DeMille D *et al* 1999 *Phys. Rev. Lett.* **83** 3978.

levels that would be missing as a result of applying the rules of spin statistics (see section 9.3). This test could be made for bosons or for fermions. The $C^{16}O_2$ molecule contains two ^{16}O boson nuclei and, as a result, the rotation–vibration transition $(v_1 v_2^l v_3, J) = (00^0 1, 26) \leftarrow (00^0 0, 25)$ is between missing levels; the observation of this 'missing' transition in the absorption spectrum would imply a violation of spin statistics. The constants in the effective rotational Hamiltonians of the $(00^0 0)$ and $(00^0 1)$ states are known very precisely so the frequency of the transition can be very accurately predicted. The failed attempt to see the transition[26] leads to an upper limit of 1.7×10^{-11} for the probability of finding the exchange-antisymmetric state $(00^0 0, 25)$. It would be desirable to increase the sensitivity of such a direct spectroscopic measurement. The highest sensitivity possible would be that obtained by attempting to count the ion(s) produced when radiation is applied at sufficient $h\nu$ energy to ionize the upper 'missing' level of a 'missing' absorption transition while tuning another radiation through the predicted frequency of the 'missing' absorption transition.

[26] Mazzotti D *et al* 2001 *Phys. Rev. Lett.* **86** 1919.

Appendix A

Answers to selected problems

Problem 2.7

By definition,

$$\chi(S) = \sum_{i=1}^{n} S_{ii} = \sum_{i=1}^{n} \sum_{j=1}^{n} \sum_{k=1}^{n} Q_{ij} R_{jk} (Q^{-1})_{ki} \tag{A.1}$$

where $(Q^{-1})_{ki}$ is the kith element of the matrix Q^{-1}. Consequently,

$$\chi(S) = \sum_{j=1}^{n} \sum_{k=1}^{n} R_{jk} \left[\sum_{i=1}^{n} (Q^{-1})_{ki} Q_{ij} \right] \tag{A.2}$$

where, relative to equation (A.1), we have changed the order of the factors and the order in which we do the summations. The term in square brackets in equation (A.2) is the kjth element of the matrix $(Q^{-1}Q)$. Since Q^{-1} is the inverse of Q, $(Q^{-1}Q)_{kj} = \delta_{kj}$, where δ_{kj} is a Kronecker delta. Thus,

$$\chi(S) = \sum_{j=1}^{n} \sum_{k=1}^{n} R_{jk} \delta_{kj} = \sum_{j=1}^{n} R_{jj} = \chi(R). \tag{A.3}$$

Problem 5.7

By cyclically permuting ξ, η and ζ in equation (5.44), we obtain

$$\hat{J}_\eta = -i\hbar \sum_{i=1}^{l} \left(\zeta_i \frac{\partial}{\partial \xi_i} - \xi_i \frac{\partial}{\partial \zeta_i} \right) \tag{A.4}$$

$$\hat{J}_\zeta = -i\hbar \sum_{i=1}^{l} \left(\xi_i \frac{\partial}{\partial \eta_i} - \eta_i \frac{\partial}{\partial \xi_i} \right). \tag{A.5}$$

For $A = \xi, \eta$ and ζ,

$$\hat{J}_A = \sum_{i=1}^{l} \hat{J}_A^{(i)} = \hat{J}_A^{(1)} + \hat{J}_A^{(2-l)} \tag{A.6}$$

where

$$\hat{J}_\xi^{(i)} = -i\hbar \left(\eta_i \frac{\partial}{\partial \zeta_i} - \zeta_i \frac{\partial}{\partial \eta_i} \right) \tag{A.7}$$

with similar expressions for $\hat{J}_\eta^{(i)}$ and $\hat{J}_\zeta^{(i)}$; $\hat{J}_A^{(2-l)} = \sum_{i=2}^{l} \hat{J}_A^{(i)}$.
From problem 5.6a,

$$[\hat{J}_\xi, \hat{J}_\eta] = [(\hat{J}_\xi^{(1)} + \hat{J}_\xi^{(2-l)}), (\hat{J}_\eta^{(1)} + \hat{J}_\eta^{(2-l)})]$$
$$= [\hat{J}_\xi^{(1)}, \hat{J}_\eta^{(1)}] + [\hat{J}_\xi^{(1)}, \hat{J}_\eta^{(2-l)}] + [\hat{J}_\xi^{(2-l)}, \hat{J}_\eta^{(1)}] + [\hat{J}_\xi^{(2-l)}, \hat{J}_\eta^{(2-l)}].$$
$$\tag{A.8}$$

Here, $[\hat{J}_\xi^{(1)}, \hat{J}_\eta^{(2-l)}] = [\hat{J}_\xi^{(2-l)}, \hat{J}_\eta^{(1)}] = 0$ because, in each of these commutators, the operator $\hat{J}_A^{(1)}$ depends only on the coordinates of nucleus 1, whereas $\hat{J}_B^{(2-l)}$ depends on the coordinates of all other nuclei. Thus,

$$[\hat{J}_\xi, \hat{J}_\eta] = [\hat{J}_\xi^{(1)}, \hat{J}_\eta^{(1)}] + [\hat{J}_\xi^{(2-l)}, \hat{J}_\eta^{(2-l)}]. \tag{A.9}$$

Obviously $\hat{J}_A^{(2-l)} = \hat{J}_A^{(2)} + \hat{J}_A^{(3-l)}$, where $\hat{J}_A^{(3-l)} = \sum_{i=3}^{l} \hat{J}_A^{(i)}$, and we can repeat the arguments just given to show that

$$[\hat{J}_\xi^{(2-l)}, \hat{J}_\eta^{(2-l)}] = [\hat{J}_\xi^{(2)}, \hat{J}_\eta^{(2)}] + [\hat{J}_\xi^{(3-l)}, \hat{J}_\eta^{(3-l)}]. \tag{A.10}$$

By repeated use of this procedure, we derive

$$[\hat{J}_\xi, \hat{J}_\eta] = \sum_{i=1}^{l} [\hat{J}_\xi^{(i)}, \hat{J}_\eta^{(i)}]. \tag{A.11}$$

Applying problem 5.6a again:

$$[\hat{J}_\xi^{(i)}, \hat{J}_\eta^{(i)}] = \left[-i\hbar \left(\eta_i \frac{\partial}{\partial \zeta_i} - \zeta_i \frac{\partial}{\partial \eta_i} \right), -i\hbar \left(\zeta_i \frac{\partial}{\partial \xi_i} - \xi_i \frac{\partial}{\partial \zeta_i} \right) \right]$$
$$= -\hbar^2 \left(\left[\eta_i \frac{\partial}{\partial \zeta_i}, \zeta_i \frac{\partial}{\partial \xi_i} \right] - \left[\eta_i \frac{\partial}{\partial \zeta_i}, \xi_i \frac{\partial}{\partial \zeta_i} \right] \right.$$
$$\left. - \left[\zeta_i \frac{\partial}{\partial \eta_i}, \zeta_i \frac{\partial}{\partial \xi_i} \right] + \left[\zeta_i \frac{\partial}{\partial \eta_i}, \xi_i \frac{\partial}{\partial \zeta_i} \right] \right). \tag{A.12}$$

It is obvious that $[\eta_i (\partial/\partial\zeta_i), \xi_i (\partial/\partial\zeta_i)] = [\zeta_i (\partial/\partial\eta_i), \zeta_i (\partial/\partial\xi_i)] = 0$. The remaining two commutators are rewritten by means of the expression from problem 5.6c,

$$[\hat{J}_\xi^{(i)}, \hat{J}_\eta^{(i)}] = -\hbar^2 \left(\eta_i \left[\frac{\partial}{\partial \zeta_i}, \zeta_i \frac{\partial}{\partial \xi_i} \right] + \left[\eta_i, \zeta_i \frac{\partial}{\partial \xi_i} \right] \frac{\partial}{\partial \zeta_i} \right.$$
$$\left. + \zeta_i \left[\frac{\partial}{\partial \eta_i}, \xi_i \frac{\partial}{\partial \zeta_i} \right] + \left[\zeta_i, \xi_i \frac{\partial}{\partial \zeta_i} \right] \frac{\partial}{\partial \eta_i} \right) \qquad (A.13)$$

where the commutators $[\eta_i, \zeta_i(\partial/\partial\xi_i)] = [(\partial/\partial\eta_i), \xi_i(\partial/\partial\zeta_i)] = 0$. By applying the expression from problem 5.6b to the remaining commutators (and removing commutators that obviously vanish), we obtain

$$[\hat{J}_\xi^{(i)}, \hat{J}_\eta^{(i)}] = -\hbar^2 \left(\eta_i \left[\frac{\partial}{\partial \zeta_i}, \zeta_i \right] \frac{\partial}{\partial \xi_i} + \xi_i \left[\zeta_i, \frac{\partial}{\partial \zeta_i} \right] \frac{\partial}{\partial \eta_i} \right). \qquad (A.14)$$

It follows from equations (2.15)–(2.17) that $[(\partial/\partial\zeta_i), \zeta_i] = 1$ and, therefore, $[\zeta_i, (\partial/\partial\zeta_i)] = -1$. Thus,

$$[\hat{J}_\xi^{(i)}, \hat{J}_\eta^{(i)}] = -\hbar^2 \left(\eta_i \frac{\partial}{\partial \xi_i} - \xi_i \frac{\partial}{\partial \eta_i} \right) = i\hbar \hat{J}_\zeta^{(i)} \qquad (A.15)$$

where we have introduced $\hat{J}_\zeta^{(i)}$ from equation (A.5). Equation (A.11) yields

$$[\hat{J}_\xi, \hat{J}_\eta] = \sum_{i=1}^l [\hat{J}_\xi^{(i)}, \hat{J}_\eta^{(i)}] = i\hbar \sum_{i=1}^l \hat{J}_\zeta^{(i)} = i\hbar \hat{J}_\zeta. \qquad (A.16)$$

Thus, $[\hat{J}_\xi, \hat{J}_\eta] = i\hbar\hat{J}_\zeta$ and, by analogous arguments, we can show that $[\hat{J}_\zeta, \hat{J}_\xi] = i\hbar\hat{J}_\eta$ and $[\hat{J}_\eta, \hat{J}_\zeta] = i\hbar\hat{J}_\xi$. These commutators have a '+' sign as do those for the XYZ components in equation (2.81).

Problem 5.8

Using problem 5.6a,

$$[\hat{J}^2, \hat{J}_\xi] = [\hat{J}_\xi^2 + \hat{J}_\eta^2 + \hat{J}_\zeta^2, \hat{J}_\xi] = [\hat{J}_\xi^2, \hat{J}_\xi] + [\hat{J}_\eta^2, \hat{J}_\xi] + [\hat{J}_\zeta^2, \hat{J}_\xi] \qquad (A.17)$$

$[\hat{J}_\xi^2, \hat{J}_\xi] = \hat{J}_\xi^3 - \hat{J}_\xi^3 = 0$ but from the expression in problem 5.6c,

$$[\hat{J}^2, \hat{J}_\xi] = [\hat{J}_\eta^2, \hat{J}_\xi] + [\hat{J}_\zeta^2, \hat{J}_\xi]$$
$$= \hat{J}_\eta[\hat{J}_\eta, \hat{J}_\xi] + [\hat{J}_\eta, \hat{J}_\xi]\hat{J}_\eta + \hat{J}_\zeta[\hat{J}_\zeta, \hat{J}_\xi] + [\hat{J}_\zeta, \hat{J}_\xi]\hat{J}_\zeta. \qquad (A.18)$$

We have just derived expressions for the commutators $[\hat{J}_A, \hat{J}_B]$ with $A, B = \xi, \eta, \zeta$ and, when we insert the expressions, we obtain

$$[\hat{J}^2, \hat{J}_\xi] = \hat{J}_\eta(-i\hbar\hat{J}_\zeta) + (-i\hbar\hat{J}_\zeta)\hat{J}_\eta + \hat{J}_\zeta(i\hbar\hat{J}_\eta) + (i\hbar\hat{J}_\eta)\hat{J}_\zeta$$
$$= i\hbar(-\hat{J}_\eta\hat{J}_\zeta - \hat{J}_\zeta\hat{J}_\eta + \hat{J}_\zeta\hat{J}_\eta + \hat{J}_\eta\hat{J}_\zeta) = 0. \qquad (A.19)$$

Similarly, $[\hat{J}^2, \hat{J}_\eta] = [\hat{J}^2, \hat{J}_\zeta] = 0$.

Problem 5.9

Using the expression from problem 5.6a, we derive that

$$[\hat{J}_\zeta, \hat{J}_s^+] = [\hat{J}_\zeta, \hat{J}_\xi + i\hat{J}_\eta] = [\hat{J}_\zeta, \hat{J}_\xi] + i[\hat{J}_\zeta, \hat{J}_\eta]. \tag{A.20}$$

The commutators on the right-hand side are known from the answer to problem 5.7 and we have

$$[\hat{J}_\zeta, \hat{J}_s^+] = i\hbar\hat{J}_\eta + i(-i\hbar\hat{J}_\xi) = \hbar(\hat{J}_\xi + i\hat{J}_\eta) = \hbar\hat{J}_s^+. \tag{A.21}$$

We show analogously that

$$[\hat{J}_\zeta, \hat{J}_s^-] = i\hbar\hat{J}_\eta - i(-i\hbar\hat{J}_\xi) = -\hbar(\hat{J}_\xi - i\hat{J}_\eta) = -\hbar\hat{J}_s^-. \tag{A.22}$$

Problem 5.10

To investigate whether $\hat{J}_s^+\psi_\lambda$ is an eigenfunction for $\hat{J}_\zeta$, we deduce that

$$\begin{aligned}
\hat{J}_\zeta(\hat{J}_s^+\psi_\lambda) &= (\hat{J}_\zeta\hat{J}_s^+)\psi_\lambda = (\hat{J}_s^+\hat{J}_\zeta + [\hat{J}_\zeta, \hat{J}_s^+])\psi_\lambda \\
&= (\hat{J}_s^+\hat{J}_\zeta + \hbar\hat{J}_s^+)\psi_\lambda = \hat{J}_s^+(\hbar m + \hbar)\psi_\lambda \\
&= \hbar\hat{J}_s^+(m + 1)\psi_\lambda = \hbar(m + 1)(\hat{J}_s^+\psi_\lambda)
\end{aligned} \tag{A.23}$$

by using equation (5.86) and the expression for $[\hat{J}_\zeta, \hat{J}_s^+]$ from the answer to problem 5.9. In summary,

$$\hat{J}_\zeta(\hat{J}_s^+\psi_\lambda) = \hbar(m + 1)(\hat{J}_s^+\psi_\lambda) \tag{A.24}$$

and, from this equation, either $\hat{J}_s^+\psi_\lambda$ vanishes (giving $0 = 0$) or it is an eigenfunction of $\hat{J}_\zeta$ with eigenvalue $\hbar(m + 1)$.

Similarly, we can show that

$$\hat{J}_\zeta(\hat{J}_s^-\psi_\lambda) = \hbar(m - 1)(\hat{J}_s^-\psi_\lambda) \tag{A.25}$$

so either $\hat{J}_s^-\psi_\lambda$ vanishes or it is an eigenfunction of $\hat{J}_\zeta$ with eigenvalue $\hbar(m - 1)$.

Problem 7.6

The operation (123) in $C_{3v}(M)$ transforms (r_1, r_2, r_3) to (r_1', r_2', r_3'). After the (123) operation, proton 1 occupies the position in space initially occupied by proton 3. Thus, $r_1' = r_3$. Similarly, $r_2' = r_1$ and $r_3' = r_2$. The effects of all other operations in $C_{3v}(M)$ on (r_1, r_2, r_3) are determined by analogous considerations.

Writing the effect of the operation R in $C_{3v}(M)$ on (r_1, r_2, r_3) in the form of equation (7.100):

$$R\begin{bmatrix} r_1 \\ r_2 \\ r_3 \end{bmatrix} = \mathbf{D}_r[R]\begin{bmatrix} r_1 \\ r_2 \\ r_3 \end{bmatrix} \tag{A.26}$$

Table A.1. The quantities Rr_1, M_{11}^R and M_{22}^R for the elements R of $C_{3v}(M)$.

R:	E	(123)	(132)	(12)*	(23)*	(31)*
Rr_1:	r_1	r_3	r_2	r_2	r_1	r_3
M_{11}^R:	1	$-1/2$	$-1/2$	1	$-1/2$	$-1/2$
M_{22}^R:	1	$-1/2$	$-1/2$	-1	$1/2$	$1/2$

where $\mathbf{D}_r[R]$ is the 3×3 representation matrix associated with R. We have

$$\mathbf{D}_r[E] = \begin{bmatrix} 1 & 0 & 0 \\ 0 & 1 & 0 \\ 0 & 0 & 1 \end{bmatrix} \qquad \mathbf{D}_r[(123)] = \begin{bmatrix} 0 & 0 & 1 \\ 1 & 0 & 0 \\ 0 & 1 & 0 \end{bmatrix}$$

$$\mathbf{D}_r[(132)] = \begin{bmatrix} 0 & 1 & 0 \\ 0 & 0 & 1 \\ 1 & 0 & 0 \end{bmatrix} \qquad \mathbf{D}_r[(12)^*] = \begin{bmatrix} 0 & 1 & 0 \\ 1 & 0 & 0 \\ 0 & 0 & 1 \end{bmatrix}$$

$$\mathbf{D}_r[(23)^*] = \begin{bmatrix} 1 & 0 & 0 \\ 0 & 0 & 1 \\ 0 & 1 & 0 \end{bmatrix} \qquad \mathbf{D}_r[(31)^*] = \begin{bmatrix} 0 & 0 & 1 \\ 0 & 1 & 0 \\ 1 & 0 & 0 \end{bmatrix}. \qquad (A.27)$$

The characters of the generated representation are determined as the traces of the matrices in equation (A.27):

$$\begin{array}{ccccccc} R: & E & (123) & (132) & (12)^* & (23)^* & (31)^* \\ \sum_{p=1}^{3}(\mathbf{D}_r[R])_{pp}: & 3 & 0 & 0 & 1 & 1 & 1 \end{array} \qquad (A.28)$$

and we use equation (7.97), together with the characters of the irreducible representations in table B.5, to reduce it to

$$\Gamma_r = A_1 \oplus E. \qquad (A.29)$$

To determine the linear combinations of r_1, r_2, r_3 that transform irreducibly, we use the projection operators given in equation (7.102). For the irreducible representation A_1, we have $D^{A_1}[R]_{11} = \chi^{A_1}[R] = 1$. For the E representation, we take $D^E[R]_{pp} = M_{pp}^R$, $p = 1$ or 2, where the 2×2 matrices $\mathbf{M}^R$ are given in equation (7.91); we have shown in chapter 7 that this group of matrices form the E irreducible representation of $C_{3v}(M)$. Table A.1 summarizes the quantities that we require to apply the two projection operators P_{11}^E and P_{22}^E [equation (7.102)] to r_1.

We derive

$$r_{A_1} = P_{11}^{A_1} r_1 = \tfrac{1}{6} \sum_R \chi^{A_1}[R]^* Rr_1 = \tfrac{1}{3}(r_1 + r_2 + r_3) \qquad (A.30)$$

$$r_{E_a} = P_{11}^E r_1 = \frac{2}{6} \sum_R (M_{11}^R)^* R r_1 = \frac{1}{6}(r_1 + r_2 - 2r_3) \qquad (A.31)$$

and

$$r_{E_b} = P_{22}^E r_1 = \frac{2}{6} \sum_R (M_{22}^R)^* R r_1 = \frac{1}{2}(r_1 - r_2). \qquad (A.32)$$

Upon normalization, we get

$$\begin{pmatrix} r_{A_1} \\ r_{E_a} \\ r_{E_b} \end{pmatrix} = \begin{pmatrix} (r_1 + r_2 + r_3)/\sqrt{3} \\ (r_1 + r_2 - 2r_3)/\sqrt{6} \\ (r_1 - r_2)/\sqrt{2} \end{pmatrix}. \qquad (A.33)$$

All operations in $C_{3v}(M)$ permute (r_1, r_2, r_3) and it is obvious that r_{A_1} in equation (A.30) has A_1 symmetry.

We investigate the effect of (123) on (r_{E_a}, r_{E_b}):

$$(123) \begin{pmatrix} r_{E_a} \\ r_{E_b} \end{pmatrix} = \begin{pmatrix} (r_3 + r_1 - 2r_2)/\sqrt{6} \\ (r_3 - r_1)/\sqrt{2} \end{pmatrix} \qquad (A.34)$$

where we have simply inserted the (123) $r_i = r_i'$ from equations (A.26) and (A.27).

We can easily verify that

$$\frac{1}{\sqrt{6}}(r_3 + r_1 - 2r_2) = -\frac{1}{2}r_{E_a} + \frac{\sqrt{3}}{2}r_{E_b} = M_{11}^{(123)} r_{E_a} + M_{12}^{(123)} r_{E_b} \qquad (A.35)$$

and

$$\frac{1}{\sqrt{2}}(r_3 - r_1) = -\frac{\sqrt{3}}{2}r_{E_a} - \frac{1}{2}r_{E_b} = M_{21}^{(123)} r_{E_a} + M_{22}^{(123)} r_{E_b} \qquad (A.36)$$

where the elements $M_{ij}^{(123)}$ of the matrix $M^{(123)}$ are given in equation (7.91). Equations (A.35) and (A.36) can be written as

$$R \begin{pmatrix} r_{E_a} \\ r_{E_b} \end{pmatrix} = M^R \begin{pmatrix} r_{E_a} \\ r_{E_b} \end{pmatrix} \qquad (A.37)$$

for $R = (123)$ and we can show that this equation is satisfied for all other R in $C_{3v}(M)$ by obtaining the left-hand side from equations (A.26) and (A.27) and checking that it is equal to the right-hand side.

Problem 11.1

From table 11.1, the equivalent rotation $R_\alpha{}^\pi$ transforms the Euler angles (θ, ϕ, χ) to $(\theta', \phi', \chi') = (\pi - \theta, \phi + \pi, 2\pi - 2\alpha - \chi)$. We obtain an expression for $R_\alpha{}^\pi |J, k, m\rangle$ by substituting $(\theta, \phi, \chi) \rightarrow (\theta', \phi', \chi')$ in equation (5.56). Inserting the expressions for (θ', ϕ', χ') and using

$$\begin{pmatrix} \cos(\theta'/2) \\ \sin(\theta'/2) \end{pmatrix} = \begin{pmatrix} \cos(\pi/2 - \theta/2) \\ \sin(\pi/2 - \theta/2) \end{pmatrix} = \begin{pmatrix} \sin(\theta/2) \\ \cos(\theta/2) \end{pmatrix} \tag{A.38}$$

we obtain

$$R_\alpha{}^\pi |J, k, m\rangle = N(-1)^m e^{-2ik\alpha} e^{im\phi} e^{-ik\chi}$$
$$\times \left\{ \sum_\sigma (-1)^\sigma \frac{(\sin \tfrac{1}{2}\theta)^{2J+k-m-2\sigma}(-\cos \tfrac{1}{2}\theta)^{m-k+2\sigma}}{\sigma!(J-m-\sigma)!(m-k+\sigma)!(J+k-\sigma)!} \right\}. \tag{A.39}$$

By substituting $\sigma = J - m - \sigma_1$ everywhere in equation (A.39), we rewrite the expression as

$$R_\alpha{}^\pi |J, k, m\rangle = N(-1)^J e^{-2ik\alpha} e^{im\phi} e^{-ik\chi}$$
$$\times \left\{ \sum_{\sigma_1} (-1)^{\sigma_1} \frac{(\cos \tfrac{1}{2}\theta)^{2J-k-m-2\sigma_1}(-\sin \tfrac{1}{2}\theta)^{m+k+2\sigma_1}}{\sigma_1!(J-m-\sigma_1)!(m+k+\sigma_1)!(J-k-\sigma_1)!} \right\}. \tag{A.40}$$

The summation index $\sigma_1 = J - m - \sigma$ has its minimum value when σ has its maximum value. The maximum value of σ is $(J - m)$ or $(J + k)$, whichever is the smaller. Thus, the minimum value of σ_1 is $[J - m - (J - m)] = 0$, or $[J - m - (J + k)] = (-k - m)$, whichever is the larger. σ_1 has its maximum value when σ has its minimum value. The minimum value of σ is 0 or $(k - m)$, whichever is the larger. Hence, the maximum value of σ_1 is $(J - m)$ or $[J - m - (k - m)] = (J - k)$, whichever is the smaller. By comparing equation (A.40) with equation (5.56), we see that we have now proved equation (11.9).

Problem 11.2

To prove the asymmetric top symmetry rule, we introduce the set of equivalent rotations

$$V = \{E, R_a{}^\pi, R_b{}^\pi, R_c{}^\pi\}. \tag{A.41}$$

Since, for the H_2O molecule, the elements in V are the equivalent rotations of the elements in $C_{2v}(M)$, the MS group for H_2O, V is isomorphic to $C_{2v}(M)$; it is called the *Vierergruppe*. Thus, the irreducible representations of V are identical to those of $C_{2v}(M)$ but we repeat them in table A.2. We use here a labelling where A is the totally symmetric representation and B_γ, $\gamma = a, b$, or c, is the irreducible representation that has a character of $+1$ under $R_\gamma{}^\pi$ (and character -1 under the two other equivalent rotations).

Table A.2. The character table for V.

V	E	R_a^π	R_b^π	R_c^π
A:	1	1	1	1
B_a:	1	1	−1	−1
B_b:	1	−1	1	−1
B_c:	1	−1	−1	1

We determine the symmetries of the asymmetric top rotational wavefunctions by *correlation* between the prolate top and oblate top limits, i.e. by following the ideas discussed for the water molecule in connection with figure 5.6. We reach the prolate limit by changing the value of the rotational constant B_e to be equal to C_e and the oblate limit by changing the value of B_e to be equal to A_e.

In the prolate limit, we use an xyz axis system defined according to the type I^r convention, i.e. with $xyz = bca$, and we consider symmetric top rotational functions $|J, k_a, m\rangle$ where k_a defines the projection of the total angular momentum on the $a = z$ axis. Since now $R_a^\pi = R_z^\pi$, we can use equations (11.8) and (11.9) to symmetry classify the $|J, k_a, m\rangle$ functions in V. We distinguish two cases: For $K_a = |k_a| > 0$, the two functions $|J, K_a, m\rangle$ and $|J, -K_a, m\rangle$ span a two-dimensional reducible representation; and, for $K_a = 0$, the single function $|J, 0, m\rangle$ spans an irreducible representation. The characters are

$$
\begin{array}{ccccc}
V: & E & R_a^\pi & R_b^\pi & R_c^\pi \\
K_a > 0: & 2 & 2(-1)^{K_a} & 0 & 0 \\
K_a = 0: & 1 & 1 & (-1)^J & (-1)^J
\end{array}
\tag{A.42}
$$

In the oblate limit, the xyz axes are defined according to the type III^r convention, i.e. with $xyz = abc$, and we consider symmetric top rotational functions $|J, k_c, m\rangle$ where k_c defines the projection of the total angular momentum on the $c = z$ axis. For $K_c = |k_c| > 0$, we determine the characters associated with the two functions $|J, K_c, m\rangle$ and $|J, -K_c, m\rangle$, and, for $K_c = 0$, we determine those generated by the function $|J, 0, m\rangle$. The characters obtained in the oblate limit are as follows:

$$
\begin{array}{ccccc}
V: & E & R_a^\pi & R_b^\pi & R_c^\pi \\
K_c > 0: & 2 & 0 & 0 & 2(-1)^{K_c} \\
K_c = 0: & 1 & (-1)^J & (-1)^J & 1.
\end{array}
\tag{A.43}
$$

The representations defined in equations (A.42) and (A.43) are reduced in terms of the irreducible representations of V (table A.2) as given in table A.3.

To determine the symmetries of the asymmetric top rotational wavefunctions, we now use the basic result that, as we make the correlation between the

Table A.3. Representations Γ_{rot} of the group V generated by the basis functions $|J, k_a, m\rangle$ and $|J, k_c, m\rangle^a$.

Prolate limit		Oblate limit	
K_a	Γ_{rot}	K_c	Γ_{rot}
0 J even	A	0 J even	A
J odd	B_a	J odd	B_c
odd	$B_b \oplus B_c$	odd	$B_a \oplus B_b$
even	$A \oplus B_a$	even	$A \oplus B_c$

^{a} $K_a = |k_a|$, $K_c = |k_c|$.

prolate and oblate limits (see figure 5.6), such a wavefunction keeps the same symmetry all the way from the prolate to the oblate limit, subject to the non-crossing rule (see section 10.5). For example, we determine the symmetry of an asymmetric top wavefunction for which K_a and K_c are both odd. We see from table A.3 that, in the prolate limit, this symmetry is contained in $B_b \oplus B_c$, whereas, in the oblate limit, it is contained in $B_a \oplus B_b$. Thus, the symmetry we seek is B_b since only this irreducible representation is common for the two limits. If K_a and K_c are both even, the only possible symmetry is A; when (K_a, K_c) is (even, odd), we get B_a, and for (odd, even) we get B_c. These results are also true if one or both of K_a and K_c are zero (zero being even). We have now proved the asymmetric top symmetry rule.

Problem 12.3

The operation (123) transforms (Q_4, α_4) to $(Q_4', \alpha_4') = (Q_4, \alpha_4 - 2\pi/3)$. By inserting the expressions for (Q_4', α_4') in equation (4.38), we obtain

$$(123)|1^{\pm 1}\rangle = F_{1,\pm 1}(Q_4')e^{i(\pm 1)\alpha'} = e^{\mp i\frac{2\pi}{3}}|1^{\pm 1}\rangle \qquad (A.44)$$

where the signs are correlated.

The operation $(23)^*$ transforms (Q_4, α_4) to $(Q_4', \alpha_4') = (Q_4, -\alpha_4)$ and so

$$(23)^*|1^{\pm 1}\rangle = F_{1,\pm 1}(Q_4')e^{i(\pm 1)\alpha'} = |1^{\mp 1}\rangle \qquad (A.45)$$

where the signs are correlated.

Problem 12.4

We determine from table 7.4 that

$$(132) = (123)(123)$$
$$(12)^* = (23)^*(123)(123)$$
$$(31)^* = (23)^*(123) \tag{A.46}$$

so all operations in $C_{3v}(M)$ can be written as products involving (123) and (23)*. From equations (A.44), (A.45) and (A.46) we determine that

$$(132)|1^{\pm 1}\rangle = e^{\mp i\frac{4\pi}{3}}|1^{\pm 1}\rangle$$
$$(12)^*|1^{\pm 1}\rangle = e^{\mp i\frac{4\pi}{3}}|1^{\mp 1}\rangle$$
$$(31)^*|1^{\pm 1}\rangle = e^{\mp i\frac{2\pi}{3}}|1^{\mp 1}\rangle \tag{A.47}$$

where, in each equation, the signs are correlated. With these equations, we can construct the representation matrices generated by $|1^{+1}\rangle$ and $|1^{-1}\rangle$ and verify that the traces of these matrices are the characters of the E irreducible representation of $C_{3v}(M)$. Since $|\psi_0\rangle$ is totally symmetric in $C_{3v}(M)$, then $|\psi_0\rangle|1^{-1}\rangle$ and $|\psi_0\rangle|1^{+1}\rangle$ have E symmetry.

Problem 12.5

In zero order, the intensity of a transition in the ν_4 band of CH_3F is determined by the integral [see equations (12.11) and (12.12)]

$$I_{TM}^0 = \langle \Phi_{rot,r'}|\langle \Phi_{vib,0v'}|\langle \Phi_{elec,0}|\mu_A|\Phi_{elec,0}\rangle|\Phi_{vib,0v''}\rangle|\Phi_{rot,r''}\rangle. \tag{A.48}$$

where the ground-state electronic wavefunction $\Phi_{elec,0}$, the vibrational wavefunction for the vibrational ground state

$$|\Phi_{vib,0v''}\rangle = |\psi_0\rangle|v_4^{l_4} = 0^0\rangle \tag{A.49}$$

(see problems 12.3 and 12.4) and the dipole moment component μ_A along the space-fixed A axis are all unchanged by the MS group operation (123). The effect of (123) on the upper-state vibrational wavefunction

$$|\Phi_{vib,0v'}\rangle = |\psi_0\rangle|1^{l_4}\rangle \tag{A.50}$$

where $l_4 = \pm 1$, is given by equation (A.44) and the effect on the rotational functions

$$|\Phi_{rot,r'}\rangle = |J',k',m'\rangle \quad \text{and} \quad |\Phi_{rot,r''}\rangle = |J'',k'',m''\rangle \tag{A.51}$$

is obtained from equation (11.8) and the fact that $R_z^{2\pi/3}$ is the equivalent rotation of (123).

As discussed in section 12.3.3, the value of an integral of a function depending on the molecular coordinates is unchanged when a symmetry operation is applied to the integrand. We apply (123) to the integrand of I^0_{TM} in equation (A.48):

$$I^0_{TM} = e^{-ik'\frac{2\pi}{3}} e^{+il_4\frac{2\pi}{3}} e^{+ik''\frac{2\pi}{3}} I^0_{TM} = e^{i(k''-k'+l_4)\frac{2\pi}{3}} I^0_{TM}. \tag{A.52}$$

Thus, I^0_{TM} can only be non-vanishing if

$$e^{i(k''-k'+l_4)\frac{2\pi}{3}} = 1 \tag{A.53}$$

which requires

$$k' - k'' - l_4 = 0, \pm 3, \pm 6, \pm 9, \ldots \tag{A.54}$$

and this is equivalent to equation (12.54). In zero order when $|\Delta k| \leq 1$, it reduces to equation (12.45).

Appendix B

Character tables

This appendix gives the character tables of some common molecular symmetry (MS) groups and it gives the character tables of the extended molecular symmetry (EMS) groups of linear molecules. The MS group is defined for rigid molecules in section 8.2 and the definition is generalized to include non-rigid molecules in section 13.1. The EMS group of a linear molecule is introduced in section 8.3.4.

For a nonlinear rigid molecule, the MS group is isomorphic to the molecular point group and, in such a case, the name of the MS group is taken to be that of the point group followed by (M), e.g. the MS group of CH_3F is called $C_{3v}(M)$. For a linear rigid molecule, the EMS group is isomorphic to the molecular point group and it is called $C_{\infty v}(EM)$ or $D_{\infty h}(EM)$ as appropriate; the MS group of a linear rigid molecule is called $C_{\infty v}(M)$ or $D_{\infty h}(M)$ but these are not isomorphic to the $C_{\infty v}$ or $D_{\infty h}$ point groups. For a non-rigid molecule, the MS group is called G_n, where n is the order of the group.

For rigid molecules, the irreducible representations are named in the same way as for the (isomorphic) point group. The irreducible representations are ordered in each symmetry group according to established convention; this convention is necessary to ensure a consistent numbering system for the normal vibrations. The normal vibrations are numbered according to their symmetry species and then within each symmetry species from highest to lowest wavenumber.

In each character table, one element from each class is given and the number of elements in the class is indicated underneath the element. For rigid molecules, the appropriate element of the molecular point group for each class is given; this shows the effect of the MS or EMS group element on the vibronic variables. The equivalent rotation (Equiv. rot.) of the MS or EMS group element is also given. The equivalent rotations of an asymmetric top molecule are called R_a^π, R_b^π or R_c^π to indicate rotations through π radians about the a, b or c axis, respectively. For a symmetric top molecule, the equivalent rotations are called R_z^β or R_α^π, as defined in table 11.1. Knowing the equivalent rotations, one can classify the symmetric top wavefunctions and the xyz molecule-fixed components of the rovibronic angular momentum operator $\hat{J}_\alpha$ (see table 11.1).

The species obtained for the $\hat{J}_\alpha$ are indicated by placing the $\hat{J}_\alpha$ to the right of the appropriate irreducible representation. The rotational coordinate R_α [see equation (4.12)] transforms in the same way as $\hat{J}_\alpha$ under permutations and permutation–inversions. The translational coordinate T_α [see equation (4.11)] transforms in the same way as $\hat{J}_\alpha$ under a nuclear permutation but with opposite sign under a permutation–inversion and the molecule-fixed dipole moment components μ_α transform in the same way as the T_α [see the discussion after equation (12.18)]. The species of the T_α and of the components $\alpha_{\gamma\delta}$ of the electronic polarizability, given by the species of $T_\gamma T_\delta$ (see section 12.5), are also indicated by placing them the right of the appropriate irreducible representation. Finally, $\Gamma(\mu_A)$, the species of the dipole moment operator along a space-fixed $A = X$, Y or Z direction (or, equivalently, along a ξ, η or ζ direction), is indicated; it has character $+1$ for the nuclear permutations and -1 for the permutation–inversions. Allowed rovibronic transitions are connected by this symmetry species (see section 12.1).

Table B.1

The group $C_s(M)$
Example: HN_3

$C_s(M)$:	E	E^*		
	1	1		
C_s:	E	σ_{ab}		
Equiv. rot.:	R^0	R_c^π		
A':	1	1	:	$T_a, T_b, \hat{J}_c, \alpha_{aa}, \alpha_{bb}, \alpha_{cc}, \alpha_{ab}$
A'':	1	-1	:	$T_c, \hat{J}_a, \hat{J}_b, \alpha_{ac}, \alpha_{bc}, \Gamma(\mu_A)$

Table B.2

The Group $C_i(\text{M})$

Example: *Trans* C(HIF)CHIF (without torsional tunnelling)

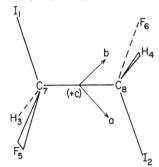

$C_i(\text{M})$:	E	$(12)(34)(56)(78)^*$		
	1	1		
C_i:	E	i		
Equiv. rot.:	R^0	R^0		
A_g:	1	1	:	$\hat{J}_a, \hat{J}_b, \hat{J}_c, \alpha$
A_u:	1	-1	:	$T_a, T_b, T_c, \Gamma(\mu_A)$

Table B.3

The group $C_2(M)$

Example: Hydrogen persulphide (without torsional tunnelling)

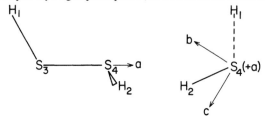

	E	$(12)(34)$	
$C_2(M)$:	1	1	
C_2:	E	C_{2b}	
Equiv. rot.:	R^0	$R_b{}^\pi$	
A:	1	1	$:\quad T_b, \hat{J}_b, \alpha_{aa}, \alpha_{bb}, \alpha_{cc}, \alpha_{ac}, \Gamma(\mu_A)$
B:	1	-1	$:\quad T_a, T_c, \hat{J}_a, \hat{J}_c, \alpha_{ab}, \alpha_{bc}$

Table B.4

The group $C_{2v}(M)$
Example: Water

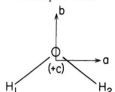

$C_{2v}(M)$:	E	(12)	E^*	$(12)^*$	
	1	1	1	1	
C_{2v}:	E	C_{2b}	σ_{ab}	σ_{bc}	
Equiv. rot.:	R^0	R_b^π	R_c^π	R_a^π	
A_1:	1	1	1	1	: $T_b, \alpha_{aa}, \alpha_{bb}, \alpha_{cc}$
A_2:	1	1	-1	-1	: $\hat{J}_b, \alpha_{ac}, \Gamma(\mu_A)$
B_1:	1	-1	-1	1	: $T_c, \hat{J}_a, \alpha_{bc}$
B_2:	1	-1	1	-1	: $T_a, \hat{J}_c, \alpha_{ab}$

Table B.5

The group $C_{3v}(M)$
Example: Methyl fluoride

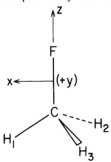

$C_{3v}(M)$:	E	(123)	$(23)^*$	
	1	2	3	
C_{3v}:	E	$2C_3$	$3\sigma_v$	
Equiv. rot.:	R^0	$R_z^{2\pi/3}$	$R_{\pi/2}^{\pi}$	
A_1:	1	1	1	$:\quad T_z, \alpha_{zz}, \alpha_{xx} + \alpha_{yy}$
A_2:	1	1	-1	$:\quad \hat{J}_z, \Gamma(\mu_A)$
E:	2	-1	0	$:\quad (T_x, T_y), (\hat{J}_x, \hat{J}_y),$
				$:\quad\quad (\alpha_{xx} - \alpha_{yy}, \alpha_{xy})$

Table B.6

The group $C_{2h}(M)$

Example: *Trans*-difluoroethylene (without torsional tunnelling)

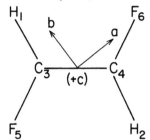

$C_{2h}(M)$:	E	(12)(34)(56)	E^*	(12)(34)(56)*	
	1	1	1	1	
C_{2h}:	E	C_{2c}	σ_{ab}	i	
Equiv. rot.:	R^0	R_c^π	R_c^π	R^0	
A_g:	1	1	1	1	: $\hat{J}_c, \alpha_{aa}, \alpha_{bb}, \alpha_{cc}, \alpha_{ab}$
A_u:	1	1	-1	-1	: $T_c, \Gamma(\mu_A)$
B_g:	1	-1	-1	1	: $\hat{J}_a, \hat{J}_b, \alpha_{ac}, \alpha_{bc}$
B_u:	1	-1	1	-1	: T_a, T_b

Table B.7

The group D_{2h}(M)
Example: Ethylene (without torsional tunnelling)

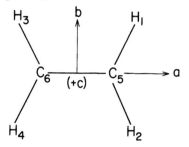

D_{2h}(M):	E	(12)(34)	(13)(24)(56)	(14)(23)(56)	E^*	(12)(34)*	(13)(24)(56)*	(14)(23)(56)*		
	1	1	1	1	1	1	1	1		
D_{2h}:	E	C_{2a}	C_{2b}	C_{2c}	σ_{ab}	σ_{ac}	σ_{bc}	i		
Equiv. rot.:	R^0	$R_a{}^\pi$	$R_b{}^\pi$	$R_c{}^\pi$	$R_c{}^\pi$	$R_b{}^\pi$	$R_a{}^\pi$	R^0		
A_g:	1	1	1	1	1	1	1	1	:	$\alpha_{aa}, \alpha_{bb}, \alpha_{cc}$
A_u:	1	1	1	1	-1	-1	-1	-1	:	$\Gamma(\mu_A)$
B_{1g}:	1	1	-1	-1	-1	-1	1	1	:	$\hat{J}_a, \alpha_{bc}$
B_{1u}:	1	1	-1	-1	1	1	-1	-1	:	T_a
B_{2g}:	1	-1	1	-1	-1	1	-1	1	:	$\hat{J}_b, \alpha_{ac}$
B_{2u}:	1	-1	1	-1	1	-1	1	-1	:	T_b
B_{3g}:	1	-1	-1	1	1	-1	-1	1	:	$\hat{J}_c, \alpha_{ab}$
B_{3u}:	1	-1	-1	1	-1	1	1	-1	:	T_c

Table B.8

The group $D_{3h}(M)$

Example: H_3^+

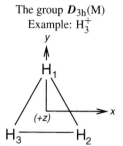

$D_{3h}(M)$:	E	(123)	(23)	E^*	(123)*	(23)*		
	1	2	3	1	2	3		

D_{3h}:	E	$2C_3$	$3C_2$	σ_h	$2S_3$	$3\sigma_v$		

Equiv. rot.:	R^0	$R_z^{2\pi/3}$	$R_{\pi/2}^{\pi}$	R_z^{π}	$R_z^{5\pi/3}$	R_0^{π}		

A_1' :	1	1	1	1	1	1	:	$\alpha_{zz}, \alpha_{xx} + \alpha_{yy}$
A_1'':	1	1	1	-1	-1	-1	:	$\Gamma(\mu_A)$
A_2' :	1	1	-1	1	1	-1	:	$\hat{J}_z$
A_2'':	1	1	-1	-1	-1	1	:	T_z
E' :	2	-1	0	2	-1	0	:	(T_x, T_y),
:							:	$(\alpha_{xx} - \alpha_{yy}, \alpha_{xy})$
E'':	2	-1	0	-2	1	0	:	$(\hat{J}_x, \hat{J}_y)$,
:							:	$(\alpha_{xz}, \alpha_{yz})$

Table B.9

The group $D_{6h}(M)$

Example: Benzene

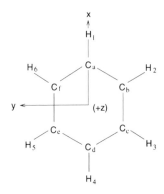

$D_{6h}(M)$:	E	(123456)(abcdef)	(135)(246)(ace)(bdf)	(14)(25)(36)(ad)(be)(cf)	(26)(35)(bf)(ce)	(14)(23)(56)(ad)(bc)(ef)	(14)(25)(36)(ad)(be)(cf)*	(135)(246)(ace)(bdf)*	(123456)(abcdef)*	E^*	(14)(23)(56)(ad)(bc)(ef)*	(26)(35)(bf)(ce)*	
	1	2	2	1	3	3	1	2	2	1	3	3	
D_{6h}:	E	$2C_6$	$2C_3$	C_2	$3C_2'$	$3C_2''$	i	$2S_3$	$2S_6$	σ_h	$3\sigma_d$	$3\sigma_v$	
Eq. rot.:	R^0	$R_z^{\pi/3}$	$R_z^{2\pi/3}$	R_z^{π}	R_0^{π}	$R_{\pi/2}^{\pi}$	R^0	$R_z^{5\pi/3}$	$R_z^{4\pi/3}$	R_z^{π}	R_0^{π}	$R_{\pi/2}^{\pi}$	
A_{1g}:	1	1	1	1	1	1	1	1	1	1	1	1	$\alpha_{xx}+\alpha_{yy},$: α_{zz}
A_{1u}:	1	1	1	1	1	1	-1	-1	-1	-1	-1	-1	: $\Gamma(\mu_A)$
A_{2g}:	1	1	1	1	-1	-1	1	1	1	1	-1	-1	: $\hat{J}_z$
A_{2u}:	1	1	1	1	-1	-1	-1	-1	-1	1	1	1	: T_z
B_{1g}:	1	-1	1	-1	1	-1	1	-1	1	-1	1	-1	
B_{1u}:	1	-1	1	-1	1	-1	-1	1	-1	1	-1	1	
B_{2g}:	1	-1	1	-1	-1	1	1	-1	1	-1	-1	1	
B_{2u}:	1	-1	1	-1	-1	1	-1	1	-1	1	1	-1	
E_{1g}:	2	1	-1	-2	0	0	2	1	-1	-2	0	0	: $(\hat{J}_x, \hat{J}_y),$ $(\alpha_{xz}, \alpha_{yz})$
E_{1u}:	2	1	-1	-2	0	0	-2	-1	1	2	0	0	: (T_x, T_y)
E_{2g}:	2	-1	-1	2	0	0	2	-1	-1	2	0	0	: $(\alpha_{xx}-\alpha_{yy},$ $\alpha_{xy})$
E_{2u}:	2	-1	-1	2	0	0	-2	1	1	-2	0	0	

Table B.10

The group $\boldsymbol{D}_{2d}(M)$
Example: Allene (without torsional tunnelling)

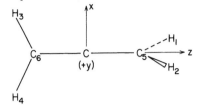

$\boldsymbol{D}_{2d}(M)$:	E	$(1423)(56)^*$	$(12)(34)$	$(13)(24)(56)$	$(34)^*$	
	1	2	1	2	2	
$\boldsymbol{D}_{2d}$:	E	$2S_4$	C_2	$2C_2'$	$2\sigma_d$	
Equiv. rot.:	R^0	$R_z^{3\pi/2}$	R_z^{π}	$R_{3\pi/4}^{\pi}$	R_0^{π}	
A_1:	1	1	1	1	1	: $\alpha_{xx} + \alpha_{yy}, \alpha_{zz}$
A_2:	1	1	1	-1	-1	: $\hat{J}_z$
B_1:	1	-1	1	1	-1	: $\alpha_{xx} - \alpha_{yy}, \Gamma(\mu_A)$
B_2:	1	-1	1	-1	1	: T_z, α_{xy}
E:	2	0	-2	0	0	: $(T_x, T_y), (\hat{J}_x, \hat{J}_y),$
						: $(\alpha_{xz}, \alpha_{yz})$

Table B.11

The group $\boldsymbol{D}_{3d}(M)$

Example: Ethane (without torsional tunnelling in the staggered conformation)

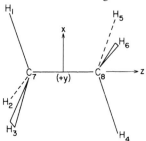

	E	$(123)(465)$	$(16)(24)(35)(78)$	$(14)(26)(35)(78)^*$	$(163425)(78)^*$	$(12)(46)^*$	
$\boldsymbol{D}_{3d}(M)$:	1	2	3	1	2	3	
$\boldsymbol{D}_{3d}$:	E	$2C_3$	$3C_2$	i	$2S_6$	$3\sigma_d$	
Equiv. rot.:	R^0	$R_z^{2\pi/3}$	$R_{\pi/6}^{\pi}$	R^0	$R_z^{2\pi/3}$	$R_{\pi/6}^{\pi}$	
A_{1g}:	1	1	1	1	1	1	$\alpha_{zz}, \alpha_{xx} + \alpha_{yy}$
A_{1u}:	1	1	1	-1	-1	-1	$\Gamma(\mu_A)$
A_{2g}:	1	1	-1	1	1	-1	$\hat{J}_z$
A_{2u}:	1	1	-1	-1	-1	1	T_z
E_g:	2	-1	0	2	-1	0	$(\hat{J}_x, \hat{J}_y), (\alpha_{xz}, \alpha_{yz}),$
							$(\alpha_{xx} - \alpha_{yy}, \alpha_{xy})$
E_u:	2	-1	0	-2	1	0	(T_x, T_y)

Table B.12

The group $T_d(M)$
Example: Methane

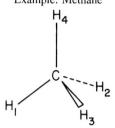

$T_d(M)$:	E	(123)	(14)(23)	(1423)*	(23)*		
	1	8	3	6	6		
T_d:	E	$8C_3$	$3C_2$	$6S_4$	$6\sigma_d$		
A_1:	1	1	1	1	1	:	$\alpha_{xx} + \alpha_{yy} + \alpha_{zz}$
A_2:	1	1	1	-1	-1	:	$\Gamma(\mu_A)$
E:	2	-1	2	0	0	:	$(\alpha_{xx} + \alpha_{yy} - 2\alpha_{zz},$
						:	$\alpha_{xx} - \alpha_{yy})$
F_1:	3	0	-1	1	-1	:	$(\hat{J}_x, \hat{J}_y, \hat{J}_z)$
F_2:	3	0	-1	-1	1	:	$(T_x, T_y, T_z),$
						:	$(\alpha_{xy}, \alpha_{yz}, \alpha_{xz})$

Table B.13

The group O_h(M)

Example: Sulphur hexafluoride

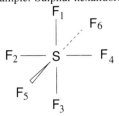

O_h(M):	E	(145)(263)	(13)(26)(45)	(1234)	(13)(24)	(13)(24)(56)*	(125346)*	(25)(46)*	(1432)(56)*	(56)*	
	1	8	6	6	3	1	8	6	6	3	
O_h:	E	$8C_3$	$6C_2$	$6C_4$	$3C_2$	i	$8S_6$	$6\sigma_d$	$6S_4$	$3\sigma_h$	
A_{1g}:	1	1	1	1	1	1	1	1	1	1	: $\alpha_{zz} + \alpha_{xx} + \alpha_{yy}$
A_{1u}:	1	1	1	1	1	−1	−1	−1	−1	−1	: $\Gamma(\mu_A)$
A_{2g}:	1	1	−1	−1	1	1	1	−1	−1	1	
A_{2u}:	1	1	−1	−1	1	−1	−1	1	1	−1	
E_g:	2	−1	0	0	2	2	−1	0	0	2	: $(2\alpha_{zz} − \alpha_{xx} − \alpha_{yy},$ $\alpha_{xx} − \alpha_{yy})$
E_u:	2	−1	0	0	2	−2	1	0	0	−2	
F_{1g}:	3	0	−1	1	−1	3	0	−1	1	−1	: $(\hat{J}_x, \hat{J}_y, \hat{J}_z)$
F_{1u}:	3	0	−1	1	−1	−3	0	1	−1	1	: (T_x, T_y, T_z)
F_{2g}:	3	0	1	−1	−1	3	0	1	−1	−1	: $(\alpha_{xz}, \alpha_{yz}, \alpha_{xy})$
F_{2u}:	3	0	1	−1	−1	−3	0	−1	1	1	

Table B.14

The group $C_{\infty v}(M)$
Example: Hydrogen cyanide

H ———— C ———— N

$C_{\infty v}(M)$:	E	E^*	
(+), Σ^+:	1	1	
(−), Σ^-:	1	−1	: $\Gamma(\mu_A)$

Table B.15

The group $D_{\infty h}(M)$
Example: Carbon dioxide

O_1 ——— C ——— O_2

$D_{\infty h}(M)$:	E	(12)	E^*	(12)*	
(+s), Σ_g^+:	1	1	1	1	
(+a), Σ_u^+:	1	−1	1	−1	
(−a), Σ_g^-:	1	−1	−1	1	
(+s), Σ_u^-:	1	1	−1	−1	: $\Gamma(\mu_A)$

Table B.16

The group $C_{\infty v}(\text{EM})$

Example: Hydrogen cyanide

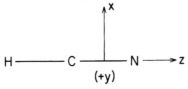

$C_{\infty v}(\text{EM})$	E_0	E_ε	$\cdots$	$\infty E_\varepsilon{}^*$	
	1	2	$\cdots$	∞	
$C_{\infty v}$:	E	$2C_\infty{}^\varepsilon$	$\cdots$	$\infty\sigma_v^{(\varepsilon/2)}$	
Equiv. rot.:	R^0	$R_z^{-\varepsilon}$	$\cdots$	$R^\pi_{(\pi+\varepsilon)/2}$	
$(+), \Sigma^+$:	1	1	$\cdots$	1	: $T_z, \alpha_{xx} + \alpha_{yy}, \alpha_{zz}$
$(-), \Sigma^-$:	1	1	$\cdots$	-1	: $\hat{J}_z, \Gamma(\mu_A)$
Π:	2	$2\cos\varepsilon$	$\cdots$	0	: $(T_x, T_y), (\hat{J}_x, \hat{J}_y), (\alpha_{xz}, \alpha_{yz})$
Δ:	2	$2\cos 2\varepsilon$	$\cdots$	0	: $(\alpha_{xx} - \alpha_{yy}, \alpha_{xy})$
$\vdots$	$\vdots$	$\vdots$	$\cdots$	$\vdots$	

Printed and bound by CPI Group (UK) Ltd, Croydon, CR0 4YY

17/10/2024

01775686-0015

Table B.17

The group $D_{\infty h}(\text{EM})$

Example: Carbon dioxide

$$O_1 \!-\!\!\! \underset{(+y)}{\overset{x}{C}} \!\!\! -\! O_2 \longrightarrow z$$

$D_{\infty h}(\text{EM})$:	E_0	E_ε	$\cdots$	$\infty E_\varepsilon{}^*$	$(12)_\pi{}^*$	$(12)^*_{\pi+\varepsilon}$	$\cdots$	$\infty(12)_\varepsilon$		
	1	2	$\cdots$	∞	1	2	$\cdots$	∞		
$D_{\infty h}$:	E	$2C_\infty{}^\varepsilon$	$\cdots$	$\infty\sigma_v^{(\varepsilon/2)}$	i	$2S_\infty^{\pi+\varepsilon}$	$\cdots$	$\infty C_2^{(\varepsilon/2)}$		
Equiv. rot.:	R^0	$R_z^{-\varepsilon}$	$\cdots$	$R_{(\pi+\varepsilon)/2}^\pi$	R^0	$R_z^{-\varepsilon}$	$\cdots$	$R_{\varepsilon/2}^\pi$		
$(+s), \Sigma_g{}^+$:	1	1	$\cdots$	1	1	1	$\cdots$	1	:	$\alpha_{xx}+\alpha_{yy},\ \alpha_{zz}$
$(+a), \Sigma_u{}^+$:	1	1	$\cdots$	1	-1	-1	$\cdots$	-1	:	T_z
$(-a), \Sigma_g{}^-$:	1	1	$\cdots$	-1	1	1	$\cdots$	-1	:	$\hat{J}_z$
$(-s), \Sigma_u{}^-$:	1	1	$\cdots$	-1	-1	-1	$\cdots$	1	:	$\Gamma(\mu_A)$
Π_g:	2	$2\cos\varepsilon$	$\cdots$	0	2	$2\cos\varepsilon$	$\cdots$	0	:	$(\hat{J}_x,\hat{J}_y),\ (\alpha_{xz},\alpha_{yz})$
Π_u:	2	$2\cos\varepsilon$	$\cdots$	0	-2	$-2\cos\varepsilon$	$\cdots$	0	:	(T_x,T_y)
Δ_g:	2	$2\cos2\varepsilon$	$\cdots$	0	2	$2\cos2\varepsilon$	$\cdots$	0	:	$(\alpha_{xx}-\alpha_{yy},\ \alpha_{xy})$
Δ_u:	2	$2\cos2\varepsilon$	$\cdots$	0	-2	$-2\cos2\varepsilon$	$\cdots$	0	:	
$\cdots$	$\cdots$	$\cdots$		$\cdots$	$\cdots$	$\cdots$		$\cdots$		$\cdots$

Table B.18

The group G_4

Example: Hydrogen peroxide
(with torsional tunnelling)

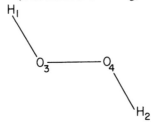

G_4 :	E	$(12)(34)$	E^*	$(12)(34)^*$
$\Gamma_1{}^a$ $\Gamma_2{}^b$ $\Gamma_3{}^c$: 1	1	1	1	

			E	$(12)(34)$	E^*	$(12)(34)^*$		
A^+	A_g	A_1 :	1	1	1	1		
A^-	A_u	A_2 :	1	1	-1	-1	:	$\Gamma(\mu_A)$
B^-	B_g	B_1 :	1	-1	-1	1		
B^+	B_u	B_2 :	1	-1	1	-1		

[a] Γ_1 notation based on effects of $(12)(34)$ and E^*.

[b] Γ_2 notation based on C_{2h} notation.

[c] Γ_3 notation based on C_{2v} notation.

Table B.19

The group G_6

Example: Methanol (with torsional tunnelling)

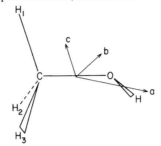

	E	(123)	$(23)^*$
G_6:	1	2	3

Equiv. rot.: R^0 R^0 R_c^π

	E	(123)	$(23)^*$		
A_1 :	1	1	1	:	$T_a, T_b, \hat{J}_c, \alpha_{aa}, \alpha_{bb}, \alpha_{cc}, \alpha_{ab}$
A_2 :	1	1	-1	:	$T_c, \hat{J}_a, \hat{J}_b, \alpha_{ac}, \alpha_{bc}, \Gamma(\mu_A)$
E :	2	-1	0		

Table B.20

The group G_{12}

Example: Nitromethane (with torsional tunnelling)

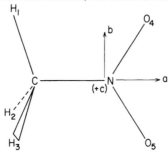

G_{12} :	E	(123)	(23)*	(45)	(123)(45)	(23)(45)*		
	1	2	3	1	2	3		
Equiv. rot.:	R^0	R^0	$R_c{}^\pi$	$R_a{}^\pi$	$R_a{}^\pi$	$R_b{}^\pi$		
A_1' :	1	1	1	1	1	1	:	$T_a, \alpha_{aa},$
							:	α_{bb}, α_{cc}
A_1'' :	1	1	1	−1	−1	−1	:	$T_b, \hat{J}_c, \alpha_{ab}$
A_2' :	1	1	−1	1	1	−1	:	$\hat{J}_a, \alpha_{bc}, \Gamma(\mu_A)$
A_2'' :	1	1	−1	−1	−1	1	:	$T_c, \hat{J}_b, \alpha_{ac}$
E' :	2	−1	0	2	−1	0		
E'' :	2	−1	0	−2	1	0		

Table B.21

The group G_{16}

Example: Ethylene (with torsional tunnelling)

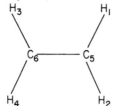

		E	$(1423)(56)^*$	$(12)(34)$	$(13)(24)(56)$	$(34)^*$	E^*	$(1423)(56)$	$(12)(34)^*$	$(13)(24)(56)^*$	(34)
G_{16}:		1	2	1	2	2	1	2	1	2	2
MW[a]	LH[b]										
A_1^+	A_1^+ :	1	1	1	1	1	1	1	1	1	1
A_2^+	A_2^+ :	1	1	1	-1	-1	1	1	1	-1	-1
B_1^+	B_2^+ :	1	-1	1	1	-1	1	-1	1	1	-1
B_2^+	B_1^+ :	1	-1	1	-1	1	1	-1	1	-1	1
E^+	E^+ :	2	0	-2	0	0	2	0	-2	0	0
A_1^-	B_2^- :	1	1	1	1	1	-1	-1	-1	-1	-1
A_2^-	B_1^- :	1	1	1	-1	-1	-1	-1	-1	1	1
B_1^-	A_1^- :	1	-1	1	1	-1	-1	1	-1	-1	1
B_2^-	A_2^- :	1	-1	1	-1	1	-1	1	-1	1	-1
E^-	E^- :	2	0	-2	0	0	-2	0	2	0	0

B_1^- row : $\Gamma(\mu_A)$

[a] Notation from Merer A J and Watson J K G 1973 *J. Mol. Spectrosc.* **47** 499. This correlates appropriately to $D_{2d}(M)$ (see table B.7).

[b] Notation from Longuet-Higgins H C 1963 *Mol. Phys.* **6** 445.

Appendix C

Books for further reading

Spectroscopy

Banwell C N and McCash E M 1995 *Fundamentals of Molecular Spectroscopy* 4th edn (New York: McGraw-Hill)

Bernath P F 1995 *Spectra of Atoms and Molecules* (Oxford: Oxford University Press)

Brown J M 1998 *Molecular Spectroscopy* (Oxford: Oxford University Press)

Brown J M and Carrington A 2003 *Rotational Spectroscopy of Diatomic Molecules* (Cambridge: Cambridge University Press)

Demtröder W 2002 *Laser Spectroscopy* 3rd edn (Berlin: Springer)

Duxbury G 2001 *Infrared Vibration–Rotation Spectroscopy: From Free Radicals to the Infrared Sky* (New York: Wiley)

Harris D C and Bertolucci M D 1990 *Symmetry and Spectroscopy: An Introduction to Vibrational and Electronic Spectroscopy* (New York: Dover)

Herzberg G 1989 *Molecular Spectra and Molecular Structure, I. Spectra of Diatomic Molecules* (Melbourne, FL: Krieger)

——1991 *Molecular Spectra and Molecular Structure, II. Infrared and Raman Spectra of Polyatomic Molecules* (Melbourne, FL: Krieger)

——1991 *Molecular Spectra and Molecular Structure, III. Electronic Spectra and Electronic Structure of Polyatomic Molecules* (Melbourne, FL: Krieger)

Hollas J M 2002 *Basic Atomic and Molecular Spectroscopy* (New York: Wiley)

——2004 *Modern Spectroscopy* 4th edn (New York: Wiley)

Jensen P and Bunker P R (ed) 2000 *Computational Molecular Spectroscopy* (New York: Wiley)

Kroto H W 2003 *Molecular Rotation Spectra* (New York: Dover)

Lefebvre-Brion H and Field R W 2004 *The Spectra and Dynamics of Diatomic Molecules: Revised and Enlarged Edition* (New York: Academic)

Wilson E B Jr, Decius J C and Cross P C 1980 *Molecular Vibrations* (New York: Dover)

Quantum mechanics

Atkins P W and Friedman R S 1999 *Molecular Quantum Mechanics* 3rd edn (Oxford: Oxford University Press)

Cohen-Tannoudji C, Diu B and Laloë F 1996 *Quantum Mechanics* (2 vol set) (New York: Wiley)

Levine I N 1999 *Quantum Chemistry* 5th edn (Englewood Cliffs, NJ: Prentice-Hall)

Pauling L and Wilson E B Jr 1985 *Introduction to Quantum Mechanics with Applications to Chemistry* (New York: Dover)

Pilar F L 2001 *Elementary Quantum Chemistry* 2nd edn (New York: Dover)

Schatz G C and Ratner M A 2002 *Quantum Mechanics in Chemistry* (New York: Dover)

Szabo A and Ostlund N S 1996 *Modern Quantum Chemistry: Introduction to Advanced Electronic Structure Theory* (New York: Dover)

Zare R N 1988 *Angular Momentum* (New York: Wiley)

Symmetry and group theory

Bishop D M 1993 *Group Theory and Chemistry* (New York: Dover)

Bunker P R and Jensen P 1998 *Molecular Symmetry and Spectroscopy* 2nd edn (Ottawa: NRC Research Press)

Cotton F A 1990 *Chemical Applications of Group Theory* 3nd edn (New York: Wiley)

Hamermesh M 1989 *Group Theory and its Application to Physical Problems* (New York: Dover)

Tinkham M 2003 *Group Theory and Quantum Mechanics* (New York: Dover)

Index